HORIZONTAL, HIGH-ANGLE, AND EXTENDED-REACH DRILLING

Engineering, Geology, Technology, and Operations

A N

A N N O T A T E D A N D I N D E X E D

B I B L I O G R A P H Y

by

David S. Pettus

SOUTHWEST GEOSERVICES

Horizontal, High-Angle, and Extended-Reach Drilling

ISBN# 0-89896-440-7

Library of Congress Cataloging in Publication Data

LARKSDALE

Printed in the United States of America

PREFACE

Horizontal, high-angle, and extended-reach drilling have become, if not commonplace, at least a recognized tool in exploration and production of oil and gas. There is some indication that these methods are also beginning to be used in hazardous waste management and remediation.

As is common when new technology is introduced in the petroleum industry, horizontal drilling went through a *panacea* stage, and now, is being more rationally examined and evaluated for its potential usefulness in finding and producing oil and gas. Horizontal, high-angle, and extended-reach drilling will not solve the basic E & P problem of where to drill, but they do provide methods of reaching previously unreachable targets and in some situations can reduce costs and increase productivity. I hope that this book will aid managers, engineers, and geologists in better utilization of these methods.

This book originated in the summer of 1990, when I was learning about horizontal drilling while working in the Austin Chalk horizontal drilling play in South Texas. After reading most of the available literature on the subject, I compiled and annotated this bibliography for my own convenience. I hope that others will find it as useful as I have. The wording of the annotations reflects my understanding of the meaning of the articles. In some instances direct quotations from the original have been made and are so indicated.

Most of the work for this book was done in the Houston Public Library, the M. D. Anderson Library at the University of Houston, the Fondren Library of Rice University, and the Texas A&M University Library. The interlibrary loan department of the Houston Public Library was especially helpful in obtaining otherwise unavailable books and articles. Julia Cosgrove of Access Information Associates, Inc., arranged for me to use the Mobil E & P technical library in Houston. The Texaco E & P technical library, also in Houston, furnished two articles that were otherwise unavailable. Irene Carlson was most helpful with the editing of this book, and contributed many useful suggestions.

Special thanks are due my mother, Mrs. Ruby Pettus, for moral and financial support during the preparation of this book.

David S. Pettus,

Houston, Texas

INTRODUCTION

Horizontal, high-angle, and extended-reach drilling are grouped together in this bibliography because they are different aspects of the same technology. These subjects may be combined under the general heading of unconventional drilling.

Unconventional drilling includes:

(1) Horizontal drilling: term used for drilling a borehole deviated from the vertical by 87 degrees or more; drilling boreholes parallel to formation boundaries in gently dipping beds.

(2) High-angle drilling: term used for drilling boreholes inclined more than 56 degrees from the vertical.

(3) Extended-reach refers to wells with horizontal displacements of more than that of "normally" deviated wells.

Horizontal wells are a special case of high-angle wells, and frequently, are referred to as drainholes.

A special case of unconventional drilling is slanthole drilling; drilling a "slanthole" by tilting the drilling rig at some angle (frequently 45 degrees) from the vertical. A slanthole may be further deviated to make it a high-angle or horizontal well. In some cases slantholes may be more productive than a horizontal drainhole, i.e., in a thick, homogeneous reservoir. Examples of slanthole drilling can be found in Morcambe Bay field, U.K. Irish Sea, and Cold Lake field, Canada.

A variation of the horizontal well is the inverted well, a well in which the distal end is elevated relative to the horizontal section of borehole. Inverted wells have been used in Prudhoe Bay field, Alaska, for example.

High-angle and horizontal drilling have as their purpose extending the borehole from the surface location, laterally, to reach horizontally displaced targets, and/or to place a longer length of borehole into the reservoir.

Variations of drainhole drilling include: multiple-drainhole drilling, and radial-drainhole drilling. Multiple-drainholes may be drilled from a single wellbore. The drainholes may be in the same reservoir, or they may be in different reservoirs. Commonly, two drainholes may be drilled in the same reservoir, oriented 180 degrees apart; this a called a dual-drainhole. Multiple-drainholes may also be drilled into several vertically separated reservoirs, such as stacked sands. Radial-drainhole drilling involves drilling several (more than 40 in one case) drainholes in a radial array from a vertical wellbore. This method is sometimes used in recovering heavy oil by steamflooding or hot-water flooding.

High-angle, horizontal, and slanthole wells all allow a larger area to be covered from a surface location than can be covered by conventional deviated wells. Therefore, if several such wells are drilled from a single surface location (drilling platform or pad), fewer locations are needed to develop a reservoir. Several North Sea and Australian fields have been developed using extended-reach wells to reduce the number of platforms normally required for field development. Extended-reach drilling has also been used in the Paris basin, to reach beneath villages, and to minimize damage to farmland. Extended-reach drilling was used in Norman Wells field, Northwest Territories, Canada, to drill development wells under the MacKenzie River.

ENGINEERING

High-angle and horizontal wells present many engineering problems, which include: cuttings transport (hole cleaning); borehole stability; directional control; completion and stimulation; and casing centralization. These and other problems are addressed by many of the papers and articles in this bibliography.

Horizontal drainholes, generally, will have higher productivity, up to 10 times, than vertical wells. Horizontal drainholes may recover more oil or gas than vertical wells. If drilling costs can be optimized, horizontal wells may improve project economics significantly, by recovering more oil or gas faster, at a relatively small incremental increase in drilling cost.

The three basic varieties of horizontal wells are: short-radius, medium-radius, and long-radius wells. The distinctions are based on the radius-of-curvature of the angle-build section of the borehole. Short-radius wells have radii of 20-40 ft (angle build rate of 1.5-3 degrees/ft drilled). A special case of the short-radius well is the ultra-short-radius well, in which the radii are 1-2 ft (angle build rate of 45-90 degrees/ft drilled). Medium-radius wells have radii of 300-700 ft (angle build rate of 8-20 degrees/ ft drilled). Long-radius wells have radii of 1,000-3,000 ft (angle build rate of 2-6 degrees/ft drilled). Geometry of the prospect, type of reservoir, completion methods, etc. determine the type of drainhole best for the project.

As high-angle and horizontal drilling practices have developed, engineers have overcome such problems as; centralization and cementing of casing; casing buckling; zonal isolation; multiple hydraulic fracturing in horizontal boreholes; and analysis of fluid behavior in horizontal boreholes.

GEOLOGY

Reservoir geology may be the most important factor in determining whether or not to drill a horizontal well. Horizontal wells have proven effective in: fractured reservoirs; tight reservoirs; thin reservoirs, in some cases; anisotropic reservoirs; discontinuous reservoirs; and multiple reservoirs. Reservoir characterization is very important in determining the placement of the borehole, completion methods used, and stimulation methods used.

Several companies have attributed their success in horizontal drilling to multidisciplinary teams of engineers and geologists. The teams work together to plan the projects, evaluate the reservoir, and execute the drilling program. In many, if not most, cases, a wellsite geologist or competent mudlogger is necessary to monitor drilling and geological data, and to make onsite decisions concerning directional drilling, in order to keep the wellbore within the desired reservoir zone.

TECHNOLOGY

Technological improvements that have made high-angle and, particularly, horizontal drilling economically attractive include: measuring-while-drilling (MWD); steerable downhole motors; and improved bits, particularly PDC bits.

In the late 1970's, when "modern" drainhole drilling began, well-logging companies developed methods of logging wells inclined at angles high enough so that gravity could not overcome frictional forces in the borehole, to pull wireline logging and completion tools to the bottom. Drillpipe-conveyed and coiled-tubing-conveyed logging and completion systems were used to push the tools to the bottom of the borehole; an alternative method uses mud pressure to pump the tools downhole. The first such system was a drillpipe-conveyed system called SIMPHOR, developed by Elf Aquitaine and Institut Francais du Petrole, for use in their early drainhole drilling experiments in the Lacq field, France. Since that time all of the major logging companies have developed their own systems.

OPERATIONS

A myriad of companies and governmental agencies are now engaged in unconventional drilling, in a variety of project types, and for an assortment of reasons. Most companies, whether operators *sensu-stricto,* or non-operating partners, are listed in the index to this bibliography, along with oil or gas field and project names. The projects discussed include: oil and gas production; enhanced oil recovery projects; gravity drainage projects; environmental remediation, in one case; and reduction or elimination of oil or gas coning.

Abbreviations:

BOPD .... barrels of oil per day
BWPD .... barrels of water per day
MCFGPD .... thousand cubic feet of gas per day
MD .... measured depth
TD .... total depth
TVD .... true vertical depth

Notes:

Abstract .... at the beginning of an entry indicates that only an abstract of the article was read.
Not examined .... indicates that the article or book was unavailable.
[ ] in the title .... indicates a translation of the title from its original language.

Bibliography of Horizontal, High-Angle, and

Extended-Reach Drilling

1. Aadnoy, B. S. and M. E. Chenevert (1987). Stability of highly inclined boreholes. SPE Drilling Engineering 2 (4), 364-374. SPE/IADC paper 16052. Also in Proceedings 1987 SPE/IADC Drilling Conference, March 15-18, New Orleans, pp 25-41. Abstract in The Engineering Index Annual (1988), 87, Part IV, 6007. Critical discussion (see Simison) and author's reply in SPE Drilling Engineering (December, 1988) 3 (4), p. 418.

   Collapse of a borehole or reduction in fracture-initiation pressure may result from alterations in the stress state around a borehole caused by inclination of the borehole. Two types of borehole instability are borehole collapse and fracturing of the borehole. Borehole collapse at low pressures is caused by shear failure combined with tensile failure. The higher the borehole inclination, the more likely collapse will occur in isotropic rocks. Weak chalks are more likely to be a problem. Collapse stability can be improved by inclining the borehole in the same direction as the least in-situ stress. Borehole fracturing is mainly tensile failure. Also see: Simison, H. (1988), Discussion of stability of highly inclined boreholes, SPE Drilling Engineering 3 (4), 418; and Aadnoy, B. S. and M. E. Chenevert (1988), Author's reply to discussion of stability of highly inclined boreholes, SPE Drilling Engineering 3 (4), 418.

2. Aadnoy, B. S. and M. E. Chenevert (1988). Author's reply to Discussion of stability of highly inclined boreholes. SPE Drilling Engineering 3 (4), 418. SPE paper 18736.

   Reply to critical discussion by Simison, H. (1988), SPE Drilling Engineering 3 (4), 418, of Aadnoy and Chenevert (1987), 2 (4), 364-374.

3. Aadnoy, Bernt S. (1988). Modeling of the stability of highly inclined boreholes in anisotropic rock formations. SPE Drilling Engineering 3 (3), 259-268. SPE paper 16526 presented at the SPE 1988 Offshore Europe Conference, September 8-11, 1987, Aberdeen. Critical discussions and author's reply: da Fontoura and dos Santos (1989), SPE Drilling Engineering 4 (2), 187; Aadnoy (1989) 4 (2) ,188.

   An anisotropic stress model has been used to simulate the fracture and collapse behavior of boreholes. Parameters considered in the model are: anisotropic elastic properties, directional shear, and directional tensile strength. Borehole orientation, *in-situ* stresses, and the bedding plane can be arbitrarily related to each other to model actual field conditions.

4. Aadnoy, B. S. and K. Larsen (1989). Method for fracture-gradient prediction for vertical and inclined boreholes. SPE Drilling Engineering 4 (2), 97-103. Also in Proceedings 1987 SPE Technical Conference & Exhibition, Drilling, September 27-30, Dallas, pp. 411-420. SPE paper 16695.

   A new method, using an overburden-stress curve equation, has been developed to predict fracture gradients in inclined boreholes. The fracture pressure can be calculated using only well depth, pore pressure, borehole inclination, and lithology as input data. The Statfjord field, Norwegian North Sea is used as an example.

5. Aadnoy, B. S. (1989). Author's reply to discussion of modeling of the stability of highly inclined boreholes in anisotropic rock formations. SPE Drilling Engineering 4 (2), 188. SPE paper 1986.

Reply to critical discussion by da Fortuna and dos Santos (1989) SPE Drilling Engineering 4 (2), 188, of original paper by Aadnoy (1988) SPE Drilling Engineering 3 (3), 259-268.

6. Abraham, Kurt S. (April, 1990). Wytch Farm development stresses environmental responsibility. World Oil 210 (4), 85-86, 88-92, 94.

The giant Wytch Farm field is being developed with environmental considerations in the forefront. The field is located in the environmentally sensitive area under and offshore of the Purbeck Peninsula, Dorset, England. High-angle (45-65 degrees) wells and extended-reach wells, horizontal displacements of up to 6,398 ft, are being used to limit offshore drilling and the number of onshore drill sites. The development plans were made with full public review and approval of local interests.

7. Adamache, I., F. McIntyre, M. Pow, D. Lewis, R. Davis, K. Kuhme, G. Bloy, N. Van Reger, G. Sherpard, and J. Green (1990). Horizontal well application in a vertical miscible flood. Part I - Pre-planning and initial results. in Preprints CIM/SPE International Technical Meeting, June 10-13, 1990, Calgary, v.III, Paper 125. The Petroleum Society of CIM, Montreal. CIM/SPE paper 90-125. Abstract in The Journal of Canadian Petroleum Technology (1990) 29 (2), 50.

A horizontal well was used in the Rainbow field miscible flood EOR project. The horizontal wellbore was placed in the reservoir just above the oil/water contact. The well placement was determined using 3-D seismic data. Production from the horizontal well is three times higher than that of the best vertical well in the field. The horizontal well was used to correct problems caused by solvent breakthrough and channeling in vertical wells.

8. Aguilera, Roberto (1980). *Naturally Fractured Reservoirs.* Petroleum Publishing Company, Tulsa.

This book reviews much of the engineering and geological literature on naturally fractured reservoirs. It is mainly concerned with reservoir engineering. Naturally fractured reservoirs currently are the most frequent horizontal drilling target.

9. Aguillera, R. and M. C. Ng (1989). Transient pressure analysis of horizontal wells in anisotropic naturally fractured reservoirs. Proceedings 1989 SPE Joint Rocky Mountain Region/Low Permeability Reservoirs Symposium & Exhibition, March 6-8, 1989, Denver, pp. 679-693. Society of Petroleum Engineers, Richardson, TX. SPE paper 19002. Abstract in The Engineering Index Monthly (January, 1990), 28 (1), 663. Also printed in SPE Formation Evaluation (1991), 6 (1), 95-100.

Analytical solutions for pressure drawdown and buildup analysis of horizontal wells in naturally fractured reservoirs have led to identification of six flow periods: (1) first radial flow period; (2) transition period; (3) second radial flow period; (4) linear flow period; (5) pseudo-radial flow period; and (6) linear flow period. Analytical solutions and descriptions of each flow period are given in the article.

10. Ahner, P. F. and A. H. Sufi (1990). Physical model steamflood studies using horizontal wells. Proceedings SPE/DOE Seventh Symposium on Enhanced Oil Recovery, April 22-25, 1990, Tulsa, pp. 717-728. Society of Petroleum Engineers, Richardson, TX. SPE/DOE paper 20247.

Scaled, physical-model steamflood tests were made to study drainage areas created by parallel horizontal wells. The model was scaled for viscous and gravitational forces. The model used vertical wells at each corner (injection and production wells) with two horizontal wells placed between the injector/producer pairs. The horizontal wells were used to heat the surrounding formation to provide mobility to the fluids. Steamflood operations were simulated for both 1.37 and 2.75 acre patterns and yielded recoveries of 60% to 80% of the oil in place.

11. Ackert, David, Manfred Boetel, Toni Marszalek, Christian Clavier, Peter Goode, Michael Tambynayagam, and Ted Stagg (1988). Looking sideways for oil. The Technical Review 36 (1), 22-36.

A brief review of the history of horizontal drilling.

12. Allison, M. Lee and Susan Juch Lutz (1991). Faulted shoreline and tidal deposits in the Moenkopi formation of the Grassy Trail Creek field, Utah. A.A.P.G. Bulletin 75 (6), 1122.

Abstract. The Moenkopi reservoir of the Grassy Trail Creek field, Utah, is a combination of tidal channel sandstones and shoreline deposits. The sandstones are dolomite cemented. Fracture porosity enhances production in parts of the field. The trapping mechanism is stratigraphic with some minor faulting. Horizontal drilling in the field has improved oil recovery.

13. American Oil & Gas Reporter (January, 1990). Horizontal well tops Texas production charts. American Oil & Gas Reporter, p. 33.

Winn Exploration Inc. - Leta Glasscock No. 10 well tested at 5,472 BOPD. The well is located in Zavala County, Texas, and produces from the Austin Chalk formation in the Pearsall field.

14. Aminian, K. and S. Ameri (1989). Predicting horizontal well production performance using type curves. in Proceedings 1989 SPE Eastern Regional Conference, October 24-27, 1989, Morgantown, pp. 321-328. Society of Petroleum Engineers, Richardson, TX. SPE paper 19342.

Production type curves may be used to predict the productivity of horizontal wells in low-permeability reservoirs such as those in the Appalachian basin. Type curves developed so far are influenced by the dimensionless well length and the dimensionless wellbore radius. The type curves are also influenced by the horizontal well penetration and the geometry of the drainage area. Horizontal well performance is also affected by the permeability anisotropy of the reservoir. Type curves are illustrated and compared with a horizontal well in West Virginia.

15. Aminian, K., J. R. Duda, S. Ameri (1989). Type curve development and application for predicting the production performance of horizontal wells in low permeability gas reservoirs. Paper presented at the 1989 SPE Annual Technical Conference & Exhibition, October, 1989, San Antonio. Society of Petroleum Engineers, Richardson, TX. SPE paper 19808. Not printed in the Proceedings of the meeting.

Type curves for predicting the productivity of horizontal wells in low-permeability gas reservoirs have been developed. The methodology used for curve development and their applications are given in the article.

16. Andersen, S. A., S. A. Hansen, and K. Fjeldgaard (1988). Horizontal drilling and completion: Denmark. Proceedings European Petroleum Conference, October 17-19, 1988, London, pp. 155-165. Society of Petroleum Engineers, Richardson, TX. SPE paper 18349. Abstract in The Engineering Index Monthly (February, 1990), 28 (2), 634.

Maersk Oil & Gas A/S has successfully drilled three horizontal wells in the Dan field North Sea, Denmark. The wells were completed in a tight, chalk reservoir using multiple hydraulic fracturing in the horizontal sections of the wells. Details of the project are given in the article.

17. Andrich, Vicente (April, 1991). Pozo horizontal da magnifico resultado. Petroleo Internacional 49 (51), 16, 18. In Spanish.

Yacimientos Petroleos Fiscales (YPF) drilled a horizontal well in the Palmar Largo field, Formosa Province, Argentina. The well was drilled to 4,319 meters MD (3,850 meters TVD). The well was completed in the Palmar Largo Formation; testing 474.6 cubic meters of oil per day through a 12mm choke.

18. Anna, Lawrence O. (1990). Analyzing fractures and lineaments to determine the direction of a horizontal well, Sand Wash basin, northwest Colorado. A.A.P.G. Bulletin 74 (8), 1314.

Abstract. Analysis of faults, stream patterns, fracture systems, and shears in the Sand Wash basin of Colorado shows that open fractures in the basin should trend northeast/southwest. Horizontal wells drilled northwest/southeast should intersect more open fractures than wells drilled northeast/southwest.

19. Anonymous (n.d.). Horizontal Wells. Dowell Schlumberger, Houston. Dowell Schlumberger brochure.

Review of horizontal well performance, cementing, sand control, hydraulic fracturing including multiple fractures, stimulation, and coiled-tubing conveyed logging and completion tools.

20. Anonymous (December, 1980). Nido reef complex oil field, Northwest Palawan, Philippines. Petroleum Geology, Taiwan 17, 133-139. Abstract in The Engineering Index Annual (1983) 85 Part IV, 4910.

Abstract. Nido reef complex is Lower Miocene-Oligocene reef. Oil production began in 1977. The discovery well flowed 7,343 BOPD; the second commercial well flowed 9,880 BOPD.

21. Anonymous (December, 1984). Deviated test well planned in Appalachia. Eastern Devonian Shales Technology Review, pp. 103-108. Published by U.S. Department of Energy.

Not examined.

22. Armessen, Pierre, Andre P. Jourdan and Christian Mariotti (May 23, 1988). Horizontal well operations - 5. Horizontal drilling has negative and positive factors. Oil & Gas Journal 86 (21), 37-40.

Horizontal drilling has been surprisingly successful in recent years but still has some negative factors. To date, horizontal drilling has been relatively free of the normal drilling problems such as fishing jobs and twist-offs. Penetration rates have been comparable to that of vertical wells and there have been few problems in trajectory control on horizontal wells. Two problems encountered in horizontal drilling are high level of friction and difficulty in removal of cuttings from the borehole. Friction is high in horizontal wells because of the greater contact of the drillstring with the side of the wellbore in the horizontal section. The friction can be overcome by using the proper mud and by rotating the drillstring during drilling. Cuttings removal is a problem because cuttings accumulate on the low-side of the wellbore in the horizontal section, and are crushed and finely-ground by the weight of the drillstring. A high performance solids removal system is required to keep the solids content of the mud low. Favorable aspects of horizontal drilling include the reduction of drillpipe buckling, reduction of sticking, and ease of drilling parallel to bedding in the reservoir.

23. Andre, Paul and Alain Doulcet (1987). Rospo Mare karstified limestone oil field. A.A.P.G. Bulletin 71 (5), 525.

Abstract. Rospo Mare heavy oil field, located in 1975, is a fractured, karstified limestone reservoir. The field is in the Adriatic Sea, 20 miles offshore from Italy. The trap is a hydrodynamic tilt caused by the low contrast between the specific gravity of oil and water. Development of the field includes four horizontal wells. Well locations are made using three-dimensional seismic data.

24. Artindale, Jim (1990). Analysis of horizontal Bakken development. in Hansen, William B. (editor), Bakken Formation Short Course, p. 16. Montana Geological Society, Billings. Abstract also printed in 1991 Guidebook to Geology and Horizontal Drilling of the Bakken Formation, p. 111, edited by William B. Hansen, Montana Geological Society, Billings, Montana.

Abstract. The Bakken has generated 100 to 150 billion barrels of oil much of which is contained in overpressured fractured reservoirs. The paper as presented gave data on reservoirs, fracture systems, and other data, although this information is not included in the abstract.

25. Artindale, Jim (1990). Analysis of horizontal Bakken development. A.A.P.G. Bulletin 74 (8), 1314.

Abstract. A large portion of the 100 to 150 billion barrels of oil generated in the Bakken formation is trapped in overpressured, fractured reservoirs.

26. Asgarpour, Soheil, Selwyn Springer, Philip Pantella, and Ashok Singhai (1990). Enhanced heavy-oil production by horizontal drilling - Case study. in Preprints CIM/SPE International Technical Meeting, June 10-13, 1990, Calgary, v. III, Paper 126. Petroleum Society of CIM, Montreal. CIM/SPE paper 90-126.

Discussion of a horizontal drilling project conducted by Gulf Canada Resources Ltd. in the West Edam field Sparky sand reservoir, a heavy-oil field. Numerical simulation and analytical models were used to predict the productivity of the horizontal well.

27. Ashton, J. P., J. Liput, R. Lemons, and J. Summerlin (1989). Gravel packing horizontal and highly deviated openhole completions using a single-screen prepacked liner in offshore California fields. Proceedings 64th Annual SPE Technical Conference & Exhibition, October 8-11, 1989, San Antonio, Texas, pp. 165-178. SPE paper 19718.

Two high-angle (87 degrees and 84 degrees) wells have been successfully gravel-packed in Giant Beaver field and Santa Clara field, offshore California. The horizontal sections of the wells are 576 ft (176m) and 1,475 ft (450m) respectively. The boreholes are 11 inches in diameter and have single screen prepacked liners. Details of the project are given along with effective shear rate calculations used.

28. Astier, B., A. Jourdan and G. Baron (1981). Elf well turns 90 degrees - and stays there. Petroleum Engineer International 53 (1), 40. Abstract in The Engineering Index Annual (1982) 81, Part IV, 5223.

Elf Aquitaine drilled the first European horizontal well in Lacq field, southwestern France, in May and June, 1980. The well was spudded conventionally and deviated to drill 330 feet (100m) horizontally (89 degrees to 92 degrees) at 2,198 feet (670m) below surface.

29. Austin, Carl, Chris Zimmerman, Bob Sullaway, and Fred Sabins (1988). Fundamentals of horizontal well completions. Drilling 50 (3), 28-31.

This article reviews horizontal well completion methods.

30. Austin, C. E., R. E. Rose, and F. J. Schuh (1988). Simultaneous multiple entry hydraulic fracture treatments of horizontally drilled wells. Proceedings 1988 SPE Annual Technical Conference & Exhibition, October 2-5, 1988, Houston, pp. 811-825. Society of Petroleum Engineers, richardson, TX. SPE paper 18623. Abstract in The Engineering Monthly Index (November, 1989), 27 (11), 773. Also in Proceedings European Petroleum Conference, London, October 17-19, 1988, pp. 31-45.

A review of the techniques of multiple entry hydraulic fracturing from horizontal wellbores.

31. Ayler, Maynard F. and Charles Brechtel (1987). Steam flooding from mine workings, a viable alternative. A.A.P.G. Bulletin 71 (5), 527.

Abstract. Steam injection into a reservoir from multiple drainholes drilled upward into a reservoir from mine workings may be an efficient way to develop heavy oil fields. Production from such a system would be by gravity drainage into the mine tunnel under the reservoir.

32. Azari, Mehdi (1991). Discussion of Using finite-system buildup analysis to investigate fractured, vugular, stimulated, and horizontal wells. Journal of Petroleum Technology 43 (1), 103-104. SPE paper 22059.

Critical discussion of article by H.N. Mead, Journal of Petroleum Technology (October, 1988), 1361-1371.

33. Azgarpour, Soheilm, Selwyn Springer, Philip Pantella, and Ashok Singhal (1990). Enhanced heavy-oil production by horizontal drilling - Case study. in Preprints of CIM/SPE International Technical Meeting, June 10-13, 1990, Calgary, v. III, Paper 126. The Petroleum Society of CIM, Montreal. CIM/SPE paper 90-126.

Not examined.

34. Babu, D. K. and A. S. Odeh (1988). Productivity of a horizontal well. Proceedings of the 1988 SPE Annual Technical Conference and Exhibition, Reservoir Engineering volume, October 2-5, 1988, Houston, pp. 373-382. Society of Petroleum Engineers, Richardson, TX. SPE paper 18298. Also in SPE Reservoir Engineering (1989), 4 (4), 417-421; Proceedings SPE Production Technology Symposium, Hobbs, New Mexico, November 7-8, 1988. Abstract in The Engineering Index Monthly (April, 1990), 28 (4), 625. Critical discussions and author's replies: Peaceman SPE Reservoir Engineering (1990), 5 (2), 252-253; Babu and Odeh SPE Reservoir Engineering (1990), 5 (2), 253.

Use of the equation for calculating the pseudo-steady state flow (productivity) of a horizontal well, presented in this article, requires the determination of two factors. The first factor is the geometric factor which accounts for the effect of permeability anisotropy, location of the well, and relative dimensions of the drainage volume. The second factor is the skin due to restricted entry accounting for the effect of the well length. The equation given is applicable for any drainage volume dimensions, any permeability anisotropy, and any length and location of the wellbore. The major variable affecting horizontal well productivity is the length of the well and its penetration of the reservoir.

35. Babu, D. K. (1988). Productivity of a horizontal well: Appendices A and B. SPE paper 18334. Unsolicited manuscript received for review by SPE Editorial Review Committee.

This paper is a supplement to "Productivity of a horizontal well" (see preceding entry) by D. K. Babu. The paper contains two appendices giving the derivations of the equations given in the original paper.

36. Babu, D. Krishna and Aziz S. Odeh (1989). Flow capabilities of horizontal wells. Journal of Petroleum Technology 41 (9), 914-915. SPE paper 19117. Abstract in The Engineering Index Monthly (February, 1990), 28 (2), 634.

Flow characteristics of the horizontal well are affected by length of borehole in the reservoir, cross-sectional flow area, and vertical permeability. Vertical permeability, including that from vertical fractures, is critical in determining the amount of advantage a horizontal well may have over a vertical well.

37. Babu, D. Krishna and Aziz S. Odeh (1990). Author's reply to discussion of productivity of a horizontal well. SPE Reservoir Engineering 5 (2), 256.

Reply to Brigham, William E. (1990), SPE Reservoir Engineering 5 (2), 254-255.

38. Babu, D. K. and A. S. Odeh (1991). Author's reply to discussion of productivity of a horizontal well. SPE Reservoir Engineering 6 (1), 147-148. SPE paper 21623.

Author's reply to critical discussion of Babu and Odeh (1989), SPE Reservoir Engineering 4 (4), 417-421.

39. Babu, D. K. and A. S. Odeh (1991). Author's reply to further discussion of productivity of a horizontal well. SPE Reservoir Engineering 6 (1), 151-152. SPE paper 21624.

Reply to a critical discussion by Peaceman (1991), SPE Reservoir Engineering 6 (1), 149-150, of Babu and Odeh (1989), SPE Reservoir Engineering 4 (4), 417-421.

40. Badry, Rob (1991). Production logs optimize horizontal tests. World Oil 212 (3), 59, 62-66.

Pressure transient analysis and production logging may be done on horizontal wells to determine flow regimes and effective wellbore length as is done with vertical wells. This information will allow identification of the potentially most productive zone in the well. A discussion and the methods of analysis and a field example are given.

41. Bacon, Jeffrey R. and Robert H. Kempthorne (1984). Waterflood oil recovery in fractured reservoirs with directionally drilled wells. Society of Petroleum Engineers Journal 24 (4), 375-381. SPE paper 11024 presented at the 1982 SPE Annual Technical Conference and Exhibition, New Orleans, September 26-29.

Anisotropically fractured reservoirs have sharp discontinuities in porosity and permeability which significantly impact fluid flow and waterflood strategies when directionally drilled wells are used. When the fracture orientations and length are known and the surface locations of wells are restricted by terrain, careful plans are needed to optimize ultimate recovery from a field. Norman Wells field in the Northwest Territories, Canada, is in the McKenzie River delta with surface locations of wells restricted to islands and the river bank. Oriented cores showed the orientations of natural fractures. Directionally-drilled wells were used to take advantage of the fracture orientations when the waterflood was planned for the field.

42. Bailey, Trey (1991). Perforated pipe: Mobile liner perforating units. Popular Horizontal 1 (2), 32-33.

Kenny Morgan, a Luling , Texas, entrepreneur, has designed a mobile pipe/liner perforating unit. The unit has the capability to perforate three joints of 2 3/8-inch pipe at a time and can be modified to perforate 7-inch casing.

43. Bakke, Steiner (1986). Directional drilling: The state of the art. Petroleum Review 40 (470), 48-68.

Not examined.

44. Baldwin, D. D., R. W. Royal, and H. S. Gill (1984). Drilling high angle directional wells. Proceedings of the 11th World Petroleum Congress, v. 3, pp. 15-24. John Wiley & Sons Ltd., London. Paper PD5 (2).

Esso Resources Canada Ltd. successfully drilled horizontal wells in the Cold Lake and Norman Wells fields, Canada. The company plans to drill an additional 154 horizontal wells in the Norman Wells field. The Cold Lake wells were drilled as part of a steamflood pilot project in the heavy oil sands of Alberta. The Cold Lake wells were drilled on pads from which up to 32 horizontal wells may be drilled. Norman Wells field produces from a shallow, naturally fractured, low-permeability reservoir. The development of the field was complicated by being located under the MacKenzie River. Current plans are to construct artificial islands in the river and to drill extended-reach horizontal wells from the islands. Details of both projects are given in the article.

45. Bardin, C. A. (January 30, 1989). Remote-controlled bent sub aids directional drilling by allowing bend-angle change. Oil & Gas Journal 87 (5), 76, 78-80.

IFP (Institut Francais du Petrole) and SMFI (Societe Materiel de Forage International) have developed a bent sub that can have its angle changed by remote control from the surface during drilling. The system is compatible with MWD tools. The sub's angles can be changed from 0 degrees (straight) to 2, 2.5, or 3 degrees. The bent sub is actuated by increasing the mud flow above a predetermined threshold value. This tool has been successfully used in Rospo Mare field, Morcambe Bay field, North Alwyn field, Beryl field, and in offshore China and the United Arab Emirates.

46. Baret, Francois and T. J. Griffin (1989). Cementing horizontal wells. Drilling 50 (3), 16-17.

Because the purpose in drilling a horizontal well is usually to 'drill as long a hole in the pay zone as possible', casing and cementing is usually not desirable. Isolation of certain zones may become necessary under some circumstances, such as to prevent coning of gas or water, or when the hole deviates out of the target horizon into incompetent rock. Problems in cementing horizontal holes are that the casing needs centralization against the force of gravity, solids (cuttings and mud filtrate) collect at the bottom of the hole, and cement slurries behave differently than in vertical holes. One effect of gravity on cement slurries, in horizontal holes, is that particle sedimentation causes thin slurries at the top of the wellbore and thick slurries at the bottom. Proper selection of cementing materials and mixtures will allow good cementing in horizontal wellbore.

47. Barnes, Pamela J. (1990). The Prudhoe Bay well planning system, an application used to evaluate potential well locations. A.A.P.G. Bulletin 74 (5), 605.

Abstract. The Prudhoe Bay Well Planning System is a software application that uses a numerical description of the Ivishak reservoir linked to a three dimensional model to estimate the hydrocarbons in place and geology of any given proposed well location in the field.

48. Baron, G. and C. Wittrisch (December, 1982). Une nouvelle methode pour la realisation des diagraphies dans les puits horizontaux: <SIMPHOR>. [A new logging method for horizontal wells: SIMPHOR]. Petrole et Technique No. 294, 42-44. In French.

Article indexed but not annotated.

49. Barrett, S. L. and Raul Lyon (1988). Navigation drilling effective in horizontal wells in the Java sea. Drilling 49 (3), 8-17.

Of 18 horizontal wells drilled in Bima field, Java Sea, by Atlantic Richfield, five were drilled with the Eastman Christensen Co. steerable navigation drilling system. The Navigation Drilling System was used in wells that required a steeper angle and longer horizontal section in the reservoir. The system is composed of a tri-cone bit, a positive displacement low-speed mud motor, a double tilted U-joint sub, and an MWD system. The use of the Navigation Drilling System established that build rates as high as 6.75 degrees/100 ft. were achievable; continuous curves of irregular paths can be drilled without changes of bottom hole assemblies; sidetracks can be made on the low side without coming out of the hole; geological objectives can be reached with a high degree of accuracy; significant drilling cost reductions can be achieved using this system.

50. Barrett, S. L. and R. G. Lyon (1988). The navigation drilling system proves effective in drilling horizontal wells in the Java Sea. Proceedings IADC/SPE 1988 Drilling Conference, February 28-March 2, 1988, Dallas, pp. 503- 510. Society of Petroleum Engineers, Richardson, TX. IADC/SPE paper 17238.

Atlantic Richfield Indonesia Inc. has drilled eighteen horizontal or high-angle wells in the development of Bima field, Java Sea, Indonesia, since 1986. The wells were drilled using standard rotary drilling equipment, mud motors, bent subs, and the Navigation Drilling System of Eastman Christensen. The Navigation Drilling System is composed of a positive displacement mud motor (PDM), double tilted U-joint, and an MWD system. Wells were drilled with inclinations of 86 degrees. Horizontal sections as long as 2,351 ft have been drilled, with total horizontal distance for the eighteen wells of 24,909 ft. The Navigation Drilling System has been used to drill 7,133 ft of horizontal hole with targets as near as 839 horizontally from the surface location. The Navigation Drilling System provides flexibility to drill a continuous curve or irregular path without changing the bottom hole assembly (BHA). The system allows 'low-side' sidetracks of the wellbore without changing the BHA or cement plugs.

51. Barry, P. M., J. C. Troncoso, and A. N. E. Sumantri (1988). High angle development wells in the N.W. Java Sea Bima field: An update. Paper presented at the 7th Offshore South East Asia Conference, February 2-5, 1988, Singapore. OSEA paper 88130.

Not examined.

52. Baturin, A. I., V. I. Khalyavkin, and A. A. Musatkin (1979). [Experimental drilling of horizontal well 196 at the Novouzybashevskiy field of Bashkiriya]. Bashk. Gos. Nauchno-Issled. Proektn. Inst. Neft. Prom-sti., 34-40. In Russian.

Not examined.

53. Bayoud, B. B. and L. W. Staiger (1989). Downhole motors increase ROP and reduce cost per foot in the Austin Chalk trend. Proceedings 1989 SPE/IADC Drilling Conference, February 28-March 3, 1989, New Orleans, pp. 139-151. Society of Petroleum Engineers, Richardson, TX. SPE/IADC paper 18631.

Downhole motors used with PDC bits allowed one drilling contractor to achieve 50% to 100% higher penetration rates in the Midway shale and Austin Chalk formations in Texas. Engineering details are given in the article.

54. Becker, T. E., J. J. Azar, and S. S. Okrajni (1991). Correlations of mud rheological properties with cuttings-transport performance in directional wells. SPE Drilling Engineering 6 (1), 16-24. SPE paper 19535.

Experiments were conducted to examine rheological parameters of drilling mud. Parameters examined included mud yield point (YP), plastic viscosity (PV), YP/PV ratio, power-law exponent, consistency index, effective viscosity, initial and 10-minute gel strength, and Fann V- G Meter dial readings at six different speeds (revolutions/minute). Mud rheology as it affected cuttings transport performance was affected by annular shear rate, hole angle, initial gel strength, and mud shear stress. Details are given in the article.

55. Beckett, Tony (1989). Test off Philippines boosts horizontal drilling technology. Petroleum Engineering International 61 (11), 24-26, 28, 30, 32. Abstract in The Engineering Index Monthly (June, 1990), 28 (6), 671.

Alcorn International Inc. drilled the Galoc 1A horizontal well in the South China Sea by reentering the Galoc 1 well and drilling a 2,000 ft horizontal extension. This well is only the second horizontal well drilled in deep water (1,005 ft) and the third from a semisubmersible rig. Details of the drilling and testing are given in the article. The well tested 4,000 BOPD of oil instead of the expected 7,000 BOPD. This disappointing performance may have been caused by using an incorrect reservoir model. Recommendations based on the Galoc 1A well are: use adequate pump capacity so that ample fluid velocity can be maintained; careful selection of the drilling motor and assembly is needed to ensure rapid angle builds after sidetracking; all BHA components should be capable of trouble-free operation in horizontal drilling; drilling fluid programs should be designed to reduce torque and drag as well as being consistent with the target formation; and use the largest possible liner or casing at least to the top of the producing formation.

56. Belaskie, J. P., M. D. Dunn, and D. K. Choo (1990). Distinct applications of MWD weight-on-bit and torque. Proceedings 1990 IADC/SPE Drilling Conference, February 27-March 2, 1990, Houston, pp. 467-476. Society of Petroleum Engineers, Richardson, TX. IADC/SPE paper 19968.

Proper use of MWD weight-on-bit and torque data can improve drilling performance. The cases of slant-hole, horizontal, and high-angle wells are given in the article.

57. Bendakhlia, H. and K. Aziz (1989). Inflow performance relationships for solution-gas-drive horizontal wells. Proceedings 1989 SPE Annual Technical Conference & Exhibition, Reservoir Engineering volume, March 8-11, 1989, San Antonio, pp. 551-560. Society of Petroleum Engineers, Richardson, TX. SPE paper 19823. Abstract in The Engineering Index Monthly January, 1990), 28 (1), 721.

Commercial oil simulators were used to predict inflow performance relationships (IPR) of horizontal wells in solution-gas drive reservoirs. The developed curves show high initial oil productivity that declines rapidly. The high productivity is caused by the 'large contact area' of the horizontal borehole in the reservoir. The productivity of thc wcll will vary with variations in the pressure drawdown. The IPR curves are compared with the equations of Vogel and Fetkovich with which they generally agree. Details of the analysis and example IPR curves are given in the article.

58. Bernard, Francois (1985). Les Nouvelles Methodes de Recuperation: La Recuperation Amelioree; Le Fouragc Horizontal. [New recovery methods: Improved Enhanced Recovery; Horizontal Drilling]. Ann. Mines 192 (3-4), 71-83. In French.

Not examined.

59. Best, D. A., G. M. Cordell and J. A. Haston (1987). Underground test facility: Shaft/tunnel laboratory for horizontal drilling technology. Journal of Petroleum Technology 39 (5), 603-612. SPE paper 14333 presented at the SPE Annual Technical Conference and Exhibition, Las Vegas, September 22-25. For critical discussion and reply see: Doscher, T. M. (1988), Journal of Petroleum Technology 40 (2), 242; Best, D. A. (1988), Journal of Petroleum Technology 40 (2), 242. Abstract in The Engineering Index Annual (1987), 86, Part IV, 6486.

Horizontal drilling and horizontal drilling/shaft/tunnel systems for recovery of oil, gas, and tar or bitumen have been used in many areas of the world. The Alberta Oil Sands Technology and Research Authority (AOSTRA) has initiated the Underground Test Facility (UTF) project to experiment with horizontal drilling from a shaft/tunnel system located near Fort McMurray, Alberta, Canada. The project is designed to operate with horizontal drilling spatially near the target reservoir to minimize costs. The project will have 15 km (9 miles) of tunnel and 356 horizontal wells will be drilled over the life of the project.

60. Best, D. A. (1988). Author's reply to "Discussion of Underground Test Facility: Shaft/Tunnel Laboratory for horizontal well technology." Journal of Petroleum Technology 40 (2), 242. SPE paper 17469.

Reply to critical discussion by T. M. Doscher.

61. Best, D. A., R. P. Lesage, and J. E. Arthur (1990). Steam circulation in horizontal wellbores. Proceedings 1990 SPE/DOE Seventh Symposium on Enhanced Oil Recovery. EOR in the 1990's: Global Prospects and Perspectives, April 22-25, 1990, Tulsa, pp. 299-312. Society of Petroleum Engineers, Richardson, TX. SPE/DOE paper 20203.

A model developed to evaluate heat loss, steam quality, and pressure profiles during steam circulation has the ability to simulate countercurrent flow streams while accounting for heat transfer interactions, fluid flow, phase changes, slippage between steam and water, and flow regime changes. Details of the model are given in the article. The model was tested on a Mobil Canada horizontal steam injection well in the Cold Lake area. Field tests confirmed that the effects of steam injectivity were negligible in the early life of the projects, as predicted. The tests confirmed that the model accurately describes the heat transfer and fluid flow in the wells, and can be used to assess the effectiveness of the steam circulation for low injectivity horizontal wells.

62. Bezaire, G. E. and I. A. Markiw (1979). Esso Resources horizontal hole project at Cold Lake. Presented at the 30th Annual Technical Meeting of the Petroleum Society of CIM, May 8-11, 1979, Banff. CIM, Montreal. Petroleum Society of CIM paper 79-30-10. Abstract in The Journal of Canadian Petroleum Technology (1979), 18 (1), 12. Abstract in The Engineering Index Annual (1980) 79, Part III, 4482.

No data given in the JCPT abstract.

63. Bierschwable, H. and R. Ridley (May 2, 1988). Steerable system adds precision to Gulf drilling. Oil & Gas Journal 86 (18), 42-44, 47-48, 50.

Steerable directional drilling systems have made the drilling of replacement wells that intersect existing wells much easier. The steerable systems are being more frequently used offshore. Details of a Gulf of Mexico project are given in the article.

64. Blanco, Eduardo R. (December 31, 1990). Horizontal wells - Conclusion. Hydraulic fracturing requires extensive disciplinary interaction. Oil & Gas Journal 88 (53), 112-117.

Hydraulic fracturing of horizontal wells must be custom planned for each well. The geometry of the horizontal segment of the wellbore affects the fluid movement in the reservoir. Design of a hydraulic fracturing job should consider the following aspects: reservoir properties, rock mechanics, material transport, and completion and treatment methodology. Important reservoir properties are: permeability anisotropy, reservoir thickness, and proximity to gas and/or water contact(s). Important aspects of rock mechanics for hydraulic fracturing are: desired orientation of the fracture, aspect of the fracture, wellbore/fracture intersection, multiple fracture generation (fracture density), and presence of natural fractures. Materials transport must be carefully considered in horizontal wellbore segments so as to avoid bridging and screen-outs. This article concisely discusses these aspects of fracturing of horizontal wells and addresses such matters as fracturing fluids, fluid-loss additives, propping agents, and diverting agents.

65. Bleakley, W. B. (1980). Barber tackles heavy oil. Petroleum Engineer International 53 (13), 132-134.

Not examined.

66. Bleakley, W. B. (1983). IPF and Elf Aquitaine solve horizontal well logging problem. Petroleum Engineer International 55 (14), 22-24. Abstract in The Engineering Index Annual (1985), 84, Part IV, 6195.

Elf Aquitaine and IFP developed a system for logging horizontal drain holes. The system, called SIMPHOR (Systeme d'Instrumentation et du Mesures en Ouits Horizontaux), uses drill pipe to push logging tools to the bottom. Logs are run both going into the hole and pulling out of the hole. The system is composed of the tools in protective housings, a connector housing containing an eight-contact electrical plug, and a lateral-window sub which allows the cable to pass inside of the drill pipe. This system has been successfully used in three wells in Lacq field and one well in Rospo Mare field. Logging tools used were gamma ray, neutron, sonic, dual-induction, dual-laterolog, four-arm caliper, dipmeter and cement bond log. The system will also handle a 4-inch perforating gun.

67. Bleakley, W. B. (1985). British Gas meets Morecambe field challenges. Petroleum Engineer International 57 (10), 30, 32, 34, 36.

Local conditions (shallow reservoir, soft shallow sediments) required slant wells to develop the Morcambe field in the Irish Sea. The rig used to drill the slant wells was designed so that the drilling mast can be tilted 30 degrees from the vertical. The wells are drilled with a buildup rate of 3 degrees/100 ft to reach an inclination of 60 degrees permitting a 5,200 ft displacement at a depth of 3,000 ft. Holes are drilled with an MWD directional tool with gamma ray logging capabilities.

68. Bloy, Graeham and Richard Davis (1990). Horizontal hole drilling increases recovery and production in a pinnacle reef. A.A.P.G. Bulletin 74 (5), 613.

Abstract. A horizontal well was drilled in 1989 into the Keg River G pool of the Rainbow member limestone pinnacle reef. A horizontal well was used to try to prevent the solvent coning that had reduced productivity in vertical wells. The horizontal well was drilled 178 ft horizontally into the reef and completed open hole. The well is capable of producing 2,515 BOPD with one-fifth of the drawdown of a vertical well. The incremental oil recovery is estimated at 723,270 barrels of oil.

69. Bobigian, Richard A. and William S. McDonald (1991). The controversy: Where to turn horizontal in the Austin Chalk, Pearsall field, Texas. in Chuber, Stewart (editor), Abstracts and Short Papers of the South Texas Geological Society Austin Chalk Exploration Symposium, February 24-26, 1991, San Antonio, pp. 1-2. South Texas Geological Society, San Antonio.

Abstract. Horizontal wells in the Pearsall field must be planned to intersect natural fractures in the Austin Chalk. The conventional azimuth for the wellbore is N60W or its reciprocal S60E. The difficult parameter to choose for horizontal wells is the depth of the target zones in the Austin Chalk. The selection is usually made based on pressured rock properties and their relation to possible fracturing.

70. Bordelon, Sam (1991). Safety first: Hydrogen sulfide safety during horizontal drilling operations. Popular Horizontal 1 (1), 34, 36.

Brief discussion of hydrogen sulfide and the relevant Texas Railroad Commission rules for operators.

71. Borisov, J. P. ( ). [Oil production using horizontal and multiple-deviation wells]. Nedra. In Russian. Partial translation by J. Combe of IFP, August, 1980. Translated into English by J. Strauss, edited by S. D. Joshi, Phillips Petroleum Co., The R & D Library Translation, Bartlesville, Oklahoma (1984).

Not examined.

72. Borkowski, Rick, Linda Hand, Danny Dickerson, and Shane Bird (1990). Horizontal exploration of the Upper Cretaceous Austin Chalk of South Texas. A.A.P.G. Bulletin 74 (5), 615.

Abstract. Horizontal wells drilled in the Austin Chalk of the Pearsall field, South Texas, are designed to intersect several fracture swarms in different stratigraphic horizons. Interdisciplinary study of the Austin and Eagleford shale allow determination of the regional fracture system, preferential flow directions, and stress field.

73. Borkowski, Rick (1991). Horizontal exploration of the Upper Cretaceous Austin Chalk of South Texas. A.A.P.G. Bulletin 75 (1), 198.

Abstract. Production of oil from the Austin Chalk in South Texas is controlled by regional fractures. Wellbores must be oriented to intersect the fracture 'swarms' which form the anisotropic and laterally partitioned reservoirs.

74. Borner, D. L., H. H. Haldorsen, P. F. Harrison, A. P. Hopwood, and C. J. MacDonald (1988). Performance of vertical and horizontal wells in reservoirs containing discontinuous shales: Numerical studies of two North Sea fields. Proceedings International Meeting on Petroleum Engineering, November 1-4, 1988, Tianjin, China, pp. 505- 513. Society of Petroleum Engineers, Richardson, TX. SPE paper 17600.

Two unidentified named North Sea fields are used to study the effects of discontinuous shales on the performance of vertical and horizontal wells. Discontinuous shales adversely affect the vertical permeability of a reservoir. The two fields studied are both in Paleocene submarine fans. The oil column in the first is 63 meters thick and in the second 30 meters thick. Field No. 1 has a gas cap in the central part and is underlain by an active aquifer. Field No. 2 has no gas cap and does have an active aquifer below the oil column. Both reservoirs have horizontal permeabilities in the oil zone of 200 md. Coning models were used to predict the coning of wells in various parts of the fields. Shales were added to the models to study their effect on coning and productivity. Conclusions reached in this study include: (1) vertical permeability from core plugs should not be used in coning studies; (2) explicit stochastic shales should be used for vertical wells; (3) foremost purposes only shales cutting the wellbore need be used; (4) when many small shales are present horizontal wells have a productivity advantage but do not improve water cut. Details of the modeling are given in the article.

75. Bosio, J. (December 20, 1984). Horizontal drilling: A fourth well successfully completed by Elf Aquitaine: What is the next step? Petrole Informations International No. 1606, 4-5, 7.

Brief discussion of the four successful horizontal wells drilled by Elf Aquitaine (Lacq field; Rospo Mare field; Castera Lou field). Directional services for these wells was provided by Top Services, a French company.

76. Bosio, Jacques (1986). Horizontal wells are now used for industrial developments. Proceedings Sixth SPE Offshore South East Asia Conference & Exhibition, January 29-31, 1986, Singapore, pp. 77-82. Society of Petroleum Engineers, Richardson, TX. SPE paper 14614.

Elf Aquitaine and IFP (French Institute of Petroleum) initiated, in 1978, a project to solve the technical problems of drilling and completion of horizontal wells. The first wells were drilled in Lacq field, France, and Rospo Mare field, Italy. Details of the successful projects are given in this article.

77. Bosio, J. C., R. W. Fincher, J. F. Giannesini, and J. L. Hatten (1987). Horizontal drilling - A new production method. Proceedings of the Twelfth World Petroleum Congress, Houston, v. 3, pp. 187-199. John Wiley & Sons Ltd., London.

Horizontal wells can be kept in a reservoir for longer intervals than vertical wells. The technology for drilling short-radius, medium-radius, and long-radius horizontal wells is discussed in the article. Also covered are the appropriate types of targets for horizontal drilling and such problems as gas and water coning which may be alleviated by horizontal wells. Several horizontal drilling projects are illustrated with a very effective chart. Editor's Note: This article is one of the best introductions to horizontal drilling covered in this bibliography.

78. Bosio, J. and L. Reis (March 21, 1988). Horizontal well operations - 2. Site selection remains key to success in horizontal well operations. Oil & Gas Journal 86 (12), 71-74, 76. Abstract in The Engineering Index Monthly (1988), 87, Part III, 6000.

In the 1950's, 43 horizontal wells were drilled in the U.S.S.R. on an experimental basis. These wells were determined to be non-economic. Two horizontal wells were drilled in China in the 1960's, again with poor results. In the late 1970's, Elf Aquitaine and other companies started the current successful horizontal drilling activity. Horizontal wells have been drilled in the tar sands at Cold Lake, Canada, Norman Wells, Canada, in the Netherlands, France, Italy, Indonesia, and other countries. Horizontal drilling has proven to be economical. The cost of horizontal wells usually is about twice that of a vertical well although wells at Rospo Mare field are only 30% more than a vertical well and at Prudhoe Bay costs on one well were only 10% higher than a vertical well. Horizontal wells, when properly utilized, can produce more oil at lower cost than vertical wells.

79. Bosio, Jaques (1988). Horizontal drilling, a new production method. The APEA Journal 28 (1), 41-46.

Horizontal wells have the capability of remaining in a reservoir for several hundred meters to enhance productivity. The three types of horizontal drilling, short-, medium-, and long-radius drilling, are described along with the applications.

80. Boulet, J. G. and P. E. Morin (1986). The remote-controlled, multi-angle bent sub for drilling- trajectory control and corrections. SPE Drilling Engineering 1 (5), 390-394. SPE paper 11384 presented at the 1983 IADC/SPE Drilling Conference, February 20-23, 1983, New Orleans.

TELEPILOTE is a downhole, adjustable, multi-angle bent sub developed for remote control of the borehole path in real-time without changing the BHA. The angle of the sub can be changed in six increments from 0 degrees to 2, 2.5, or 3 degrees depending on size. The sub is available in 6.75-, 7.75-, 8-, and 9.5-inches nominal sizes.

TELEPILOTE can be directly connected to any kind of survey instrument.

81. Bourke, Lawrence, Pierre Delfiner, Jaen-Claude Trouller, Tom Fett, Mike Grace, Stefan Luthi, Oberto Serra, and Eric Standen (1989). Using Formation MicroScanner images. The Technical Review 37 (1), 16-40.

Review of the interpretation of fractures, sedimentary structures and other features in boreholes using the Formation MicroScanner (Schlumberger trademark) log.

82. Branagan, P. T., C. L. Cipolla, S. J. Lee, and R. H. Wilmer (1985). Comprehensive well testing and modeling of pre- and post-fracture well performance of the MWX lenticular tight gas sands. Proceedings of the 1985 SPE/DOE Symposium on Low-Permeability Reservoirs, May 19-

22, 1985, Denver, pp. 191-201. Society of Petroleum Engineers, Richardson, TX. SPE paper 13867.

The U.S. Department of Energy Multiwell Experiment Site (MWX) drilled and tested a well in Rulison field, Colorado. The MWX well was completed in two tight, lenticular, naturally-fractured sands in the Mesaverde Group. Both sands were hydraulically fractured. The post-fracture production/interference testing of the reservoirs showed a decline in productivity from pre-fracture testing. The hydraulic fracturing degraded the ability of the natural fracture system to flow natural gas. The experiment also showed that the presence of highly conductive natural fractures may not be detected in using standard analytical methods.

83. Brannin, C. S., L. Velser, and M. P. Williams (1990). Drilling a record horizontal well: A case history. Proceedings 1990 IADC/SPE Drilling Conference, February 27 - March 2, 1990, Houston, pp. 627-636. Society of Petroleum Engineers, Richardson, TX. IADC/SPE paper 19986. Abstract in The Engineering Index Monthly (May, 1990), 28 (5), 559.

The first offshore horizontal well in Denmark was drilled in 1989 by Maersk Oil & Gas as operator for the Danish Underground Consortium. The well was drilled in the Tyra field, North Sea, and had a measured depth of 13,660 ft with a 3,276 ft horizontal interval. The well was logged with drillpipe-conveyed logging assembly composed of DIL, LSS, LDL, CNL, GR, and CAL tools. Completion was in the Danian chalk. Details of the planning, drilling, and completion of the well are given in the article.

84. Brewer, Herbert L. (January 7, 1985). Improved economics prompt another look at France's Paris basin. Oil & Gas Journal 83 (1), 105-107.

The petroleum laws and regulations of France are among the most reasonable of any country in the world. Exploration permits are of five year terms with provision of two renewals of five years each. There are no rental payments. Each permit has a specified work obligation and non-compliance will cancel the permit. Exploitation permits for fields of less than 2.1 million barrels of oil are for five years with two five year renewals allowed. There are no royalties or rentals. Exploitation concessions for fields of greater that 2.1 million barrels of oil have up to a fifty year term with no rentals with a graduated royalty ranging from 0% to 12% for oil dependent on the daily production rate. Royalty for gas production over 30 MMcfd is 5%. The Paris basin is a relatively unexplored basin of 51,000 square miles with a sedimentary section of up to 11,000 feet thick. Several fields produce oil and gas at present. Editor's Note: The Paris basin now has several extended-reach or horizontal wells which were drilled to reach reservoirs under towns and villages, and are used to minimize the amount of farm land required for drilling locations.

85. Briggs, Gary M. (1989). How to design a medium-radius horizontal well. Petroleum Engineer International 61 (9), 26, 30-32, 36-37. Abstract in The Engineering Index Monthly (June, 1990), 28 (6), 671.

Medium-radius horizontal wells require careful planning in which build rate, sidetrack methods, and motor configurations are major considerations. These issues are addressed in this article.

86. Brigham, William E. (1990). Discussion of productivity of horizontal wells. SPE Reservoir Engineering 5 (2), 254-255.

Critical discussion of Babu and Odeh (November, 1989), SPE Reservoir Engineering 5 (2), 417-421.

87. Broman, W. H., T. O. Stagg, J. J. Rosenzweig (1990). Horizontal well performance evaluation at Prudhoe Bay. Preprints CIM/SPE International Technical Meeting, June 10-13, 1990, Calgary,

Canada, v. III, Paper 124. Petroleum Society of CIM, Montreal. CIM/SPE paper 90-124. Abstract in The Journal of Canadian Petroleum Technology (1990), 29 (2), 50.

Horizontal wells have been used at Prudhoe Bay field to reduce gas coning and to produce oil from very thin oil reservoirs. Twenty-seven horizontal production wells and two horizontal gas injection wells have been drilled to date in the field. Horizontal wells have been completed at angles of up to 98 degrees with horizontal displacements of 2100 ft. Details of the project are given in the article.

88. Brown, C. A., J. W. Crafton and J. G. Golson (1982). The Niobrara gas play: Exploration and development of a low-pressure, low-permeability gas reservoir. Journal of Petroleum Technology 34 (12), 2863-2870. SPE paper 10304 presented at the 1981 SPE Annual Technical Conference and Exhibition, San Antonio, October 5-7.

The Niobrara formation chalks produce gas from more than 60 fields in Colorado, Kansas, and Nebraska. The Niobrara is a low-permeability, low-pressure reservoir rock which must be fractured to produce gas in commercial amounts. The Niobrara is its own source rock with the gas being biogenic. The individual traps are highly faulted, low-relief structures. Editor's Note: The Niobrara is now a target for horizontal drilling.

89. Brown, David (1990). Horizontal prospects wait in line. A.A.P.G. Explorer 11 (11), 8, 10.

The boom in horizontal drilling caused by the Austin Chalk play has revealed several problems for the industry: a shortage of MWD tools; lack of experienced personnel; need for new completion methods; need for better reservoir analysis; need for better drilling-at-depth technology. The current increase in demand for horizontal drilling equipment may exceed the supply in the near term.

90. Brown, David (1991). Horizontal drilling reaches out. A.A.P.G. Explorer 12 (5), 5.

Report of interviews with the staff of Joshi Technologies, a horizontal drilling consulting firm.

91. Brown, N. P. and P. A.Bern (1989). Cleaning deviated holes: New experimental and theoretical studies. Proceedings 1989 SPE/IADC Drilling Conference, February 28-March 2, 1989, New Orleans, pp. 171-180. Society of Petroleum Engineers, Richardson, TX. SPE/IADC paper 18636.

Removal of cuttings from deviated boreholes is necessary to prevent high torque and drag during drilling, prevent stuck pipe, improve casing and cementing operations, and to keep the hole in good condition. The experimental work reported in this article was designed to improve cuttings transport and removal. A Cuttings Transport Flow Loop facility was built to simulate hole cleaning on a field scale. Recommendations are made for efficient hole cleaning.

92. Bruckert, Louis (December 18, 1989). Horizontal well improves oil recovery from polymer flood. Oil & Gas Journal 87 (5), 35-39.

Horizontal wells in the Chateaurenard field in the Paris basin, France, have demonstrated that the technique can be used successfully in thin unconsolidated sand. The target sand was 7 meters thick at a depth of 580 meters and located in a polymer-flood project. The horizontal well was efficient in enhancing oil recovery because of its ability to overcome permeability anisotropy. Production was increased 7-fold over that of vertical wells.

93. Burton, Bill (1988). The market impact of steerable motor systems in the Gulf of Mexico. Drilling 49 (3), 20-24.

Downhole steerable motors were introduced into the Gulf of Mexico in 1985, where they improved drilling efficiency in directional and horizontal drilling. The article gives a brief history of steerable systems and downhole motors.

94. Butler, R. M. and D. J. Stephens (1980). Gravity drainage of steam-heated heavy oil to parallel horizontal wells. Preprints, 31st Annual Technical Meeting of the Petroleum Society of CIM, May 25-28, 1980, Calgary, Alberta. Petroleum Society of CIM, Montreal. CIM paper No. 80-31-31. Abstract in The Engineering Index Annual (1982) 81, Part IV, 5256. Also in Journal of Canadian Petroleum Technology, (April-June, 1981), pp. 90-96.

Equations are given for gravity drainage of steam-heated heavy oil reservoirs with oil drainage into several parallel horizontal drainholes.

95. Butler, R. M., G. S. McNab, and H. Y. Lo (1981). Theoretical studies on the gravity drainage of heavy oil during in-situ steam heating. Canadian Journal of Chemical Engineering 59 (4), 455-460. Abstract in The Engineering Index Annual (1982) 81, Part IV, 5199.

Theoretical treatment of in-situ recovery of heavy-oil/bitumen using steam-assisted gravity drainage.

96. Butler, R. M. (1984). The potential of horizontal wells - Some introductory comments. Fifth Annual AOSTRA Advances in Petroleum Recovery & Upgrading Technology Conference, June 14-15, 1984, Calgary.

A review of horizontal drilling activity and its applicability to Canadian projects. Projects are indexed.

97. Butler, R. M. (1985). A new approach to modeling of steam- assisted gravity drainage. Journal of Canadian Petroleum Technology 24 (3), 42-51.

A new method is presented to predict the progress of steam-assisted gravity drainage and drainage rates in systems using horizontal drains. The spacing of horizontal wells with respect to steam injection wells is addressed.

98. Butler, R. M. and G. Petela (1987). Theoretical estimation of breakthrough time during a steamflood between long parallel horizontal wells. University of Calgary, Calgary. Internal Report for Saskatchewan Energy and Mines, Contract No. 7348-1.

Not examined.

99. Butler, R. M. (1989). The Application of Horizontal Wells to Petroleum Production - A Short Course. University of Calgary, Calgary.

Not examined.

100. Butler, Roger M. (1989). The potential for horizontal wells for petroleum production. Journal of Canadian Petroleum Technology 28 (3), 39-47. CIM paper 88-39-22 presented at the 39th Annual Technical Meeting, Petroleum society of CIM, June 12-16, 1988.

Horizontal wells may be used to advantage in situations where conventional wells have low productivity, in reservoirs with vertical fractures, in reservoirs where water or gas coning is a problem, and in heavy oil reservoirs using steam flooding and gravity drainage. Horizontal wells are now suitable in thick reservoirs with high horizontal and low vertical permeability.

101. Butler, R. M. and I. J. Mokrhys (1990). A new process (VAPEX) for recovering heavy oils using hot water and hydrocarbon vapor. Paper presented at the International Technical Meeting. The Petroleum Society of CIM and the Society of Petroleum Engineers, June 10-13, 1990, Calgary. The Petroleum Society of CIM, Montreal. CIM/SPE paper 90-133. Abstract in The Journal of Canadian Petroleum Technology (1990), 29 (2), 51.

Heavy oil may be produced through horizontal drainholes by injecting a hydrocarbon vapor into a chamber above the production well. The vapor will dissolve the heavy oil allowing it to drain into the production well. At the top of the reservoir, hot water and a solvent should be injected in a horizontal wellbore to raise the reservoir temperature and mobilize the bitumen.

102. Buza, J. W. and A. Unneberg (July 21, 1986). Statfjord field - 1. Statfjord geology, reservoir performance. Oil & Gas Journal 84 (29), 68-72. Abstract in The Engineering Index Annual (1986), 85, Part IV, 6499.

Statfjord field is the largest field in Europe with recoverable reserves of 3 billion barrel of oil and 3 trillion cubic-feet of gas. The field is in the Norwegian sector of the North Sea, East Shetland basin. Production is from two reservoirs: Statfjord formation and Brent formation. Reservoir pressures are maintained in the Brent reservoir by water injection and in the Statfjord reservoir by gas injection. Both reservoirs are in sandstones with porosities of 20-36%. A history of exploration in the area is given along with a summary of the geology of the field.

103. Buza, John W. and Arild Unneberg (July 28, 1986). Statfjord field - 2. Statfjord field development strategies. Oil & Gas Journal 84 (30), 110-114.

Discussion of the development strategy of the Statfjord field, the largest oil field in Europe. Reservoir characteristics of the Brent reservoir are described.

104. Byrd, Bob and Mario Zamora (1988). Fluids are key in drilling highly deviated wells. Petroleum Engineer International 60 (2), 24-26.

Proper drilling fluid selection in high-angle wells can reduce the problems of poor hole cleaning, excessive torque and drag, well bore instability, etc. Although no one fluid will meet all needs, invert-emulsion oil-base mud comes closest to meeting requirements. High-angle holes have been successfully drilled with gel-water, polymer, saturated-salt, and sized-salt water-base fluids.

105. Camilleri, D. and S. A. Mobarak (1987). Application of a numerical simulator to investigate the use of horizontal wells in steamfloods. Paper presented at the II Simposio Internacional Sobre Recuperacion Mejorada De Crudo, February 24-27, 1987, Maracaibo, Venezuela.

Not examined.

106. Campbell, Michael J. (1991). Sleeping giant: Union Pacific Resources horizontal activity awakens sleeping giant in the Rockies. Popular Horizontal 1 (1), 16-17.

Recent successful horizontal wells completed in the Niobrara chalk of the Silo field, Wyoming, may have 'awakened a sleeping giant', the Niobrara play. Union Pacific Resources Company has completed two horizontal wells in Silo field with flow rates of 1,525 BOPD and 767 BOPD. A third well, drilled by another operator and now on pump, produces 400 BOPD and 300 CFGPD. The Niobrara production is from naturally fractured reservoirs. A potential Niobrara fairway is proposed for northeastern Colorado and southern Wyoming. The Niobrara is compared to the Austin Chalk of Texas.

107. Campbell, Michael J. and Roc E. Saint (1991). New understanding of Niobrara reservoir characteristics based on horizontal drilling in Silo field, southeastern Wyoming. A.A.P.G. Bulletin 75 (6), 1123.

Abstract. Logging of horizontal wells in Silo field reveals that fractures in the Niobrara reservoir parallel and trend northwest. Fracture apertures are 1- 3mm. The Niobrara is an open-marine limestone deposited in deep water.

108. Canter, D. B. and J.-M. Simon (1991). A decade of drilling innovation and progress in the Paris basin. SPE Drilling Engineering 6 (1), 65-70. SPE paper 19573. This paper was originally presented at the 1989 SPE Annual Technical Conference & Exhibition, October 8-11, 1989, San Antonio, Texas.

Over the ten-year period from 1979 to 1989, Esso Rep has drilled more than 150 exploration and development wells in the Paris basin. Most of the development wells are directional (high-angle) wells drilled from multiwell surface clusters for environmental and economic reasons. Drilling time per well has decreased from 52 days to 20 days over the course of the decade. Details of the drilling methods, formation evaluation, drilling mud, drilling contracting strategy, and other factors are discussed in the article.

109. Carden, Richard S. (December 26, 1988). Equations determine maximum horizontal wellbore length. Oil & Gas Journal 86 (52), 121-125. Abstract in The Engineering Monthly Index (May, 1989), 27 (5), 433.

A simple program for Hewlett Packard HP-41C calculators is given to calculate drag and torque in horizontal well drilling. Down drag is the principal limiting factor in the length of a horizontal well. In a clean hole, torque should not be a problem.

110. Carden, Richard S. (April 8, 1991). Air drilling has some pluses for horizontal wells. Oil & Gas Journal 89 (14), 76-78. Based on a paper presented at the 14th Annual ASME Energy-Sources Technology Conference & Exhibition, January 20-23, 1991, Houston, Texas.

Air-drilling has several advantages in drilling horizontal wells. These advantages are: increased rate- of-penetration, elimination of lost circulation problems, continual drill stem test, and minimization of formation damage. Disadvantages of air-drilling include: shorter downhole motor life; MWD will not work, hole cleaning is a problem in high-angle wells; increased friction reduces the possible length of horizontal boreholes; and targets are limited to certain lithologies.

111. Cardwell, W. T. and R. L. Parsons (1949). Gravity drainage theory. Transactions of AIME - Petroleum Branch 179, 199-215.

Theoretical discussion of the gravity drainage of liquids from a sand column accounting for the permeability variation as the upper part of the column becomes unsaturated.

112. Carlisle, Joe (1990). The Bakken formation of the Williston basin: Deposition, maturation, and fracturing. in Hansen, William B. (editor), Bakken Formation Short Course, p. 16. Montana Geological Society, Billings. Study Notes No. 1.

Abstract. The Bakken formation of the Williston basin is a marine sequence divided into three parts: two organic rich shales separated by a silty-sandy middle member. Hydrocarbon generation began in the early Cretaceous. Overpressuring was caused by the maturation process and, in turn, fractured the Bakken. Tectonic fracturing occurred during the Laramide orogeny.

113. Carlisle, W. J. (1990). The Bakken formation of the Williston basin: Deposition, maturation, and fracturing. A.A.P.G. Bulletin 74 (8), 1318. Abstract also printed in Hansen, William B (editor) (1991), 1991 Guidebook to Geology and Horizontal Drilling of the Bakken Formation, p. 89, Montana Geological Society, Billings, Montana. The paper was presented at the Montana Geological Society Bakken Workshop, Billings, March 27, 1990.

Abstract. The Mississippian-Devonian Bakken formation of the Williston basin is one of the richest hydrocarbon source rocks in the world. The Bakken is a marine clastic sequence composed of a deep-marine shale deposited under anoxic conditions, and a shallow-marine deposit.

Hydrocarbon generation began in the Cretaceous and caused overpressuring which fractured the rocks. The Laramide orogeny caused tectonic fracturing.

114. Carlson, Clarence G. (1990). The horizontal Bakken play in North Dakota. A.A.P.G. Bulletin 74 (8), 1318.

Abstract. The Meridian Oil-MOI Elkhorn Ranch #33-11 well started the horizontal drilling play in the Bakken formation in North Dakota. The medium-radius horizontal well penetrated 1,288 ft of the Bakken and has produced over 216,123 barrels of oil. Thirty-two additional Bakken wells had initial potentials of 11 to 439 BOPD with an average of 198 BOPD. Horizontal wellbore penetrate from 187 to 3,544 ft into the Bakken.

115. Carlson, Eric S. and James C. Mercer (1991). Devonian shale gas production: Mechanisms and simple models. Journal of Petroleum Technology 43 (4), 476-482.

Devonian shales of the eastern United States are special cases of dual-porosity gas reservoirs. Devonian shales are producible only where extensive networks of natural fractures exist. Unlike other dual-porosity gas reservoirs, the Devonian shales have very little matrix porosity in which to store gas. The shales store enormous amounts of gas in a sorbed state. The Devonian shale matrix behaves as a semi-infinite medium, i.e. portions of the matrix remain at the original reservoir pressure during production.

116. Carlson, N. R. and M. J. Davarzani (1990). Profiling horizontal oil-water production. Proceedings 1990 SPE Annual Technical Conference & Exhibition, Formation Evaluation and Reservoir Geology volume, September 23-26, 1990, New Orleans, pp. 653-663. Society of Petroleum Engineers, Richardson, TX. SPE paper 20591. Also in Journal of Petroleum Technology (1991) 43 (7), 780-785.

Report of experiments conducted to evaluate the effectiveness of production logs in determining total flow and water cut in horizontal wellbores. The fluids used in the experiment were kerosene and water. In segregated flow, a centralized continuous spinner flowmeter was found to be ineffective in measuring total flow. A diverting flowmeter was better able to measure both mixed flow and segregated flow. The unfocused density tool was found to respond to the full range of water cuts in both mixed and segregated flow. Fluid capacitance and focused density tools were unresponsive at water cuts above 40%.

117. Carlson, Roy F. (September 1, 1952). Special directional drilling in west Texas. Oil & Gas Journal 51 (34), 99-101.

Humble Oil & Refining Co. drilled a directional well to intersect vertical fractures in the Spraberry in the Pembrook area. The bottomhole location was offset 1,200 ft with an average hole angle of 45 degrees below the kickoff point. An oriented core was taken to determine the orientation of the natural fractures in the Spraberry. The well was deviated on an azimuth determined to intersect the maximum number of fractures (N15W). The equipment and drilling methods are described in the article. Editor's Note: This well was not horizontal but is an early illustration of the thinking that has led to the present practice of drilling horizontal wells in naturally fractured reservoirs.

118. Carnes, P. S., Jr. (1966). Effects of natural fractures on directional permeability in flooding. Paper presented at the 1966 SPE Secondary Recovery Symposium, May 2-3, Wichita Falls, Texas. Society of Petroleum Engineers, Richardson, TX. SPE paper 1423.

Directional movement of injected water in waterflood projects may be caused by natural fractures or permeability anisotropy in the reservoir. Injection well patterns and rate of injection must be designed for the reservoir to obtain maximum recovery of oil. Details of a Calvin sand waterflood project in Oklahoma are given in the paper.

119. Carrigy, Maurice A. (1988). Thermal recovery from tar sands. Proceedings 1988 California Regional Meeting, March 23-25, 1988, Long Beach, California, pp. 25-34. Society of Petroleum Engineers, Richardson, TX. SPE paper 12556.

Review of the recovery of bitumen from tar sands including the horizontal drilling at Cold Lake, Alberta, Canada, by Esso Resources Canada Ltd. and Athabasca by Texaco. Topics include steamflood, fireflood, electrical preheating, and oxygen injection. Most of the major projects are mentioned.

120. Carvalho, R. S. (1988). Compartamento de Presao em Pocas Horizontais. M.S. Thesis. Universidade Federal de Ouro Preto, Brazil. In Portuguese.

Not examined.

121. Carvalho, R. de S. and A. J. Rosa (1988). Transient pressure behavior for horizontal wells in naturally fractured reservoir. Proceedings 1988 SPE Annual Technical Conference & Exhibition, Reservoir Engineering volume, October 2-5, 1988, Houston, pp. 399-410. Society of Petroleum Engineers, Richardson, TX. SPE paper 18302.

A mathematical model is presented to describe the pressure behavior of horizontal wells in double-porosity reservoirs (naturally fractured reservoirs). The model shows the three typical flow regimes of heterogeneous systems. Conventional and/or type-curve matching techniques can be applied to estimate well and reservoir parameters in cases where wellbore storage effects are not dominant.

122. Cawvey, J. L. (July/August, 1981). ARCO's Empire Abo Unit drainhole program advances. Well Servicing 21 (4), 56, 58.

Arco Oil & Gas Co. drilled three horizontal drainholes in the Empire Abo Unit, New Mexico, in the period 1979-1981. The wells were drilled to combat serious gas coning problems. Two of the three wells were drilled using a well servicing rig which was cost effective even though more time was required for the drilling. The Empire Abo Unit produces from the Permian Abo Reef dolomite.

123. Celier. G. C. M. R., P. Jouault, and O. A. M. C. de Mortigny (1989). Zuidwal: A gas field development with horizontal wells. Proceedings 1989 SPE Annual Technical Conference & Exhibition, Reservoir Engineering volume, March 8-11, 1989, San Antonio, pp. 591-606. Society of Petroleum Engineers, Richardson, TX. SPE paper 19826.

Horizontal wells were chosen over vertical wells to develop the Zuidwal gas field because horizontal wells will produce twice as much gas as deviated wells at only 20% greater cost. The difference in production is accounted for by the lower pressure drop in the vicinity of the horizontal wellbore. Three horizontal wells have been drilled instead of six deviated conventional wells. The horizontal wells are drilled with two horizontal sections 100 to 200 meters in length in the two separate reservoirs in the field. The well profiles chosen will reduce water coning. Zuidwal field is a structural trap situated at 1900 meters depth.

124. Chambers, Mike and Jim Hanson (1982). Australian well breaks horizontal displacement record. World Oil 194 (4), 123-124, 126. Abstract in The Engineering Index Annual (1982) 81, Part IV, 5226.

Esso Australia Mackerel Well A-16 set a horizontal displacement record by reaching 15,085 ft horizontally with a TVD of 7,974 ft The well had a measured depth of 18,153 with an angle of 63.75 degrees and a heading of S 37 W. Operations and engineering details included in the article.

125. Chang, H. L., S. M. Farouq Ali, and A. E. George (1990). Performance of horizontal-vertical well combinations for steamflood bottom water formations. in Preprints CIM/SPE International Technical Meeting, June 10-13, 1990, Calgary, Canada. Petroleum Society of CIM, Montreal. CIM/SPE paper 90-86. Abstract in The Journal of Canadian Petroleum Technology (1990) 29 (2), 42-43.

Abstract. Scaled model studies of steamfloods were conducted using a heavy oil deposit in Saskatchewan as the prototype. Various combinations of vertical and horizontal wells were modeled. The case of the horizontal injection well-horizontal producing well performed best with oil recoveries as high as 60%.

126. Chang, M. M. (1988). Simulation of production from wells with horizontal/slanted laterals. Department of Energy Bartlesville Project Office, Bartlesville, Oklahoma. Department of Energy Report No. NIPER-326.

Not examined.

127. Chang, Ming-Ming and Bill Linville (March 27, 1989). Simulator predicts production from nonvertical wells. Oil & Gas Journal 87 (13) 53-55, 58-59.

The National Institute for Petroleum Energy Research has developed BOAST, a 'black oil' simulator and model for predicting the production of nonvertical wells. The three-dimensional, three-phase (gas, oil, water) model predicts the Productivity Index for each reservoir grid block penetrated by the wellbore. The model and examples are discussed in the article.

128. Chaperon, I. (1986). Theoretical study of coning toward horizontal and vertical wells in anisotropic formations: Subcritical and critical rates. Paper presented at the 1986 SPE Annual Technical Conference & Exhibition, October 5-8, 1986, New Orleans. Society of Petroleum Engineers, Richardson, TX. SPE paper 15377.

A simple equation is given for estimating the critical production rate of a horizontal well in an anisotropic reservoir. This equation and its consequences are discussed in the article.

129. Chen, H. Y., S. W. Poston, and C. H. Wu (1986). Characterization of the Austin Chalk producing trend. Paper presented at the 1986 SPE Annual Technical Conference & Exhibition, October 5-8, 1986, New Orleans. Society of Petroleum Engineers, Richardson, TX. SPE paper 15533.

The Austin Chalk of the Giddings field, Texas, is a low-permeability, low porosity chalk that produces oil from natural fractures. This study used production data from 1,235 wells in the Giddings field to study well performance. Based on decline curve analysis two types of wells were observed. Type 1 wells have an initial rapid decline followed by a gradual change to a slow decline exhibited by a hyperbolic decline curve or double-exponential decline. Type 2 wells have single exponential decline. 56% of the wells studied are Type 1 wells. Of the 1,235 wells studied only 63 were non- productive. Ultimate recovery of oil is dependent on the extensiveness of the fracture systems encountered by the wellbore. Most of the oil production comes from fracture porosity with a small component from the matrix porosity. Details of the decline curve analysis are given in the article. Editor's Note: This paper suggests that gravity drainage may have occurred in some parts of the Pearsall field to explain the lack of updip production where fractures are present. This phenomenon is attributed to areas of more intense fracturing.

130. Chen, S. M. and J. Olynyk (1985). Sweep efficiency improvements using horizontal wells or tilted horizontal wells in miscible floods. Journal of Canadian Petroleum Technology 24 (2), 24. CIM paper 85-36-62. Paper presented at the Joint Annual Meeting of The Petroleum Society of CIM and the Canadian Society of Petroleum Geologists, June 2-5, 1985, Edmonton.

Abstract. Horizontal or tilted horizontal wells may improve sweep efficiency of miscible floods and oil recovery. Results of simulation studies reported.

131. Chen, T. (1988). Pressure buildup analysis for wells influenced by fractures and other heterogeneities. Journal of Canadian Petroleum Technology 27 (2), 61-71.

A modified Horner analysis is proposed for analyzing pressure buildup data in naturally fractured and heterogeneous reservoirs.

132. Chen, Yu-Che, Yu-Hsu Lin, and John B. Cheatham (1990). Tubing and casing buckling in horizontal wells. Journal of Petroleum Technology 42 (2), 140-141, 191. SPE paper 19176 presented at the 1989 Offshore Technology Conference, Houston, May 1-4, 1989. Abstract in The Engineering Index Monthly (June, 1990), 28 (6), 667.

Pipe subjected to axial loading in a horizontal borehole is subject to both sinusoidal and helical buckling. This behavior of the pipe can be predicted and can be calculated using equations given in the article.

133. Chen, Yu-Che, Yu-Hsu Lin and John B. Cheatham (1990). Author's reply to Discussion of tubing and casing buckling in horizontal wells. Journal of Petroleum Technology 42 (8), 1063. See Wu, Jiang and Hans C. Juvkam-Wold (1990), Journal of Petroleum Technology 42 (8), 1062-1063, for critical discussion.

Reply to discussion of original paper (Journal of Petroleum Technology (1990), 42 (2), 140-141, 191).

134. Cheng, A. M. (1990). Development of an inflow performance relationship (IPR) for a slanted/horizontal well. U.S. Department of Energy, Bartlesville, Oklahoma. D.O.E. Report NIPER-458.

Not examined.

135. Cheng, A. M. (1990). Inflow performance relationships for solution-gas-drive slanted/horizontal wells. Proceedings 1990 SPE Annual Technical Conference & Exhibition, Reservoir Engineering volume, September 23-26, 1990, New Orleans, pp. 77-84. Society of Petroleum Engineers, Richardson, TX. SPE paper 20720.

Inflow performance relationships (IFR) of inclined wells increase at inclinations above 45 degrees. A wellbore inclined 60 degrees may be twice as productive as a wellbore inclined 45 degrees. The IPR's for inclined wells are similar to the parabolic behavior of the Vogel IPR curve. The Vogel curve was shown in this work to be accurate at borehole inclinations of less than 45 degrees.

136. Chidsey, Thomas C. and Craig D. Morgan (1991). Horizontal drilling potential in Utah. A.A.P.G. Bulletin 75 (3), 553.

Abstract. At least 18 reservoirs in Utah have potential as horizontal drilling targets. These reservoirs include the Twin Creek, Nugget, Phosphoria, Dinwoody, Madison, Green River, Ferron, Dakota, Kaibab, and Paradox formations. The Blackhawk and Straight Cliffs formations have coalbed methane with horizontal well potential. Grassy Trail Creek field, which produces from the Moenkopi formation, has been rejuvenated by horizontal drilling.

137. Chidsey, Thomas C., Craig D. Morgan, and Michael D. Lane (1991). The fractured Green River and Wasatch formations of the Uinta basin, Utah: Ideal targets for horizontal drilling. A.A.P.G. Bulletin 75 (6), 1123-1124.

Abstract. The Eocene Green River and Wasatch formations in the Uinta basin contain ideal horizontal drilling targets. These targets are low-permeability sandstones and siltstones with predictable fracture orientations controlling production. Some of the fields which are potentially good horizontal drilling targets are: Altamont-Bluebell, Duchesne, Star Flat, and Cottonwood Wash.

138. Childers, Ronald D. (1990). Application of MWD (Measurement- While-Drilling) gamma ray logging to horizontal Bakken wells. in Hansen, William B. (editor), Bakken Formation Short Course, p. 17. Montana Geological Society, Billings. Study Notes No. 1.

Abstract. Since 1987, 45 horizontal wells have been drilled in the Bakken formation. The MWD-gamma ray log may allow real-time correlation of a well with offsetting wells.

139. Chin, Wilson C. (1990). Annular flow model explains Conoco's borehole cleaning success. Offshore 50 (10). 41-42.

Conoco successfully used an oil-based mud to remove cuttings from high-angle wells in Green Canyon block 184, Gulf of Mexico. The oil-based mud was successful because annular flow on the high and low sides of the borehole was more nearly equal than when other mud were used. The article describes the model used to calculate the annular flow.

140. Chisholm, Pat and Kevin Dunn (1991). Chalk stimulation: Stimulating the Austin Chalk. Popular Horizontal 1 (2), 24-31.

Review of stimulation techniques used in the Giddings and Pearsall field for the Austin Chalk, Buda, and Georgetown formations. Waterflood and sand fracturing are the most effective stimulation techniques in the Giddings field. The Pearsall field is best stimulated using closed fracture acidizing (CFA). Details are given in the article.

141. Chopra, A. K., M. H. Stein and J. C. Ader (1989). Development of reservoir descriptions to aid in design of EOR projects. SPE Reservoir Engineering 4 (2), 143-150. SPE paper 16370 presented at the SPE California Regional Meeting, April 8-10, 1987, Ventura.

Enhanced Oil Recovery (EOR) projects require detailed models of the reservoir(s) for maximum effectiveness. This paper presents an approach to producing a reservoir description using geological, petrophysical, and pressure-transient data.

142. Christian, D. B. (1988). Planning and operational requirements for a shallow-objective high-angle well in the Gulf of Mexico. SPE Drilling Engineering 3 (3), 241-247. SPE/IADC paper 16168. Also in Proceedings 1987 SPE/IADC Drilling Conference, March 15-18, New Orleans, pp. 1035-1046. Critical comments are in Flak, Larry H. (1989), SPE Drilling Engineering 4 (3), 76.

This paper is a case history of a highly deviated exploration well drilled from a semisubmersible rig in East Cameron Block 378 in the Gulf of Mexico. The target sand was 6,160 ft (1878 m) horizontally from the surface location at a depth of 3,520 ft (1073 m). The kickoff point was 270 ft (82 m) below the mudline. Drilling was done with a steering-tool/mud motor combination. Three combinations of MWD tools were used for different parts of the well: MWD directional, MWD directional and logging with real-time data, and MWD directional and logging with stored data only. Conclusions and recommendations for drilling are: (1) minimum circulating rate needs to be established to clean the hole and reduce washout, (2) a limber hole finder may need to be run during hole opening, (3) during tripping in the hole, care must be used to avoid unintentional sidetracking.

143. Chuber, Stewart (editor) (1991). Abstracts and Short Papers of the South Texas Geological Society Austin Chalk Exploration Symposium, February 24-26, 1991, San Antonio. South Texas Geological Society, San Antonio.

The South Texas Geological Society Austin Chalk Symposium included papers on exploration, geophysics and formation evaluation of the Austin Chalk and related plays and two papers on the Niobrara play in the Rocky Mountain region and horizontal drilling in the Mid-Continent (Oklahoma). The papers and abstracts are separately considered in this bibliography. Taken together, these papers and abstracts represent the current knowledge and confusion concerning the horizontal drilling plays in the Austin Chalk.

144. Chugbo, A. I., G. D. Roux, and J. C. Bosio (1989). Thin oil columns: Most people think horizontal wells, Obagi field case suggests the contrary. Proceedings 1989 SPE Technical Conference & Exhibition, Formation Evaluation and Reservoir Geology volume, March 8-11, 1989, San Antonio, pp. 297-306. Society of Petroleum Engineers, Richardson, TX. SPE paper 19599.

The Obagi field in Nigeria produces oil from a 36 ft thick unconsolidated deltaic sand that has permeability of 2 darcies and oil viscosity of 0.8 cp. The field has natural geological barriers that limit water and gas coning. These barriers are discontinuous shale interbeds or laminations which limit vertical permeability. Vertical wells have recovered 19% of the oil in place (21 million barrels) and will ultimately recover 30% of the oil. Horizontal wells would probably have lower recoveries because of the limited vertical permeability.

145. Cinco-Ley, H. (1984). Unsteady-state pressure distribution created by a slanted well or a well with a slanted fracture. M.S. Thesis, Stanford University, Stanford, California.

Not examined.

146. Claridge, Elmond L. and Shan Qin (1989). Sweep efficiency comparisons of horizontal and vertical wells. Journal of Canadian Petroleum Technology 28 (2), 38. CIM paper 89-40-55. Paper presented at the 40th Annual Technical Meeting of The Petroleum Society of CIM.

Abstract. Well productivity or injectivity is higher in horizontal wells than in vertical wells resulting in lower pressure drawdown in the case of production.

147. Clark, A. C. and D. A. Cocking (1989). The planning and drilling of the World's first horizontal well from a semisubmersible rig. Proceedings 1989 SPE/IADC Drilling Conference, February 28-March 3, 1989, New Orleans, pp. 773-790. Society of Petroleum Engineers, Richardson, TX. SPE/IADC paper 18712. Abstract in The Engineering Monthly Index (November, 1989), 27 (11), 762.

In June 1988, the first horizontal well to be drilled from a semi-submersible rig was spudded in Cyrus field, U.K., North Sea. The well was drilled with a 565 m (2,854 ft) horizontal section in the Paleocene Andrew formation, the main producing horizon. Details of the planning and drilling of the well are given in the paper.

148. Clary, M. M. and T. W. Stafford, III (1987). MWD performance and economic benefits in the Zu horizontal drilling program. Proceedings 1987 SPE/IADC Drilling Conference, March 15-18, 1987, New Orleans, pp. 1065-1076. Society of Petroleum Engineers, SPE/IADC paper 16171.

Between June and December, 1986, Atlantic Richfield Indonesia Inc. drilled twelve high-angle and horizontal wells to develop the Zu field, Java Sea, Indonesia. The target formation was the Batu

Raja formation. The principal MWD tool used was a resistivity-gamma-ray directional tool. Conventional rotary drilling assemblies and navigation motors were used to drill the well. The assemblies are described in the article. Use of the MWD directional and logging tools replaced wireline logging on the last four wells and saved $500,000 for the project. MWD logging allowed real-time modifications of the well plans. The MWD directional surveys require less than three minutes of rig time independently of depth and are definitive. MWD logging confirmed the top of the target formation with better bed resolution than wireline logs, and allowed picking the gas sand without having to rely on well cuttings alone.

149. Clavier, C. (1991). The challenge of logging horizontal wells. The Log Analyst 32 (2), 63-84.

A comprehensive review of horizontal well and extended- reach well logging. Problems considered include conveyance of tools into the borehole, formation dissymmetry, gravity effects on borehole fluids and cuttings, tool eccentricity effects, and new logging problems (fractures, gas and water coning, tight reservoirs). The final sections of the article cover interpretation of logs of horizontal wells and predictions of future of well logging, considering the increase in the number of horizontal wells being drilled.

150. Claycombe, E. (February 28, 1983). Pressure buildup characteristics in Austin Chalk wells. Oil & Gas Journal 81 (9), 71-77.

Four examples of pressure build up data are given and analyzed for Austin Chalk wells in the Giddings field, Texas. The Giddings field is 100 miles long and 15 miles wide and contains 2,500+ wells. Pressure build up data, if carefully obtained, can be of use in predicting well performance in the Austin Chalk.

151. Claytor, S. B., Jr., R. King, and J. Speed (1989). Steerable systems drilling; the right angle for horizontal drilling. Proceedings of the 1989 SPE Asia-Pacific Conference, September 13-15, 1989, Sydney, Australia, pp. 7-16. Society of Petroleum Engineers, Richardson, TX. SPE paper 19466. Abstract in The Engineering Index Monthly (October, 1990), 28 (10), 611.

A well with a horizontal section of 1,270 ft was drilled in the South Pepper field, offshore Australia, using a steerable drilling system. The well was drilled to improve productivity from a relatively thin oil zone between an aquifer and a gas cap. The well TD was 4,826 ft with the vertical section TVD held between 4,037 and 4,041 ft. The horizontal well was kicked off from a pilot well designed to locate the oil/water contact. The well flowed 4,230 barrels of oil per day with a 550 GOR.

152. Clements, William R. and Henk H. Jelsma (1989). Horizontal wells pose special hydraulic design considerations. Petroleum Engineer International 61 (11), 34, 36, 39. Abstract in The Engineering Index Monthly (June, 1990), 28 (6), 671.

Hydraulic considerations when drilling horizontal wells include: drilling fluids design, pressure design, fluid flow design, and hydraulic diameter design. Drilling fluids are particularly important in horizontal wells because of the greater difficulty in cleaning the hole, the need to drive the downhole motor, and hole stability problems.

153. Clerke, E. A. and T. J. Van Akkeren (1988). Borehole Televiewer improves completion results in a Permian basin San Andres reservoir. SPE Production Engineering 3 (1), 89-95. SPE paper 15033 presented at the 1986 SPE Permian Basin Oil and Gas Recovery Conference, March 13-14, Midland.

The borehole televiewer was used in the Hobbs field, New Mexico, to detect vuggy zones which supplied water to producing intervals. The field produces from dolomitized Grayburg/San Andres. The vuggy interval was not detected by other logging tools. Identification of the porous, vuggy zones changed the completion strategy in the field. Editor's Note: The borehole televiewer and similar acoustic tools can be very helpful in identifying fractured zones, vuggy zones, and fine-scale features in the borehole that may not be detectable in other logs. These tools can be useful in areas where horizontal drilling is to be done in fractured reservoirs.

154. Clonts, M. D. and H. J. Ramey, Jr. (1986). Pressure-transient analysis for wells with horizontal drainholes. Proceedings 1986 SPE California Regional Meeting, April 2-4, 1986, Oakland, pp. 215-230. Society of Petroleum Engineers, Richardson, TX. SPE paper 15116.

This article presents analytical solutions for transient pressure response of horizontal drainholes in an anisotropic reservoir of finite thickness and can be applied to multiple drainholes in a vertical array. If the drainhole is short relative to the reservoir thickness, initial radial flow perpendicular to the drainhole axis is followed by transition to a pseudo-radial flow period. If the drainhole length is long relative to reservoir thickness, initial radial flow period ends rapidly and transient pressure behavior becomes identical to that of a uniform flux vertical fracture.

155. Coates, R. (1984). Physical simulator for horizontal well. Paper presented at the First WRI-DOE Tar-Sand Symposium, June 26-29, 1984, Vail, Colorado.

Not examined.

156. Coffin, Ph. (1989). Pelican Lake and Winter field horizontal well projects. 1989 Horizontal Well Seminar, April 20, 1989, Calgary.

Not examined.

157. Coffman, Charles A. (1982). How to drill and complete Austin Chalk wells. Petroleum Engineer International 54 (8), 116, 120, 122.

U.S. Operating Inc. has 160 wells producing 9,748 BOPD in the Giddings field of Texas. Their preferred method of completing an Austin Chalk is: (1) set 7 inch casing 50 to 100 feet above the top of the chalk; (2) 5 inch liner is set in the chalk and cemented at the top, middle and bottom of the formation to isolate the potential producing zones; (3) if lost circulation material is required, an acid soluble material is used; (4) if circulation is lost, the hole is dry-drilled in 10 foot intervals until TD is reached; (5) after the liner is set the appropriate intervals are perforated and tested in sequence. Treatment decisions are frequently made from the mud log which indicates the fractured zones, fluorescence and gas shows, and from drilling troubles. These methods are typical of pre-horizontal drilling practice in the Austin Chalk.

158. Colamasi, C. and A. delle Canne (March, 1963). Drainhole drilling successful in Sicily. World Oil 156 (4), 88-89.

Lateral drainhole drilling was used in the Ragusa field, Sicily, to reduce the water coning problem. Of the three wells drilled, two successfully eliminated the water cut; the third well did not eliminate the water. The Ragusa field produces from the naturally fractured Taormina formation, a Triassic dolomite. The reservoir is underlain by a thick aquifer. Details of the project are given in the article.

159. Collins, David, Nghien Long, Ravi Sharma, Yau-Kun Li, and Kamal Jha (1990). Field-scale simulation of horizontal wells. in Preprints of CIM/SPE International Technical Meeting, June 10-13, 1990, Calgary, v III, Paper 121. CIM, Montreal. CIM/SPE paper 90-121.

A field-scale simulation of reservoir and horizontal wells with the wells treated as a 'separate porosity' as in some fractured-reservoir simulations. The test uses a thin reservoir with a bottom-water drive and shows the effect of wellbore length and production rate on recovery.

160. Combe, J., J. Burger, G. Renard, and E. Valentin (1988). New technologies in thermal recovery: Contribution of horizontal wells. in 4th UNITAR/UNDP Conference on Heavy Crude and Tar Sands, August 7-12, 1988, Edmonton, Paper 117.

A detailed and systematic numerical simulation study was made of the relative recovery efficiencies of heavy oil/tar sands using three cases: (1) horizontal well patterns; (2) vertical well patterns; (3) combined horizontal and vertical well patterns. Other factors considered were: (1) reservoir thickness; (2) simultaneous steamflooding of separated, superimposed reservoirs; (3) oil viscosity; (4) initial mobile water saturation; (5) effective horizontal wellbore length; (6) distance between wells; and economics of three cases.

161. Comer, John B. and Henry H. Hinch (1987). Recognizing and quantifying expulsion of oil from the Woodford formation and age-equivalent rocks in Oklahoma and Arkansas. A.A.P.G. Bulletin 71 (7), 844-858.

Discussion of oil in the Woodford formation and equivalent rocks in Oklahoma and Arkansas. Oil was generated in, and migrated within the Woodford before being expelled into adjacent formations. Oil has accumulated in fractures, stylolites, burrows, and sandstone lenses within the Woodford. Editor's Note: The Woodford is now being discussed as a possible horizontal drilling target comparable to the Bakken formation. To date little horizontal drilling has been done in the Woodford.

162. Comer, John B. (1991). Potential for producing oil and gas from Woodford shale (Devonian-Mississippian) in the southern Mid-Continent, USA. A.A.P.G. Bulletin 75 (3), 555.

Abstract. The Woodford Shale is a potentially prolific oil and gas producer from fractured reservoirs. The Woodford extends through large areas of Oklahoma, Texas, New Mexico, and Arkansas. Horizontal drilling may be the proper technology to exploit this formation.

163. Conti, P. E. (1989). Controlled horizontal drilling. Proceedings 1989 SPE/IADC Drilling Conference, February 28-March 3, 1989, New Orleans, pp. 749-754. Society of Petroleum Engineers, Richardson, TX. SPE/IADC paper 18708. Abstract in The Engineering Monthly Index (November, 1989), 27 (11), 722.

The first successful horizontal well drilled in Michigan was completed in the Antrim formation using MWD and steerable systems. The average Antrim gas well has a daily productivity of 70-80 Mcf of dry gas and 40-80 barrels of salt water. The well was drilled vertically to 309 where it was kicked-off with an average build of 5 degrees/100 ft. The well was drilled to 2,428 ft MD (1,455 ft TVD). The well was later nitrogen-foam fractured in stages. Tests resulted in 100MCFGPD range with the final stages unfractured. When the remaining stages have been fractured, the production estimates are 130-140 MCFGPD. The horizontal well took seven days to drill compared with three days for a vertical well. The production obtained was potentially 30-40% higher than that of the average vertical well.

164. Cooper, R. E. (1988). Coiled tubing in horizontal wells. Proceedings 1988 International Petroleum Engineering Meeting, November, 1-4, 1988, Tianjin, China, pp. 323-334. Society of Petroleum Engineers, Richardson, TX. SPE paper 17581.

Coiled-tubing conveyed logging and completion tools are effective in horizontal wells because they do not depend on gravity to place the tools in the proper portion of the borehole. This paper addresses coiled-tubing for a variety of uses including: drillstem testing, stimulation, zone isolation, perforating, setting bridge plugs, gravel packing downhole televiewer, multi-zone completion systems, and other conventional and non-conventional uses.

165. Cooper, R. E. and J. C. Troncoso (1988). An overview of horizontal well completion technology. Proceedings International Meeting on Petroleum Engineering, November 1-4, 1988, Tianjin, China, pp. 335-350. Society of Petroleum Engineers, Richardson, TX. SPE paper 17582.

Horizontal wells are useful in a number of circumstances, such as reducing coning, increasing production from thin reservoirs, reducing the number of wells need to develop a field, in fractured-reservoirs, and in waterfloods in thin reservoirs. These and other factors are discussed and case histories are given to illustrate their use. Proper completion methods must be used in order to achieve the full potential productivity of horizontal wells.

166. Cooper, Robert E. and George Birch (1990). Chapter 15. Horizontal well cementing. in Nelson, Erik B. (editor), *Well Cementing*, pp 15-01 - 15-15. Schlumberger Educational Services, Houston.

A thorough treatment of cementing horizontal wells and the problems encountered.

167. Corbett, Kevin, Melvin Freidman and John Spang (1987). Fracture development and mechanical stratigraphy of Austin Chalk, Texas. A.A.P.G. Bulletin 71 (1), 17-28. See also Nolen-Hoeksema, Richard C. (1987), Fracture development and mechanical stratigraphy of Austin Chalk, Texas: Discussion. A.A.P.G. Bulletin 71 (6), 792-793, and Corbett, Kevin *et. al.* (1987). A.A.P.G. Bulletin 71-(6), 794-795, for discussion of this paper.

The mechanical stratigraphy of the Austin Chalk, Texas, was determined by study of fracture intensity in outcrop and subsurface. Three mechanical/stratigraphic units were identified: a massive, fractured upper Austin unit corresponding to the Big House Chalk member; a ductile Middle Austin unit, corresponding to the Dessau Chalk and Burditt Marl members; massive, fractured Lower Austin unit corresponding to the Atco Chalk member. Experimental results and linear regression analysis show that porosity is the most strongly correlated factor and smectite-content is the second most strongly correlated factor in fracture strength. Editor's Note: This paper and the discussions bear on the fractured reservoirs of the Austin Chalk which are targets of horizontal drilling.

168. Corbett, Kevin E., Melvin Friedman and John Spang (1989). Fracture development and mechanical stratigraphy of Austin Chalk, Texas: Reply. A.A.P.G. Bulletin 73 (6), 794-795. See Corbett, Kevin et al. (1987). A.A.P.G. Bulletin 71 (1), 17-28, the original paper discussed in this reply.

This reply to Nolen-Hoeksema's paper reiterates the authors' view that strain is the more appropriate predictor of strength than confining pressure.

169. Corbett, Kevin P., M. Friedman, D. V. Wiltschke, and J. Hung (1991). Characteristics of natural fractures in the Austin Chalk outcrop trend. in Chuber, Stewart (editor), Abstracts and Short Papers of the South Texas Geological Society Austin Chalk Exploration Symposium, February 24-26, 1991, San Antonio, pp. 3-11. South Texas Geological Society, San Antonio.

Natural fractures were studied in outcrops of the Austin Chalk from near Del Rio, Texas, to a quarry in Dallas, Texas. Detailed maps and descriptions were made of each locality showing fracture spacing, length of fractures in dip direction, type of fracture surfaces, and relation to local structures. The study shows that the Austin Chalk exhibits brittle fracture development and has four distinct structural styles which cause more intense fracturing than in surrounding flat-lying beds.

170. Cornett, James C. (1984). Polymer mud improves chalk gas production. World Oil 198 (1), 89-93, 96.

Union Carbide has developed an new polymer-drilling fluid that has improved drilling and production of Austin Chalk wells near Somerville, Texas. The fluid, UCAR DVX Polymer Drilling System, allows near gauge holes to be drilled in water sensitive carbonate formations with minimal formation damage. Data from four wells are presented.

171. Corrigan, Mark (April, 1988). Horizontal wireline push broadens CT horizons. Offshore Engineer, 33-34.

Coiled-tubing conveyed tools and systems for workovers of horizontal wells have been developed by Dowell Schlumberger. Brief descriptions of their systems are given.

172. Cortes, Jaques and Alain Robbin (1982). Total-CFP's diamond turbodrilling experiments prove profitable. Petroleum Engineer International 54 (1), 28, 32, 34, 36, 40.

Development of downhole motors with high rotation rates made diamond drilling more economical in some conditions. Total-CFP conducted tests in Abu Dhabi to test the turbodrill-PCD diamond bit combination. Results indicate that this combination can be much more economical and efficient than rotary drilling with other kinds of bits. This combination had longer life thereby reducing trip time, better calibrated holes, and better well bore equilibrium. These tests are considered applicable to vertical holes and straight sections of directional holes.

173. Cottrill, Adrian (October, 1988). Bass Strait: Esso displays full spectrum of marginal solutions. Offshore Engineer, pp. 24-25, 27-28.

Description of Esso Australia's offshore operations in Bass Strait, Australia. High-angle, extended-reach drilling has been used by Esso Australia/BHP for wells in the Mackerel, Snapper, Bream and Flounder fields, Bass Strait. Wells in Bream field are deviated up to 65 degrees. Two extended-reach wells have been drilled in Mackerel field. Four wells in Snapper field have been drilled at angles greater than 71 degrees. Four high-angle, extended-reach wells have been drilled in Flounder field.

174. Cowan, John F. (1985). 1928-1985: The Devran gravity assisted drainage of petroleum project. Twenty-Fourth Annual Conference, September 29-30 - October 1, 1985, London, Ontario. Technical Paper No. 12. Ontario Petroleum Institute, London, Ontario.

Leo Ranney, in 1928, developed and patented methods for recovering oil using mining and drilling methods in combination in a gravity drain process. These methods are being used in Ontario in the Devran (Devonian and Ranney) to produce oil from the Alberta tar sands. The method involves sinking a shaft through the shallow oil sands, digging drifts under the sands, and then drilling boreholes upward into the reservoir. The oil drains into the 'mine' and can be removed by pumping. Horizontal boreholes can be used to link horizontally separated 'sweet spots.' Details of such a project in Sarnia-London Road field, Ontario, are given in the paper.

175. Cramer, D. D. (1986). Reservoir characteristics and stimulation techniques in the Bakken formation and adjacent beds, Billings Nose area, Williston basin. Proceedings 1986 SPE Rocky Mountain Rendezvous: Rocky Mountain Regional Technical Program, May 18-21, 1986, Billings, Montana, pp. 331-344. Society of Petroleum Engineers, Richardson, TX. SPE paper 15166.

The Bakken formation of the Williston basin has three subdivisions: upper black shale, middle calcareous siltstone or silty dolostone, and lower black shale. The Bakken is a source rock that generated oil at depths greater than 9,000 ft. The reservoirs are all naturally fractured with very low matrix-permeability. The Bakken produces from 35 fields in the U.S. portion of the Williston basin. Sixty percent of the production comes from three fields (Buckhorn, Devils Pass, and Elkhorn Ranch) on the 'Billings Nose.' Well stimulations commonly used in the Bakken are oil/surfactant breakdown and gelled oil fracture treatments with small quantities of fluid loss additives and high concentrations of proppant. Details of the reservoir characteristics and stimulation techniques are given in the paper.

176. Cramer, D. D. (March 27, 1989). Guides exist for fracture treatment in horizontal wells. Oil & Gas Journal 87 (13), 41, 44, 46, 48-49, 51.

Discussion of hydraulic fracturing of horizontal wells and how to select candidates for fracturing.

177. Cramer, David D. (1990). Stimulation treatments in the Bakken formation. in Hansen, William B. (editor), Bakken Formation Short Course, p. 17. Montana Geological Society, Billings. Study Notes No. 1. Complete paper is printed in Hansen, William B. (editor) (1991), 1991 Guidebook to the Geology and Horizontal Drilling of the Bakken Formation, pp. 117-140, Montana Geological Society Billings, Montana.

This paper presents a detailed analysis of stimulation treatments used in the Bakken formation of the Williston basin. Examples are given of treatments in Antelope field, Bicentennial field, Buckhorn field, Devils Pass field, Elkhorn Ranch field, and Squaw Gap field.

178. Cramer, David D. (April 22, 1991). Stimulating shales - 1. Understanding the reservoir important to successful stimulation. Oil & Gas Journal 89 (16), 53-56, 58-61.

The Bakken formation is one of the few oil producing shales in the world. Bakken production comes from anisotropic, naturally fractured reservoirs. Bakken fractures are believed to be caused by in-situ hydrocarbon generation which causes overpressuring of the rock. A thorough understanding of the nature of the reservoir is required for planning production and stimulation programs. Studies have been made of the in-situ stress, temperature logs, drill stem tests, acid solubilities, permeability anisotropy, fracture distribution, rock properties, and hydrocarbon shows to plan stimulation programs. Details of these studies and data on major Bakken fields are given in this article.

179. Cramer, David D. (April 29, 1991). Stimulating shales - Conclusion. Experience reveals better Bakken stimulation techniques. Oil & Gas Journal 89 (17), 56-61.

Discussion of completion methods used in the Bakken formation horizontal drilling play, Williston basin. Examples are given from Antelope, Bicentennial, Devil's Pass, Buckhorn, Elkhorn Ranch, and Squaw Gap fields, North Dakota. Methods discussed include hydraulic fracturing with water and acids, proppants, and post-treatment surveys. Scale removal and gel breakers are also covered.

180. Cranfield, J. (October, 1984). French refine horizontal drilling technique. Oil and Gas Digest 6, 17.

Not examined.

181. Crook, R. J., S. R. Keller and M. A. Wilson (1987). Deviated wellbore cementing: Part 2 - Solutions. Journal of Petroleum Technology 39 (8), 961-966. SPE paper 14198 presented at the 1985 SPE Annual Technical Conference and Exhibition, Las Vegas, September 22-25.

Channeling may occur during the cementing of highly deviated wellbores. Proper rheological control of drilling fluids can prevent the accumulation of solids on the low side of a deviated wellbore. Maintaining the value of the mud yield-point at a high enough level can prevent the settling of solids. Very-low-viscosity preflushes can improve the displacement of settling-type mud. Casing centralizers may improve mud displacement. Pipe rotation and reciprocation can improve mud displacement in high-angle wellbores. Wall cleaners can also improve mud removal in permeable zones.

182. Crossey, Laura J. and Daniel Larsen (1991). Integrating diagenesis and burial history of the Mesaverde group, Piceance basin, CO. 1991 Abstracts with Programs, Rocky Mountain and South-Central Sections, Geological Society of America, Meeting, April 22-24, 1991, Albuquerque, New Mexico, p. 14.

Abstract. Studies of organic maturation of Mesaverde group sandstones and shales have been made on core samples from the Department of Energy Multiwell Experiment Site, Colorado. The vitrinite reflectance of these rocks ranges from 1.0 to 2.2. Sandstone diagenesis at the site is put into a time-temperature framework.

183. Crouse, Philip C. (1989). Reserve potential due to horizontal drilling is substantial. World Oil 209 (4), 47-49. Abstract in The Engineering Index Monthly (June, 1990), 28 (6), 671.

Ultimate recovery of oil can be increased in conventional reservoirs by use of horizontal wells. Horizontal wells can provide better sweep efficiency, better control of injected fluids, and tighter well- spacing. Horizontal wells are suitable for reservoirs that are thin and have high vertical permeability, naturally fractured reservoirs, offshore reservoirs, reservoirs that are difficult to fracture using conventional techniques, and areas with limited surface access.

184. Crouse, Philip C. (February, 1991). Horizontal drilling spurs optimism. World Oil 212 (2), 35-37.

In 1990, 1,000 (estimated) horizontal wells were drilled worldwide, with the Austin Chalk of Texas accounting for 65% of the total. The Bakken formation of the Williston basin was second in number of horizontal wells with 90. Horizontal drilling expenditures in 1990 were $1.2 billion, four times that of 1989. Financing of horizontal wells now includes promotions to small investors with some deals having questionable economics. Applications of horizontal drilling are increasing and drilling is expected to double in 1991.

185. Cummings, B. C. (April, 1991). Let's go after the oil we left behind. World Oil 212 (4), 73-76.

Review of the status of oil mining in the United States and its potential for adding producible oil reserves to U.S. stocks. Oil mining is here defined as recovering oil by gravity drainage into tunnels and/or horizontal drainholes. Potential "oil mines" are reservoirs less than 3,000 ft deep. Sandstone reservoirs are most likely to be economical at $10/barrel or higher. Many older fields still have 90% of their original oil in place.

186. Cunningham, R., F. B. Zelt, P. D. Snavely, S. R. Morgan, Q. R. Passey, and R. L. Webster (1990). Recognition of hydrocarbon expulsion using well logs: Bakken formation, Williston basin. A.A.P.G. Bulletin 74 (5), 637.

Abstract. The Bakken formation is one of the source rocks and reservoir rocks in the Williston basin. The Bakken source intervals can be identified on sonic and resistivity logs where the two curves separate.

187. da Fontoura, S. A. B. and H. M. R. dos Santos (1989). Discussion of modeling of the stability of highly inclined boreholes in anisotropic rock formations. SPE Drilling Engineering 4 (2), 187. SPE paper 19213.

Critical discussion of the paper by Aadnoy (September, 1988), SPE Drilling Engineering 3 (3), 259-268.

188. Damgaard, A., D. S. Bangert, R. P. Rubbo, and G. W. Stout (1989). A unique method for perforating, fracturing and completing horizontal wells. Paper presented at the 1989 SPE

European Offshore Conference: Offshore Europe '89, September 5-8, 1989, Aberdeen. Society of Petroleum Engineers, Richardson, TX. SPE paper 19282.

Horizontal drilling started in Dan field, Danish North Sea, in 1987. Horizontal wells were used to improve productivity. This article describes in detail perforating, hydraulic fracturing, and completion methods used in the Dan field horizontal wells.

189. Daniel, W. L. and W. W. Fertl (December 3, 1984). Logging high-angle, long-reach boreholes. Oil & Gas Journal 82 (49), 103-108. Abstract in The Engineering Index Annual (1985), 84, Part IV, 6196.

Dresser Atlas has developed a pipe-conveyed system for logging high-angle borehole. The system is called the 'Slant-hole Express.' Tools that can be used by this system include induction-electric, acoustic, density, neutron, and gamma ray logs.

190. Dann, Dennis and David Jetelina (1990). New logging approach detects fractures in horizontal wells. Petroleum Engineer International 62 (9), 30, 32-33, 35-36.

An acoustic wellbore imaging tool used in combination with a spectral gamma-ray tool was used to detect natural fractures in a horizontal wellbore. Logging was done with a drillpipe conveyed system. The fractured formation was a chalk. Details of the logging system are given in the article.

191. Daviau, F., G. Mouronval, P. Curutchet, and F. Giger (April, 1985). Ecoulement des fluides autour d'un puits horizontal. Paper presented at the 3rd European Meeting on EOR, Rome. In French.

Not examined.

192. Daviau, F., G. Mouronval, G. Bourdarot, and P. Crutchet (1988). Pressure analysis for horizontal wells. SPE Formation Evaluation 3 (4), 716-724. SPE paper 14251 presented at the 1985 Annual SPE Technical Conference and Exhibition, September 22-25, Las Vegas. Abstract in The Engineering Monthly Index (July, 1989), 27 (7), 578.

An analytical model was developed for horizontal-well test design and interpretation. The method is suitable for use by minicomputer. Two successive types of flow in horizontal wells showed up: vertical radial flow and horizontal pseudo-radial flow. In favorable cases, the two flows correspond to interpretable semi-log straight lines to which conventional semi-log analysis can be applied.

193. Davis, A. P., Jr. and O. M. Knight (1983). Reduce logging problems in high angle wellbores. World Oil 196 (5), 59-62.

Exxon Production Research Co. and Gearhart Industries have developed an openhole logging system using conventional logging tools run on drill pipe. The trade name for this system is TOOL-PUSHER. The system has been successfully tested in holes with angles ranging from 65 degrees to 90 degrees. Logging tools mentioned are dual induction, compensated neutron, compensated density, and gamma ray.

194. Dawson, Rapier and P. R. Pasley (1984). Drillpipe buckling in inclined holes. Journal of Petroleum Technology 36 (11), 1734-1738. SPE paper 11167 presented at the 1982 SPE Annual Technical Conference and Exhibition, New Orleans, September 26-29. Abstract in The Engineering Index Annual (1984), 83, Part IV, 6672.

In high-angle holes, drillpipe can carry high compressive axial loads without buckling. In boreholes of greater than 50 degrees inclination, use of drillpipe in compression will permit reduction in BHA eight and reduction of torque and drag on the drillstring.

195. Dawson, William C. (1988). Stylolite porosity in carbonate reservoirs. A.A.P.G. Bulletin 72 (2), 176.

Abstract. Stylolite porosity is known to exist in carbonate reservoirs in the Williston, Paradox, and Delaware basins. Stylolite-associated porosity may be exhibited as fractures, open stylolite seams, fabric-selective megaporosity halos, vugs, microporosity halos, and crystal molds. Editor's Note: This article and a few others on stylolite porosity are listed here because of my belief that some of the Austin Chalk production in the Maverick basin may be related to the features mentioned here. Stylolite porosity is also known in the Woodford shale of the Anadarko basin.

196. Deans, Stuart, Lee Scherer and John Pulley (1990). Horizontal drilling in the Bakken: Boom of bust? in Hansen, William B. (editor), Bakken Formation Short Course, p. 18. Montana Geological Society, Billings. Study Notes No. 1. The complete paper is printed in Hansen, William B. (editor) (1991), 1991 Guidebook to Geology and Horizontal Drilling of the Bakken Formation, pp. 95-104, Montana Geological Society, Billings, Montana.

This paper discusses all aspects of horizontal drilling in the Bakken formation, Williston basin, including: well spacing, economics, productivity of wells, and well interference. Horizontal wells drilled in the Bakken should be drilled perpindicular to the predominant fracture azimuth, and should be drilled with an eye to preventing drainage by nearby wells on other leases. Horizontal well costs are 50% higher than the cost of vertical wells. When the larger drainage area of a horizontal well is considered the well costs are 26% less. Reserves of horizontal wells are 83% higher than for vertical wells. Bakken horizontal wells are marginal at $18/barrel of oil.

197. Dech, J. A., David D. Hearn, Frank J. Schuh, and Bob Lenhart (July 14, 1986). New tools allow medium-radius horizontal- drilling. Oil & Gas Journal 84 (28), 95-99. Abstract in The Engineering Index Annual (1986), 85, Part IV, 6522.

ARCO has developed a system for drilling medium-radius horizontal wells. A six-inch diameter hole was drilled in the Austin Chalk in Rockwall county, Texas, with a horizontal distance of 1,340 ft and a total displacement of 1,630 ft. The horizontal hole was kept within a 30 ft vertical section. Build rate from vertical to horizontal was a constant 20 degrees/100 ft. The horizontal hole was kicked-off at 1,098 ft measured depth. Equipment used for drilling the horizontal section includes a steerable hold/correct mud motor, PDC bits, and flexible non-magnetic collars.

198. Dech, J. A. and L. Wolfson (April 7, 1987). Advances in horizontal drilling. Presented at the 7th Biannual Petroleum Congress of Turkey, April 7, 1987, Ankara. SPE paper 16868.

Not examined.

199. Decker, David (1987). Coal bed methane: Dynamic new segment of petroleum industry. A.A.P.G. Bulletin 71 (5), 547.

Abstract. Coalbed methane is a potentially major source of methane for the United States. The U.S. Department of Energy has funded two projects to study the extraction of methane from coalbeds. One project is the Multi-Seam Project, Black Warrior basin, Alabama, which deals with shallow coal beds. The other project is the Deep Coal Seam Project, Piceance basin, Colorado, operated by Resource Enterprises Inc. The latter project deals with coalbeds at depths greater than 3,000 ft. Editor's Note: The Deep Coal Seam Project has utilized horizontal drilling to extract methane from coalbeds.

200. Dees, J. M., T. G. Freet, and G. S. Hollabaugh (1990). Horizontal well stimulation results in the Austin Chalk formation, Pearsall field, Texas. Proceedings 1990 SPE Annual Technical Conference & Exhibition, Production Operations and Engineering volume, September 23-26, 1990, New Orleans, pp. 665-676. Society of Petroleum Engineers, Richardson, TX. SPE paper 20683.

Oryx Energy Co. uses three stimulation methods on horizontal wells in the Austin Chalk of the Pearsall field. The methods are: water fracturing, acid washes, and acid fracturing. All three methods have been used in completing more than 20 wells. The horizontal wellbores range from 150 to 4,800 ft in length. The most effective treatment for cleaning up formation fines and stimulation of fracture systems is acid fracturing with diverter pills. Acid washes prove less effective for deep stimulation, but costs less and assures treatment along the whole length of the drainhole. Water fracs are beneficial if the horizontal well does not have high water cuts. Details of the completion methods are given in the article.

201. de Geeter, P. J. (1988). North Herald/South Pepper oil field development. Oil and Gas in Australia, pp. 1-18.

Not examined.

202. Delafon, H. (1989). Part 1. BHA prediction software improves directional drilling. World Oil 208 (3), 45-48.

Total Oil Marine plc and Elf UK have developed bottom-hole assembly (BHA) behavior prediction software for use in the Alwyn North field, North Sea. The software and its predictions were used in development drilling and resulted in some record performances, including the longest run of a BHA with a drilling motor in a directional well. The "record run" was in the 12.25-inch section of the hole, which was drilled 7,218 ft with an average drilling rate of 72 ft/hour. The Total system, which uses rotary BHAs instead of steerable systems, has proved to have better rate-of-penetration and better holes in medium-soft formations.

203. Delafon, H. (1989). Part 2. BHA prediction software improves directional drilling. World Oil 208 (4), 51-56, 60. Abstract in The Engineering Monthly Index (August, 1980), 27 (8), 757.

Total Oil Marine plc and Elf UK have developed simulation software to model bottom-hole assembly (BHA) behavior. The simulations were done "in-house" and were used in designing and operating wells in the North Alwyn field, North Sea. The model is discussed in the article.

204. Dellinger, T. B. W. Gravley, and G. Tolle (September 15, 1980). Directional technology will extend drilling reach. Oil & Gas Journal 78 (37), 153-156, 161-162, 164, 169.

This article discusses extended-reach drilling and its advantages and problems.

205. DeLucia, F. V. (1989). Benefits, limitations, and applicability of steerable system drilling. Proceedings SPE/IADC Drilling Conference, February 28-March 3, 1989, New Orleans, pp. 333-340. Society of Petroleum Engineers, Richardson, TX. SPE/IADC paper 18656.

Steerable directional drilling systems offer several benefits over conventional drilling. The steerable system, using MWD and downhole positive displacement motors can give better directional control, reduce the number of trips, change from rotating to sliding (drilling with the downhole motor rather than rotating the drillstring), and produce a better borehole. This system can lower drilling costs in either straight or highly deviated holes.

206. de Montigny, Olivier and Jean Combe (April 11, 1988). Horizontal well operations - 3. Hole benefits, reservoir types key to profit. Oil & Gas Journal 86 (15), 50-54, 56.

Proper utilization of horizontal drilling requires careful planning. Horizontal wells are most effective in low-permeability reservoirs where the benefits of greater length of wellbore in the reservoir, intersection of more natural fractures, and control of the geometry of the wellbore in the reservoir can increase productivity significantly. Formation damage may be reduced in horizontal wells. In gas reservoirs, flow velocities are less in horizontal wells than in vertical wells, reducing the turbulence that causes pressure drops. Coning of water and gas are reduced in horizontal wells. Horizontal wells are also useful in steamfloods, allowing a greater volume of oil to be reached by the steam. Recent experience shows that when horizontal wells are planned and operated properly they are a cost-effective way to increase productivity and ultimate recovery in a wide variety of reservoir types.

207. de Montigny, Olivier, Patrick Sorriaux, Alain J. Louis, and Jacques Lessi (July 4, 1988). Horizontal well operations - conclusion. Horizontal well drilling data enhance reservoir appraisal. Oil & Gas Journal 86 (27), 40-48. Abstract in The Engineering Index Annual (1988), 87, Part IV, 6001.

Horizontal wells add a new dimension to reservoir evaluation. Lateral heterogeneities of a reservoir may be detected and measured in horizontal wells. Elf Aquitaine drilled two horizontal wells in Lacq field, France, to evaluate the drilling technique and to try to improve production from the field. The two wells encountered significant fractures and facies not intersected by the more than 100 vertical wells previously drilled in the field. Elf drilled another horizontal well in Castera Lou field to test drilling and completion technology in a deeper horizontal well and to locate the exact position of the fault forming the eastern boundary of the reservoir. This well succeeded in proving the technology as well as providing significant new and unexpected information about the reservoir. As a result of the success at Lacq and Castera Lou fields, Elf developed Rospo Mare field in the Italian Adriatic Sea using horizontal wells. Rospo Mare is a heavy oil field producing from a fractured, karstic, carbonate reservoir. Well logs were run in each of the horizontal wells, cores and well cuttings were obtained where possible, and the information was used to build a three dimensional model of the reservoirs. Several methods were used to evaluate the reservoir: well testing, well seismic surveys, acoustic measurements, and resistivity measurements. In addition to these methods, MWD gamma-ray logs were run. Horizontal boundaries of reservoirs may be determined by pressure drawdown and pressure buildup tests. Reservoir geometry may be determined, at least broadly, by well seismic surveys. Acoustic measurements made with a full wave-form tool can show the geometry of the layers in the vicinity of the wellbore.

208. Dempsey, P. (February, 1983). What's happening offshore: New drilling concept. World Oil 196 (2), 19.

A drilling system called Robot Drill has been developed by Advanced Drilling Inc. of California. The system promises better directional control in high-angle wells, lower labor costs, reduced trip time, lower capital investment in rig equipment, and enhanced safety. A brief description is given.

209. Dempsey, Paul (1987). MWD technology comes of age. World Oil 205 (1), 29-31.

Discussion of MWD directional-only tools and formation evaluation using MWD. Equipment of NL Sperry-Sun, Eastman Christensen, Gearhart, Baker, Smith International, Schlumberger, and Teleco are described.

210. Denham, Miles E. and Thomas T. Tieh (1990). Stylolites of the Upper Smackover formation of north Louisiana. A.A.P.G. Bulletin 74 (5), 640.

Abstract. Stylolites are common in the Upper Smackover of north Louisiana. Minerals present along stylolites indicate that fluid flow occurred after formation of the stylolites.

211. Dennis, B., E. Standen, D. T. Georgi, and G. O. Callow (1988). Fracture identification and productivity predictions in a carbonate reef complex. A.A.P.G. Bulletin 72 (2), 177.

Abstract. Fractures in a Devonian reef complex were identified and their orientations determined using the Borehole Televiewer, Schlumberger's Formation MicroScanner, and Array-Sonic logs. These logs were combined with core data and production flowmeter data to refine the interpretations and confirm the utility of the logs.

212. Desbrandes, Robert, Adam T. Bourgoyne, Jr. and Joseph A. Carter (1987). MWD transmission data rates can be optimized. Petroleum Engineer International 59 (6), 46-52.

Experimental results of MWD data transmission by pressure pulse in a borehole are presented. Equations are given for the prediction of pulse amplitude and velocity in water, water-based mud, and oil-based mud. Experiments were conducted in a 9,460-ft long 'mud loop' apparatus at Louisiana State University. Use of these equations may allow optimization of MWD data transmission in a borehole.

213. Desbrandes, Robert (1988). Status report: MWD technology, Part 1 - Data acquisition and downhole recording and processing. Petroleum Engineer International 60 (9), 27-30, 32-33.

MWD was initially used for directional control of directionally-deviated wells. Addition of drilling parameters and formation evaluation logs enhanced the usefulness of MWD. Currently directional data is obtained using three accelerometers to measure three components of gravity acceleration in any frame of reference. The three frames of reference in MWD are: borehole or tool axis, tool face direction, and an axis normal to both. Solid state magnetometers are used instead of compasses to determine borehole orientation. Drilling parameters measured by MWD are: weight-on-bit (WOB), torque, bending moment, temperature, and pressure. The first three parameters are measured with strain gages. Formation evaluation tools that can be used with MWD are: natural gamma-ray, resistivity, density, and neutron density. Other parameters that can be measured, or soon will be available are: mud density in the annulus; fluid inflow from the formation; and estimates of formation pressure.

214. Desbrandes, Robert (1988). Status report: MWD technology, Part 2 - Data transmission. Petroleum Engineer International 60 (10), 48-54.

MWD data may be transmitted to the surface by wireline, mud pressure waves, acoustic waves, or electromagnetic waves.

215. Desbrandes, Robert (1988). Status Report: MWD Technology. Part 3 - Processing, display, and applications. Petroleum Engineer International 60 (11), 42-44.

MWD data is transmitted to the surface from the downhole tool by means of pressure pulses transmitted through the drilling mud. These signals may be processed at the surface and may be used for directional survey and other types of data. Pump noise may interfere with the MWD data, so care must be taken to minimize this problem.

216. Detmering, T. J. (1984). Update on drainhole drilling - Empire Abo. Proceedings of the 31st Annual Southwestern Petroleum Short Course, April 25-26, 1984, Lubbock, pp. 25-41. Southwestern Petroleum Short Course Association, Lubbock, Texas.

Arco Oil & Gas Co. has drilled six horizontal drainholes in the Permian Empire Abo Unit and plans four more such wells. The horizontal sections of the wells extend 200 ft into the reservoir. The drainholes were designed to reduce a gas coning problem in the field. Four of the six wells have had superior performance over their life. The wells have been a success both in reducing gas coning and economically. The project is described in detail in the paper with supporting data.

217. Diaconescu, Peter and Larry LeBlanc (1987). Optimized drilling through use of downhole motors. Journal of Canadian Petroleum Technology 26 (2), 38-43.

Rate-of-penetration (ROP) in many formations is directly related to the rotational speed of the bit. Downhole motors can provide much higher rotational speeds (250 RPM) than drilling with the rotary table (150 RPM). PanCanadian Petroleum Ltd. tested downhole motors in the Wembly field of Alberta, and discovered that they increased ROP with acceptable bit wear. Downhole motors also proved to be economically competitive with other drilling methods.

218. Diamond, William P. and David C. Oyler (1980). Drilling long horizontal coalbed methane drainage holes from a directional surface borehole. Proceedings 1980 SPE/DOE Symposium on Unconventional Gas Recovery, May 18-21, 1980, Pittsburgh, pp. 341-346. Society of Petroleum Engineers, Richardson, TX. SPE/DOE paper 8968.

Three horizontal drainholes drilled from a surface borehole have been drilled in the Pittsburgh coalbed. The longest of the drainholes has a horizontal section of 3,207 ft. Details of the project are given in the article.

219. Dickey, Doug B. (1988). Nonconventional drilling in Prudhoe Bay field. A.A.P.G. Bulletin 72 (2), 178.

Abstract. Nonconventional wells (horizontal, drainhole, high-angle) have been used to improve productivity and recovery at Prudhoe Bay field, Alaska.

220. Dickinson, R. Wayne, R. A. Anderson, and R. Wade Dickinson (1989). The ultrashort-radius radial system. SPE Drilling Engineering 4 (3), 247-254. SPE paper 14804. Abstract in The Engineering Index Monthly (January, 1990), 28 (1), 663.

The Ultrashort-Radius Drilling System (URRS), developed over 10 years, has been used to drill more than 500 horizontal drainholes with a total of 27,000 ft of borehole. The system uses an erectable whipstock placed in an under-reamed cavity in the borehole and a hydraulic drilling head. Drilling is done by directing a water stream at 8,000-10,000 psi pressure to do the "cutting". The system can drill holes with a radius of curvature of 1-ft and can drill multiple drainholes. The system is described in detail in the article.

221. Dickinson, W. and R. W. Dickinson (1985). Horizontal radial drilling system. Proceedings 1985 55th SPE Annual California Regional Meeting, March 27-29, 1985, Bakersfield, pp. 887-892. Society of Petroleum Engineers, Richardson, TX. SPE paper 13949.

A high-pressure, water-jet system has been developed to drill multiple horizontal radial holes, 100 to 200 ft in length, from a single vertical wellbore. The system is a short-radius (10 to 12 inch radius) drilling system. The 4-inch diameter boreholes are designed for placement of 1.25-inch diameter production tubing. The water-jet drilling system operates at 8,000-10,000 psi. The system has been used in unconsolidated sands in the Kern River, Taft, and Olinda fields, California, at depths of 500-2,600 ft. The system is applicable for steam-drive or injection. Details of the system, sand control, and other problems are given in the article.

222. Dickinson, W., R. R. Anderson, R. W. Dickinson (1986). A second-generation horizontal drilling system. Proceedings 1986 SPE/IADC Drilling Conference, February 10-12, 1986, Dallas, pp. 673-682. Society of Petroleum Engineers, Richardson, TX. SPE/IADC paper 14804.

A new high pressure water-jet drilling system has been developed for drilling radial drainholes from a vertical wellbore. The new system is an improved version of a previously reported BecWell System (Dickinson, W. and R. W. Dickinson, SPE paper 13949 presented at the 1985 California Regional Meeting). The new system uses an improved whipstock which reduces underreaming from 48 inches to 24 inches, and does not require tripping between the drilling of the multiple radial holes. The system also includes the Model V Radius-of-Curvature Tool which provides directional information, and a Hydraulic Tail which provides real-time information on production-tube movement, formation drag, and control of the rate of movement during drilling. Completion technologies include: perforation of production tube for oil production or steam injection; sand control using slotted liner in the production tube; filters to keep gravel out of the end of the slotted liner; and gravel filling or packing.

223. Dickinson, W., R. D. Wilkes, R. W. Dickinson (1987). Conical water jet drilling. Paper presented at the Fourth U.S. Water Jet Conference, Berkeley, California.

Not examined.

224. Dickinson, W., R. R. Anderson, H. Dykstra (1987). Gravel packing of horizontal wells. Proceedings 1987 SPE Annual Technical Conference & Exhibition, Production Operations and Engineering volume, September 27-30, 1987, Dallas, pp. 519-528. Society of Petroleum Engineers, Richardson, TX. SPE paper 16931. Also presented at the SPE Formation Damage Control Symposium, Bakersfield, California, February 8-9, 1988.

Full scale laboratory tests show that 100% gravel packing of horizontal wells can be accomplished. The method successfully tested is a two-lift system using a gravel slurry; the first lift fills 70-90%; a second, reverse direction, lift completes the 100% fill. Hole deviation has been varied in the tests from 70-110 degrees. Details of the method are described in the article.

225. Dickinson, W. and D. F. Overdyke, III (1990). Production comparison of ultrashort-radius radials completed totally in oil and in oil plus water columns. Paper presented at the International Technical Meeting, The Petroleum Society of CIM and the Society of Petroleum Engineers, June 10-13, 1990, Calgary. The Petroleum Society of CIM, Montreal. CIM/SPE paper 90-127. Abstract in The Journal of Canadian Petroleum Technology (1990) 29 (2), 50.

Abstract. Ultrashort-radius radial drainholes were drilled in two wells in Caddo Parish, Louisiana. The drainholes in one well were completed solely in the oil zone. The radials in the other well were completed in both the oil and water zones.

226. Dickinson, Wade, Michael J. Pesavento, and R. Wayne Dickinson (1990). Data acquisition, analysis, and control while drilling with horizontal water-jet drilling systems. Preprints of CIM/SPE International Technical Meeting, June 10-13, 1990, Calgary, v. III, Paper 127. The Petroleum Society of CIM, Montreal. CIM/SPE paper 90-127.

Two water-jet, ultrashort-radius drilling systems have been developed by Petrolphysics Ltd. The two systems described in this article use "Control While Drilling" of multiple, radial drainholes. The Ultrashort Radius Radial System (URRS) uses coiled-tubing and can drill 100 to 200 ft radial drainholes from a vertical wellbore. The Remotely Piloted System uses conventional flush joint tubing and is designed "to drill for many kilometers with high three dimensional accuracy". Both systems are described in detail in the article.

227. Dietrich, J. K. (1988). The Kern River horizontal-well steam pilot. SPE Reservoir Engineering 3 (3), 935-944. SPE paper 16346 presented at 1987 SPE California Regional Meeting, April 8-10, Ventura.

The Kern River horizontal-well steam pilot project was necessary because of the sluggish oil flow from the vertical wells drilled in the field. The reservoir environment is not favorable for neither conventional oil recovery techniques nor oil-mining techniques. The project was conducted on a 25-acre test site. Eight horizontal wellbores were drilled radially from a vertical chamber (shaft) into the reservoir. The horizontal wellbores were spaced 45 degrees apart. The horizontal wells were rotary-drilled by men from the bottom of the chamber. Four of these "laterals" were 400 ft long and four were 700 ft long. The project, although not commercial, was encouraging. The tests agreed with simulator predictions. Steam override was reduced. Oil cut and OSR were increasing throughout the life of the project.

228. Dikken, B. J. (1989). Pressure drop in horizontal wells and its effect on their production performance. Proceedings 1989 SPE Annual Technical Conference & Exhibition, Reservoir Engineering volume, March 8-11, 1989, San Antonio, pp. 561-574. Society of Petroleum Engineers, Richardson, TX. SPE paper 19824. Abstract in The Engineering Index Monthly (January, 1990), 28 (1), 719.

Proper design of a horizontal well must consider the along-hole well pressure-gradients so as to understand turbulent flow in the wellbore and its effect on productivity. Practically, the drawdown is reduced at the distal end of the wellbore, thereby reducing gas or water coning at that end. A model is given in the article for analyzing flow in the horizontal wellbore.

229. Doan, Q., Ali S. M. Farouq, and A. E. George (1990). Scaling criteria and model experiments for horizontal wells. in Preprints of CIM/SPE International Technical Meeting, June 10-13, 1990, Calgary, v. III, Paper 128. Petroleum Society of CIM, Montreal. CIM/SPE paper 90-128.

Horizontal well modeling studies were done for two conditions; (a) isothermal experiments using gas and water, and (b) steam model with a concentric well. Substantial pressure drops can be experienced in the early stages of production, when the reservoir is being heated and the steam-to-oil ratio is high. A variable-diameter horizontal well may lower pressure loss, have a better steam-to-oil ratio, and have better ultimate recovery than a constant-diameter horizontal well.

230. Dobson, Jay and T. Mondshine (1990). Unique completion fluid suits horizontal wells. Petroleum Engineer International 62 (9), 42, 44, 46, 48.

Horizontal wells have been successfully drilled in unconsolidated sediments using sized-salt polymer system in Gabon and the North Sea. A brief description of the system and of two case histories (West Africa and North Sea) are given in the article.

231. Dolson, John, Dave Muller, M. J. Evetts, and J. A. Stein (1991). Regional paleogeographic trends and production, Muddy sandstone (Lower Cretaceous), central and northern Rocky Mountains. A.A.P.G. Bulletin 75 (3), 409-435.

The Cretaceous Muddy Sandstone of the Dakota Group is a major petroleum producer in the Rocky Mountain region. This paper presents a regional analysis of the petroleum geology and the paleogeography of the Muddy Sandstone.

232. Dorel, M. (1983). Horizontal drilling methods proven in three test wells. World Oil 196 (6), 127-135. Abstract in The Engineering Index Annual (1983) 85 Part IV, 4928.

Three horizontal wells drilled (2 in Lacq field, 1 in Rospo Mare field) prove the feasibility of horizontal wells. Primary applications of horizontal wells are: large but thin reservoirs, heavy-oil recovery, steam injection, and fractured reservoirs. Horizontal wells have provided up to five times the productivity of vertical wells, increased ultimate recovery, and reduced water and gas coning. Details of the three test wells are discussed in the article.

233. Doscher, T. M., A. Ying, and J. A. Kostura (1988). Improving the steam drive: Physical model studies of inert gas injection and horizontal wells. Proceedings 1988 SPE California Regional Meeting, March 23-25, 1988, Long Beach, pp. 443-448. Society of Petroleum Engineers, Richardson, TX. SPE paper 17448.

Scaled physical model studies were done to clarify optimum well configurations and operating conditions for steam injection. Model studies show that high velocity gas injection in shallow reservoirs is capable of increasing recovery of all reservoir fluids. Optimally placed wells, both vertical and horizontal, can lead to greater recovery and profit.

234. Doscher, T. M. (1988). Discussion of Underground Test Facility: shaft/tunnel laboratory for horizontal well technology. Journal of Petroleum Technology 40 (2), 242. SPE paper 17469. A critical discussion of "Best, D. A. et al. (May 1987) Journal of Petroleum Technology, 39 (5), 603-612."

Doscher believes that the evidence given by Best *et al.* does not support the argument that horizontal wells are economical in recovering heavy oil and bitumen by steamflood.

235. Drake, Gerald E. (1991). Re-development of Giddings field. in Chuber, Stewart (editor), Abstracts and Short Papers of the South Texas Geological Society Austin Chalk Exploration Symposium, February 24-26, 1991, San Antonio, pp. 13-16. South Texas Geological Society, San Antonio.

Horizontal drilling is now being used to "re-develop" the Giddings field Austin Chalk reservoir. Three case histories are given for individual wells. Horizontal wells in this field are, on average, more productive than vertical wells. The examples given are from a Union Pacific Resources Company drilling program.

236. Dravis, Jeffrey J. (1991). Regional depositional setting and porosity development of the Austin Chalk formation, South Texas. in Chuber, Stewart (editor), Abstracts and Short Papers of the South Texas Geological, Society Austin Chalk Exploration Symposium, February 24-26, 1991, San Antonio, 17-23. South Texas Geological Society, San Antonio.

The Austin Chalk is generally believed to produce oil from naturally-fractured intervals. Core studies show that, while fractures exist in the Austin, they may not provide enough permeability or enhance the porosity enough to explain the highly productive reservoirs present in parts of the Austin trend. This article suggests that preserved matrix porosity may be needed for the Austin to be highly productive. A useful discussion of the Austin Chalk regional deposition and porosity trends is given in this paper.

237. Drilling (1985). Offshore Bulletin: Horizontal drilling progress. Drilling 46 (11), 15.

Elf Aquitaine has drilled a well horizontally more than 600 meters into the reservoir in Rospo Mare field, Italy. The well cost was about three times that of a vertical well. Production from a horizontal well is expected to be four to five times that of a vertical well. Four horizontal wells are planned for Rospo Mare.

238. Drilling (1987). Downhole tool sniffs out fractures. Drilling 48 (3), 29.

Sandia National Laboratories has developed a radar tool using pulses eight billionths of a second in duration with a frequency in the VHF band. The tool can detect fractures up to 30 meters from the borehole.

239. Drilling (1987). Drainhole drilling projects under way. Drilling 48 (4), 30-31.

Drainhole drilling is becoming increasingly common as a method to increase productivity in older fields. Current projects include drainhole or horizontal drilling in Yates field, Texas, Spraberry Trend, Texas, and Prudhoe Bay field, Alaska, to increase productivity and ultimate recovery of oil. Drainhole drilling usually is a re-entry of an old well in which a short section of high-angle or horizontal borehole is drilled into the reservoir.

240. Drilling (1987). Horizontal drilling lab completed. Drilling 48 (6), 16.

British Petroleum has completed wellbore simulators that can simulate high-angle and horizontal wells. A key research goal is to improve cuttings transport and hole cleaning in high-angle/horizontal wells. Another goal of the facility is to improve casing cementing in high-angle/horizontal wells.

241. Drilling (1990). The North Sea: Activity humming despite uncertainty. Drilling 50 (2), 32.

Shell UK Exploration & Production drilled an extended-reach water-injection well in Cormorant field. The well has a displacement of 15,618 ft and took 92 days to drill. Conoco canceled plans for a 32,000 ft displacement extended-reach well when no partners could be found.

242. Drilling Contractor (1985). Lateral drilling system extends rig reach. Drilling Contractor 41 (11), 48, 50.

Texas Eastern Corp. through its subsidiary Eastman Whipstock now offers the TEDSI (Texas Eastern Drilling Systems, Inc.) horizontal drilling system. TEDSI is an ultrashort-radius drilling system. The system is briefly described in the article.

243. Drilling Contractor (1986). Extended reach debuts in Prudhoe Bay. Drilling Contractor 42 (4), 8-9.

Sohio Petroleum Co. has successfully drilled the first horizontal well in the Prudhoe Bay field, Alaska. Horizontal drilling is expected to improve recovery of the remaining 12 to 13 billion barrels of oil in place. Some details of the project are given in the article.

244. Drilling Contractor (1987). Horizontal drilling has bright future, says Elf. Drilling Contractor 43 (4), 29-30.

Brief report on horizontal drilling as presented at the World Petroleum Congress by Jacques Bosio of Elf Aquitaine.

245. Drilling Contractor (October/November, 1988). Semi drills horizontal well in North Sea. Drilling Contractor 44 (5), 14.

Not examined.

246. Drilling Contractor (December, 1988/January, 1989). Horizontal re-completion techniques refined. Drilling Contractor 44 (6), 28-29.

Not examined.

247. Druyff, Lorraine (1990). Reservoir properties of the Bakken shale. in Hansen, William B. (editor), Bakken Formation Short Course, pp. 18-19. Montana Geological Society, Billings. Study Notes No. 1. Abstract also printed in Hansen, William B. (editor) (1991), 1991 Guidebook to Geology and Horizontal Drilling of the Bakken Formation, p. 113, Montana Geological Society, Billings, Montana.

Abstract. Fractures in the Bakken are of three types: structurally related tectonic fractures, stress related regional fractures, and expulsion fractures associated with overpressuring due to hydrocarbon generation. Lithology controls the degree and type of fracturing.

248. Dubrule, Plivier and Philip H. Nelson (1987). Evaluation of directional survey errors at Prudhoe Bay. SPE Drilling Engineering 2 (3), 257-267. SPE paper 15462 presented at the 1986 Annual SPE Technical Conference and Exhibition, New Orleans, October 5-8. For critical discussion and reply see: Thorogood, J. L. (1988), SPE Drilling Engineering 3 ( ), 334-335; Dubrule, O. and P. H. Nelson (1988), SPE Drilling Engineering 3 ( ), 335-336.

Directional surveys have been made for more than 8,000 wells in the Prudhoe Bay field. The surveys are magnetic single-shot and, since 1982, MWD with fluxgate magnetometers. Post-drilling surveys are done with gyroscopic instruments. Comparisons of the different types of survey showed that at 8,000 ft the largest lateral differences can be a few hundred feet. Drillstring magnetization effects cause magnetic surveys to read north of the gyros. Vertical and radial differences may be on the order of tens of feet and are negatively correlated. Lateral and vertical errors increase faster than radial errors as the horizontal offset increases.

249. Dubrule, O. and P. H. Nelson (1988). Author's reply to 'Discussion of evaluation of directional survey errors at Prudhoe Bay.' SPE Drilling Engineering 3 ,335. SPE paper 18581.

Reply to discussion by Thorogood, J. K. (1988), SPE Drilling Engineering 3 ( ), 334-335.

250. Duda, J. R. and H. K. Kumar (1988). Production forecasts for alternative completion methods in low-permeability gas reservoirs: Simulation of high-angle and horizontal wells. Proceedings 1988 SPE Annual Technical Conference & Exhibition, Reservoir Engineering volume, March 2-5, 1988, Houston, pp. 299-308. Society of Petroleum Engineers, Richardson, TX. SPE paper 18291.

Simulation of horizontal and high-angle wells and completion methods is a cost effective way to predict well performance and analyze reservoir properties. Simulation of Upper Cretaceous Mesaverde formation reservoirs in the Piceance basin show that intermediate length horizontal wells can be expected to produce four times as much gas as a vertical well. The Mesaverde is a low-permeability reservoir. The advantages of high-angle wells in the same formation has not been quantified but may have a significant potential. Details of the simulation are given in the paper.

251. Duda, J. R., S. P. Salamy, and K. Aminian (1989). Pressure analysis of an unstimulated horizontal well using type curves. Proceedings 1989 SPE Eastern Regional Conference, October 24-27, 1989, Morgantown, pp. 329-338. Society of Petroleum Engineers, Richardson, TX. SPE paper 19343.

Type cures were used for pressure analysis of a horizontal well in the Devonian shale of West Virginia. This analysis showed that the effective length of the wellbore was less than the actual drilled length. Details of the analysis are given in the article.

252. Durand, L. J., F. A. Samhouri and D. L. Bathe (1981). Kicking off in large diameter holes. Proceedings of the SPE Middle East Technical Conference, March 9-12, 1981, Bahrain, pp. 695-708. Society of Petroleum Engineers, Richardson, TX. SPE paper 9649.

ADMA-OPCO has standardized kick off operations in large diameter holes (17.5 inches and 26 inches) using a turbine with offset stabilizer. This method is useful for both sidetracking and azimuth changes.

253. Durham, C. O., Jr. and R. A. Bobigian (1990). Horizontal drilling in the Austin Chalk: Stratigraphic factors. A.A.P.G. Bulletin 74 (5), 646.

Abstract. Horizontal drilling in the Austin Chalk of south- central Texas has increased interest in the Giddings and Pearsall fields. The Giddings field at the end of 1988 had 4,500 wells which had produced 209 million barrels of oil and 934 BCFG. Pearsall field had 1,440 wells which have produced 57 million barrels of oil and 35 BCFG.

254. Durham, C. O. and Sheila Burnette Hall (1991). The Austin Chalk - bed by bed through Central Texas. in Chuber, Stewart (editor), Abstracts and Short Papers of the South Texas Geological Society Austin Chalk Exploration Symposium, February 24-26, 1991, San Antonio, pp. 25-40. South Texas Geological Society, San Antonio.

A description of the Austin Chalk stratigraphy and the role of stratigraphy in horizontal drilling.

255. Dussert, P., G. Santoro and H. Soudet (February 29, 1988). Horizontal well operations - 1. A decade of drilling developments pays off in offshore Italian oil field. Oil & Gas Journal 86 (9), 33-34, 36-37.

Rospo Mare field, offshore Italy, is the only field in the world producing systematically through horizontal wells. Rospo Mare is in the Adriatic Sea in 200-300 ft deep water, 12 miles offshore. The reservoir is a karstic reservoir with many dolines/sinkholes and enlarged fractures forming an irregular surface. The lower zone of the reservoir has vugs and partially closed horizontal caves. The reservoir produces heavy oil. Five horizontal wells have been drilled in the field for the cost originally estimated for four wells. Flow rates of the horizontal wells range from 2,240 to 12,000 BOPD. Production results are better than predicted and costs are lower. Even with lower oil prices the project is economically viable.

256. Dykstra, H. (1977). Prediction of oil recovery by gravity drainage. Paper presented at the 1977 SPE Annual California Regional Meeting, April 13-15, 1977, Bakersfield. Society of Petroleum Engineers/AIME. SPE paper 6548. Published in Journal of Petroleum Technology (May, 1978), 29 (5), 818-830.

Theoretical treatment of oil recovery by gravity drainage. This article expands on the work of Cardwell and Parsons (1949), and others.

257. Dykstra, Herman (1988). Productivity of radials or horizontal wells as affected by their length and number. Dykstra. Privately printed monograph.

Not examined.

258. Dykstra, H. and W. R. Dickinson (1989). Oil recovery by gravity drainage into horizontal wells compared with recovery from vertical wells. in Proceedings 1989 SPE Annual Technical Conference & Exhibition, Reservoir Engineering volume, March 8-11, 1989, San Antonio, 607-618. Society of Petroleum Engineers, Richardson, TX. SPE paper 19827.

Recovery of oil by gravity drainage in horizontal wells versus vertical wells is dependent on reservoir thickness and dip. Recovery of oil by gravity drainage is not affected by time in flat formations. Reservoir thickness is important in gravity drainage, with horizontal wells in flat reservoirs being more productive than vertical wells in reservoirs with thickness less than 0.85 times the distance between wells. In dipping formations, a horizontal well will produce up to 2.5 times the rate of a vertical well in certain circumstances. Details of these relationships are given in the article.

259. Easterling, D. and W. Jones (August, 1987). Steerable system optimizes drilling time. Offshore, 120.

Not examined.

260. Eastman Well Survey Company (1952). Lateral drilling. in 19th Composite Catalog of Oil Field and Pipeline Equipment, v. I, 1460. Gulf Publishing Co., Houston.

Not examined.

261. Eastman, H. John (1954). Lateral drainhole drilling, development of equipment and results. Eastman Oil Well Survey Co., Denver.

Not examined.

262. Eastman, H. John (November, 1954). Lateral drainhole drilling. First of a series. Petroleum Engineer 36 (11), B-57—B-58, B-61—B-62, B-64, B-67—B-68, B-70, B-73.

Lateral drainhole drilling is a method of increasing the borehole length in the reservoir. An early example of oil production by gravity drainage in the United States is an oil mine in Pennsylvania where the New York Enterprise and Mining Company dug a 165 ft shaft into an oil sand. Oil mining is now used in Germany in the Pechelbronn oil sands. In 1891, a patent was awarded to J.L. Addison for a device to expose more of the reservoir to the wellbore by drilling. In 1919, Bernard Granville applied for a patent to drill lateral drainholes. By 1930, Robert E. Lee of Coleman, Texas, had developed an air-drilling device for drilling lateral drainholes. He later returned to rotary drilling for lateral drainholes. The first true lateral drainholes were drilled with Lee's new rotary equipment at Texon, Texas, for the Big Lake Oil Co. Two 5.25-inch diameter laterals were drilled 23-24 ft horizontally from a vertical wellbore. This article discusses the continued development of horizontal drilling up to the development of the whipstock for directional drilling. Projects discussed include horizontal drilling at McConnelsville, Ohio; the Franklin Heavy Oil field, Pennsylvania; and the Round Mountain field, California. Also discussed are equipment designed and used by Zublin Oil Tool Co. and Oil Well Drain Hole Drilling Co.

263. Eastman, H. John (1954). Lateral drainhole drilling. Second of a series. The Petroleum Engineer 36 (12), B-44, B-47. This article discuss

Discussion of directional well survey state-of-the-art in the early 1950's.

264. Eaton, Nick (1990). Newly developed systems meet demands of coring. World Oil 210 (3), 62-64, 66.

Currently coring systems are available for short-radius and medium-radius horizontal wells. The short-radius coring system uses 3-ft long core barrels and can be used with build rates of up to 1.45 degrees/ft. Medium-radius coring can be done with a modified conventional 30-ft core barrel. This article describes the equipment and gives case histories of its use in West Virginia (Devonian shale), Texas (Austin Chalk), North Sea (chalk), and Canada (Midale formation).

265. Eberts, R. D. and R. D. Barnett (July 19, 1976). Ultrahigh angle wells are technical and economic successes. Oil & Gas Journal 74 (29), 115-120.

Tenneco Oil Co. has used high-angle drilling to reach reserves otherwise requiring additional offshore platforms. One such well, drilled from the West Cameron 643-A platform, was drilled to 12,300 ft MD at an average angle of 82 degrees. The well tested 7.7 MMCFGPD. Details of the project are given in the article.

266. Eckles, Wesley W., Jr. (1989). New simulator for slant and horizontal wells developed by Niper. Petroleum Engineer International 61 (4), 58, 60, 62. Abstract in The Engineering Monthly Index (August, 1989), 27 (8), 758.

Brief review of a well model, Beta-VHS, developed by the National Institute for Energy Research (NIPER) for horizontal and slanted wells.

267. Economides, Michael J. (1988). Formation conditions dictate stimulation. The American Oil & Gas Reporter 31 (12), 33-34.

Important factors in horizontal drilling are: (a) horizontal wells can be directed to intersect discrete fractures or fissures; (b) horizontal wells can improve volumetric recovery of hydrocarbons; (c) horizontal wells allow lower pressure drawdown; (d) horizontal wellbores can be oriented in the direction of maximum wellbore stability; (e) horizontal wells can improve sweep efficiency; (f) horizontal wells allow better control of hydraulic fracture direction. These factors are discussed briefly in this article.

268. Economides, M., J. McLennan, and R. P. Marcinew (1989). Fracturing of highly deviated and horizontal wells. in 40th Annual Technical Meeting of The Petroleum Society of CIM, May 28-31, 1989, Banff, Alberta. The Petroleum society of CIM, Montreal. CIM paper 89-40-38.

Abstract not published.

269. Economides, M. J., K. Ben-Naceur, and R. C. Klem (1989). Matrix stimulation method for horizontal wells. Proceedings 1989 SPE Annual Technical Conference & Exhibition, Production Operations and Engineering volume, October 8-11, 1989, San Antonio, pp. 179-187. Society of Petroleum Engineers, Richardson, TX. SPE paper 19719. Abstract in The Engineering Index Monthly (January, 1990), 28 (1), 680. Also in Journal of Petroleum Technology (1991) 43 (7), 854-861.

Matrix stimulation in horizontal wells may be beneficial, even in cases where only hydraulic fracturing should be used in vertical wells. Matrix stimulation may be done in open holes, through slotted liners, or through perforated, cemented casing. Stimulation of horizontal wells using reactive fluids can best be done by using coiled tubing, which ensures complete coverage of the selected zones. The tubing is pushed to the end of the well, and, as it is slowly withdrawn, the stimulation fluid is pumped through the tubing while an inert fluid is pumped through the annulus between the tubing and the well to prevent backpressure. Criteria are given for selection of horizontal wells for matrix stimulation, techniques of the treatment, and design of the stimulation fluids.

270. Economides, M. J., John D. McLennan, Ernest Brown, and Jean-Claude Roegiers (1989). Part 1. Performance and stimulation of horizontal wells. World Oil 208 (6), 41-45. Abstract in The Engineering Index Monthly (December, 1989), 27 (12), 563.

Horizontal wells may be drilled for a number of reasons including: intersecting of natural fractures; improving hydrocarbon recovery; lowering drawdown during production; improving wellbore stability; improving sweep efficiency; and improving hydraulic fracturing. Earlier attempts to drill horizontal wells in the 1950's and 1960's proved to be disappointing. By the 1980's the method had achieved major success in Cold Lake, Canada: Rospo Mare field, Lacq field, and others. The superior performance of horizontal wells in proper circumstances and the stimulation of these wells are discussed in this article.

271. Economides, Michael J., John D. McLennan, Ernest Brown, and Jean-Claude Roegiers (1989). Part 2. Performance and stimulation of horizontal wells. World Oil 209 (1), 69-72. Abstract in The Engineering Index Monthly (December, 1989), 27 (12), 563.

Methods and performance for multiple hydraulic fracturing of horizontal wells are addressed in this article.

272. Economides, M. (editor) (1989). *Horizontal Wells.* Prentice Hall, New York.

Not examined.

273. Economides, M. J., F. X. Deimbacher, C. W. Brand, and Z. E. Heinemann (1990). Comprehensive simulation of horizontal well performance. Proceedings 1990 SPE Annual Technical Conference & Exhibition, Reservoir Engineering volume, September 23-26, 1990, New Orleans, pp. 37-48. Society of Petroleum Engineers, Richardson, TX. SPE paper 20717.

Simulation of horizontal wells using real geometries has been developed using locally-refined perpendicular gridding. Hydraulic fractures penetrating the well transversely and logitudinaly have also been simulated. Details of the simulation model are given in the paper.

274. Edlund, P. A. (1988). Application of recently developed medium-curvature horizontal-drilling technology in the Spraberry trend area. Journal of Petroleum Technology 40 (9), 1178-1182. SPE paper 16170. Also in Proceedings 1987 SPE/IADC Drilling Conference New Orleans, March 15-18, pp. 1057-1063. Reprinted in SPE Reprint Series 30, Directional Drilling (1990), pp. 178-182. Abstract in The Engineering Index Annual (1988), 87, Part IV, 6001.

A medium-radius horizontal well was drilled in the Spraberry Trend, Midland County, Texas. The well was drilled so that 891 feet of the hole was in the fractured Spraberry formation instead of the 70 feet of a vertical well. The W.M. Schrock 38 No. 7 well was drilled to a measured depth of 8,930 ft (2,722 m) and vertical depth of 7,929 ft (2,417 m). An angle-build BHA with MWD was used from the kick-off point to the top of the Spraberry at which point an angle-hold BHA was substituted. The MWD malfunctioned throughout the operation and the downhole motor broke at one point. The hole was logged with the MWD sensors after the motor was removed.

275. Edwards, Don A. (1990). Assessment of fracture-induced anisotropy in the Austin Chalk formation (Upper Cretaceous), central Texas. A.A.P.G. Bulletin 74 (5), 647.

Abstract. Horizontal drilling has been estimated to increase productivity of Austin Chalk wells in central Texas from 75,000 barrels of oil to 500,000 barrel of oil per well. Gas productivity also increases significantly. Three-component vertical seismic profiles can be used to analyze shear waves to detect the dominant fracture orientations in the wells.

276. Edwards, R. D. and George Strelkov (1988). Slant-hole drilling finds expanding role in Canada. Petroleum Engineer International 60 (2), 20, 22-23. Abstract in The Engineering Index Annual (1988), 87, Part IV, 6001.

Slant-hole drilling is a method for achieving horizontal displacement of wells to reach targets not accessible to conventional directional drilling and in multi-well programs where the wellheads are close together on a drill site. Slant-hole drilling can be used to minimize land use and to reduce costs by limiting surface gathering systems and maintenance costs. Slant-hole drilling has been used successfully at Wolf Lake, Alberta, Canada, with 10 of 23 wells being slant-holes and 13 wells conventionally directionally drilled. The slant-holes were drilled with steerable motors. The wells were drilled from a multiwell drilling pad with a central mud system which improved economics. Slant-hole drilling proved to be cost- and time-effective.

277. Ehlers, R., L. Kracht, and J. Witte (1989). Case history of horizontal wells drilled with navigation technology in European operations. Proceedings 1989 SPE/IADC Drilling Conference, February 28-March 3, 1989, New Orleans, pp. 315-324. Society of Petroleum Engineers, Richardson, TX. SPE/IADC paper 18654. Abstract in The Engineering Monthly Index (November, 1989), 27 (11), 761.

Case histories of four medium- and long-radius horizontal wells drilled in Europe using navigation drilling. The wells were drilled in Denmark and The Netherlands. The horizontal sections ranged from 1,643 ft to 2,650 ft in length.

278. Ekrann, S. E.(September 29, 1987). Production from thin oil zones. Rogaland Research Institute. Technical Report No. SPT T-6/87. SPOR Program Report.

Not examined.

279. El Rabaa, W. (1989). Experimental study of hydraulic fracture geometry initiated from horizontal wells. Proceedings 1989 SPE Eastern Regional Conference, October 24-27, 1989, Morgantown, pp. 349-364. Society of Petroleum Engineers, Richardson, TX. SPE paper 19720. Also in Proceedings 1989 SPE Annual Technical Conference & Exhibition, Production Operations and

Engineering, San Antonio, October 8-11, 1989, pp. 189-204. Abstract in The Engineering Index Monthly (January, 1990), 28 (1), 683.

Laboratory experiments using blocks of rock were done to identify stimulation problems in horizontal wells. The rock blocks were subjected to tri-axial loading with boreholes at angles varying from 0 to 90 degrees to the minimum applied stress, and hydraulic fractures were induced. The following problems were identified: multiple fractures were created from the same perforated interval; created fractures are non-planar; fluid communication between perforations and fractures can be through small channels; and interference can occur between fractures from different perforated intervals. Some recommendations given in the article are: in-situ stress field must be determined in the target zone; stresses must be determined above and below the target zone for fracture height containment; wellbore diameter should be as large as feasible; and wellbore deviation from minimum stress must be controlled.

280. Elson, T. D., R. H. Darlington and M. A. Mantooth (1983). High-angle gravel pack completion studies. Proceedings of the 53rd SPE Annual California Regional Meeting, March 23-25, 1983, Ventura, pp. 19-32. Society of Petroleum Engineers, Richardson, TX. SPE paper 11012 presented at the 57th Annual SPE Fall Technical Conference and Exhibition, September 26-29, 1982, New Orleans. Journal of Petroleum Technology (1984), 36 (1), 69-78.

Laboratory studies were made to determine the optimum gravel pack procedures in high-angle wells. The experiments were done in an 1,100-ft long tubing-string at 80 degrees inclination. After gravel slurries with satisfactory transport characteristics were developed, a full scale cased hole completion model was constructed. Results of this study show that high-viscosity carrier fluids (600-700 cp) with high gravel concentrations (15 lb/gal) provide good transport but are unsuitable for high inclination boreholes. Transport and packing were improved by using lower carrier viscosity (300-400 cp) and 4 lb/gal gravel concentration. Optimum results require special liner-tailpipe (washpipe) geometry.

281. Emery, Maurice Myron (1985). Optimizing results for bent-sub drilling applications. Drilling 46 (4), 46-48.

Use of a bent-sub in directional drilling causes several problems stemming from side-loading. Side-loading can cause unexpected changes in drilling direction, and robs power from the mud-motor. Alternatives to a bent-sub are the use of offset stabilizers or a bent-housing. Offset stabilizers also cause side-loading and will cause the hole to be slightly out of gauge. A bent-housing is a bend in the motor assembly close to the bit. This arrangement eliminates power loss to the mud motor, bit performance is improved, accurate hole orientation can be achieved, and the bent-housing assembly can be rotated to drill straight sections of hole.

282. Erdol Erdgas Kohle (1988). Interview -Erdolbergbau: Konkurrenz fur die Horizontalbohrtechnik? Erdol Erdgas Kohle 104 (2), 58-59. In German.

Indexed but not annotated.

283. Ertekin, T., W. Sung, and F. C. Schwerer (1988). Production performance analysis of horizontal drainage wells for the degasification of coal seams. Journal of Petroleum Technology 40 (5), 625-632. SPE paper 15453. Abstract in The Engineering Index Annual (1988), 87, Part IV, 6029.

Models are presented for the evaluation of the performance of one or more horizontal boreholes used for coalbed degasification. The important parameters of a horizontal borehole in degasification are: diameter of the borehole, length of penetration of the coalbed, positioning of the borehole, and orientation of the borehole. Important coalbed properties are: thickness, porosity, permeability, and sorption characteristics. All of these factors are considered in the model presented.

284. Escaron, P. C. (1983). A technique to evaluate deviated wells with standard logging tools. Preprints, 1983 SPE Annual Technical Conference & Exhibition, October 5-8, 1983, San Francisco. Society of Petroleum Engineers, Richardson, Tx. SPE paper 12180.

A new technique has been developed to log horizontal and high-angle wells using a combination of drillpipe- conveyed tools and pump-down stinger. The system involves attaching the logging tools to the drillpipe with a releasable locking mechanism. The tool is conveyed to the top of the zone to be logged, the electrical cable is then sent down on a locomotive which makes the electrical connection to the tools and releases the locking mechanism. After release, the tool and cable assembly the are pumped to the bottom of the zone to be logged, and then retrieved with the cable. Logging is done during tool retrieval. Details of the system are given in the article.

285. Esteban, Mateu (1987). Unconformities, paleokarst facies, and porosity evaluation. A.A.P.G. Bulletin 71 (11), 1440.

Abstract. Paleokarsts formed at unconformities may present a picture different from that of the modern karst model of meteoric diagenesis of Quaternary carbonates in the tropics. Facies analysis and modeling offers advantages in the study of such karsts. Certain typical log characteristics indicate paleokarsts. Editor's Note: Paleokarsts are now recognized as a potential target of horizontal drilling e.g. Rospo Mare field, Italy - see index.

286. Evans, Hilton (April, 1991). Evaluating differences between wireline and MWD systems. World Oil 212 (4), 51- 52, 54, 56, 58-60, 62, 136.

Logs run for formation evaluation with MWD may not be the same as the same log run as a wireline tool. One of the primary reasons for this difference is that the MWD log will be made before formation invasion by drilling fluids has been completed. This article reviews the reasons for the differences in the logs and the reasons for those differences.

287. Ewing, Thomas E. (1991). Tectonic factors affecting the Austin Chalk play. in Chuber, Stewart (editor), Abstracts and Short Papers of the South Texas Geological Society Austin Chalk Exploration Symposium, February 24-26, 1991, San Antonio. South Texas Geological Society, San Antonio.

Abstract. Tectonic analysis of the Austin Chalk trend of Texas indicates that much of the faulting and fracturing is associated with extension toward the Gulf of Mexico in the San Marcos Arch area and Giddings field area. Other regional stresses may be post-Miocene (between Atascosa County and Pearsall field) or Laramide (southwestern part of the Pearsall trend).

288. Falconer, I. G. and D. Normaore (October 26, 1987). MWD bit-efficiency model provides real-time answers. Oil & Gas Journal 85 (43), 40-42, 44, 47-48.

Values of torque and weight-on-bit from MWD data have been used to develop a drilling model. The model can be used to calculate the mechanical efficiency of the bit, wear of the bit teeth and strength of the formation, on a foot-by-foot basis. The interpretation output is the Mechanical Efficiency Log (MEL). The uses of the log are: to avoid unnecessary trips; to reduce time spent drilling with a worn out bit; to prevent the loss of cones; to identify tight zones while drilling; selecting coring and casing points while drilling.

289. Fantoft, S., P. K. Krogh, and S. Pollen (1988). Evaluation of oil recovery from a thin oil column by a horizontal well in the Oseberg Gamma North reservoir. Proceeding European Petroleum Conference, October 17-19, 1988, London, pp. 73-84. Society of Petroleum Engineers,

Richardson, TX. SPE paper 18341. Abstract in The Engineering Index Monthly (February, 1990), 28 (2), 636.

A horizontal well increased oil recovery 2.4 times from a thin oil zone in the Oseberg field on the Gamma North structure in the Norwegian sector of the North Sea. The thin oil zone was sandwiched between a gas cap and an aquifer. The well extended 500 meters horizontally into the Statfjord formation reservoir. Details of the project are given in the article.

290. Farouq Ali, S. M. (1989). Application of horizontal wells to bottom/top water reservoirs. Journal of Canadian Petroleum Technology 28 (2), 38. CIM paper 89-40-56. Paper presented at the 40th Annual Technical Meeting of The Petroleum Society of CIM.

Abstract not printed.

291. Feenstra, R. (1988). Status of polycrystalline-diamond-compact bits: Part 1 - Development. Journal of Petroleum Technology 40 (6), 675-684. SPE paper 17919.

Polycrystalline-diamond-compact (PDC) bits are a new generation of bits particularly suited to drilling soft-to medium-hard, non-abrasive formations. A PDC bit is one in which PDC cutters are mounted in a bit, and should be differentiated from the generic terms "shear bits" and "polycrystalline diamond." A PDC cutter is a thin PDC layer on top of a thicker tungsten carbide substrate. This article gives a detailed history of the construction, uses, limitations, and development history of PDC bits. Editor's Note: PDC bits are widely used in horizontal drilling, particularly in the Austin Chalk.

292. Feenstra, R. (1988). Status of polycrystalline-diamond-compact bits: Part 2 - Applications. Journal of Petroleum Technology 40 (7), 817-821. SPE paper 17920, Distinguished Author Series.

Polycrystalline-diamond-compact (PDC) bits can usually be used successfully only in soft, soft to medium-hard, and medium-hard, non-abrasive formations. Bit performance is affected most by bit hydraulics, mud type, and bit design. PDC bits are particularly useful in drilling depleted reservoirs, in deviation control, in directional control, in bit steering, drilling with turbine and positive displacement motors, slim-hole drilling, large-hole drilling, oversized-hole drilling, underreaming, milling, and drainhole drilling (horizontal drilling). In suitable formations PDC bits are also good for coring.

293. Fernandes, Gerson (1989). A perfuracas horizontal em aguas profundas. [Horizontal drilling in ground water]. Mineracao Metalurgia 53 (504), 44-45. In Portuguese.

Not examined.

294. Ferreira de Souza, Reinaldo and Adalberto Jose Rosa (1987). Slug test in horizontal wells. Unsolicited SPE paper. SPE paper 17136.

Physical and mathematical models are given for analysis of slug tests in horizontal wells.

295. Fertl, Walter H. (November 25, 1987). Well logging technology for highly deviated and horizontal wellbores. Journal of Petroleum Science and Engineering 1 (1), 83-90. Abstract in The Engineering Index Annual (1988), 87, Part IV, 6008.

Not examined.

296. Fertl, Walter H. and Rod F. Hotz (1987). Efficiently log and perforate 60 [degree]+ wells with coiled tubing. World Oil 205 (1), 32-33, 35.

Coiled-tubing-conveyed logging is an efficient logging method in extended-reach and horizontal wells. The advantages of this method are: continuous logging capability at optimum logging speeds; pushing and pulling capability; direct and reverse circulation capabilities; protection for wireline cable; elimination of side-entry subs and wet connectors; availability of conventional well logging and completion services; and the elimination of the need for a workover rig. Coiled-tubing logging has been used in the Berkhoepen 2001 project, West Germany; Chateaurenard field, France; and Prudhoe Bay field, Alaska.

297. Fertl, Walter H. and Sid B. Nice (1988). Well logging in extended-reach and horizontal boreholes. Proceedings of the 20th Annual Offshore Technology Conference, May 2-5, 1988, Houston, pp. 193-206. OTC paper 5828.

Three pipe-conveyed logging (PCL) systems have been developed by Atlas Wireline Services to overcome logging problems in high-angle (60 degrees) boreholes. In all three systems full-sized standard logging tools, side-entry sub, and 8-conductor wet connectors are used. The "connect" PCL system uses a fiberglass or metal housing to protect the logging tools and an orientation- sub monitors the position of the pad-type devices in the hole. The tool assembly is attached to the drillpipe. The "extended" PCL system houses the logging assembly in a rigid metal housing attached to the drillpipe until the zone to be logged is reached, where the logging assembly is extruded from the housing, and logs obtained by tripping out of the borehole. The "direct-connect" system uses unprotected logging tools directly attached to the drillpipe. In addition to the pipe-conveyed systems, a coiled-tubing-conveyed system has been developed for logging and perforating high-angle boreholes. A variety of logging tools, including dual-induction, gamma-ray, dipmeter, compensated density, and compensated neutron, can be used with these systems. Pipe-conveyed logging systems have been successfully used in a number of fields, including Prudhoe Bay and Chateaurenard.

298. Fertl, Walter H. and Sid B. Nice (1990). Well logging and completion technology for horizontal wellbores. in Kinsland, Gary and Tony Cagle (editors), Transactions of the 40th Annual Meeting, GCAGS & GCS-SEPM, October 17-19, 1990, Lafayette, Louisiana, pp. 217-226.. Gulf Coast Association of Geological Societies. Abstract in A.A.P.G. Bulletin (1990), 74 (9), 1494-1495.

Logging and completion of horizontal wellbores may be done with pipe-conveyed or coiled-tubing-conveyed logging tools. Details of Atlas Wireline Services logging and completion systems are discussed in the article. Examples are given from the Austin Chalk and Bakken formation.

299. Fett, Tom (1991). Fracture evaluation in the Austin Chalk. in Chuber, Stewart (editor), Abstracts and Short Paper of the South Texas Geological Society Austin Chalk Exploration Symposium, February 24-26, 1991, San Antonio, pp. 41-49. South Texas Geological Society, San Antonio.

The Austin Chalk produces oil from naturally fractured intervals. The fractured intervals may vary in extent from not present to several hundred feet horizontally within a specific chalk bed. Some tools that may detect the natural fractures in a wellbore are: the dipmeter used as a Fracture Identification Log; Borehole Televiewer; and Formation Microscanner. These tools are discussed in this article.

300. Finch, William C. (1969). Abnormal pressure in the Antelope field, North Dakota. Journal of Petroleum Technology 21 (7), 821-826. SPE paper 2227 presented at the 43rd Annual SPE Technical Conference & Exhibition, September 29-October 2, 1968, Houston.

The Antelope field has an overpressured reservoir, Sanish pool, of Devonian age. The Sanish reservoir is fractured and it is proposed in this article that the fracturing is natural hydraulic fracturing caused by tectonic forces. Editor's Note: Later workers have concluded that the fracturing in the Bakken was caused by the generation of hydrocarbons which hydraulically fractured the reservoir rocks by the accompanying change in volume.

301. Fincher, R. (October 15, 1985). Vertical wellbore communication via lateral boreholes - An alternative to hydrofracturing. Paper presented at the 1985 Solution Mining Research Institute Meeting, October 15, 1985, Houston.

Not examined.

302. Fincher, Roger W. (1987). Short-radius lateral drilling: A completion alternative. Petroleum Engineer International 59 (2), 29-30, 32, 34-35. Abstract in The Engineering Index Annual (1987), 86, Part IV, 6499.

Eastman Christensen has developed a system for drilling short-radius multiple-drainholes from a single wellbore. Short-radius in this context means angle build rates of 1.5-3.0 degrees/foot. The short-radius rotary drilling system can change the hole geometry from vertical to horizontal in 30 to 60 drilled feet. Multiple lateral-drainholes allow the reservoir to drained more efficiently at lower cost than conventional drilling.

303. Finley, Robert J., S. E. Laubach, Noel Tyler, and M. H. Holtz (1991). Opportunities for horizontal drilling in Texas. A.A.P.G. Bulletin 75 (3), 574.

Abstract. In addition to the Austin Chalk, Texas has several potential horizontal drilling plays. These plays include fractured reservoirs in the Buda formation, Caddo Reef, San Andres/Grayburg formation, and Spraberry/Dean sandstone. Tight gas reservoirs in Texas that may be good horizontal drilling targets include Wolfcamp formation, Lower Cretaceous-Jurassic of the Sabine Uplift, and several Paleozoic units in West Texas.

304. Finley, S. J. and J. C. Lorenz (1989). Characterization and significance of natural fractures in Mesaverde reservoirs at the Multiwell Experiment Site. Proceedings of the 1989 SPE/DOE Joint Rocky Mountain Regional/Low Permeability Reservoirs Symposium and Exhibition, March 6-8, 1989, Denver, pp. 721-728. Society of Petroleum Engineers, Richardson, TX. SPE paper 19007.

Report on the natural fractures at the U.S. Department of Energy Multiwell Experiment Site, Colorado. Fracture orientations, mineralogy of fracture fillings, and discussion of the origins of the fractures are given in the article. Editor's Note: Horizontal drilling experiments have been conducted at the Multiwell Experiment Site (see index).

305. First Boston (n.d.). Horizontal drilling: An initial assessment of the impact on the oil service business. Equity Research Report OS0676.

Not examined.

306. Fischer, David W. and Marvin E. Rygh (1989). Overview of Bakken formation in Billings, Golden Valley, McKenzie counties, N. D. Oil & Gas Journal 87 (47), 71-73.

The Bakken formation of the Williston basin is the focus of an horizontal drilling play. Meridian Oil Inc. initiated the play and gives four reasons for drilling horizontal wells in the Bakken: (1) increase the probability of success; (2) increase profit potential; (3) increase effective drainage area and maximize recoverable reserves; (4) increase productivity. The Lower Mississippian/Upper Devonian Bakken is a thin shale/limestone/siltstone sequence, characterized on logs by high gamma-ray deflections (200 API) and high-resistivity (1,000 ohms). The Bakken produces from naturally fractured reservoirs. Bakken fields are related to overpressured areas which may be the result of hydrocarbon generation which produces high pore pressures which cause the fracturing.

307. Flak, L. H. (1989). Discussion of planning and operational requirements for a shallow-objective, high-angle well in the Gulf of Mexico. SPE Drilling Engineering 4 (1), 76. SPE paper 19015.

Critical discussion of Christian, (1988), Planning and operational requirements for a shallow-objective, high- angle well in the Gulf of Mexico. SPE Drilling Engineering 3 (3), 241-247.

308. Folefac, A. N. and J. S. Archer (1987). Application of horizontal wells in the simulation of heavy oil recovery process. SPE unsolicited manuscript. SPE paper 17078. Abstract in The Engineering Index Annual (1988), 87, Part IV, 6027.

Horizontal wells may provide solutions to some of the problems in steamflood and hot waterflood displacement of heavy oil. Horizontal wells should allow better placement of injection fluids to give better downward displacement of oil toward horizontal drainholes. Conclusions reached are: (a) horizontal well productivity is dependent on well length and lateral extent of the reservoir; (b) horizontal wells should reduce breakthrough time under conditions of gravity flow; (c) sweep efficiency is enhanced if the injection-producer pair of wells are placed as far apart as possible and away from impermeable boundaries; (d) increasing the horizontal well length reduces coning and increases critical production rate.

309. Fontenot, John E. (1986). Measurement While Drilling - A new tool. Journal of Petroleum Technology 38 (2), 128-130.

Measurement-while-drilling (MWD) allows resistivity and natural gamma-ray logging during the drilling process. Data is sent to the surface using a mud-pulse or wireline telemetry system.

310. Fontenot, J. E. (1986). Author's Reply to discussion of measurement while drilling - a new tool. Journal of Petroleum Technology 38 (5), 590-591.

Reply to critical discussion by L. M. Moseley of "Measurement while drilling - A new tool," Journal of Petroleum Technology 38 (5), 590.

311. Fontenot, John E. and M. Vikran Rao (February 15, 1988). MWD poised for future - 3. MWD can improve well safety, control. Oil & Gas Journal 86 (7), 40-41, 44-47.

MWD can improve well safety by improving detection of overpressured zones and in the future detection of formation fluid ingress.

312. Fontenot, J. E. (1988). Successful high angle completions, cementing, and drilling impact. Proceedings International Meeting on Petroleum Engineering, November 1-4, 1988, Tianjin, China, pp. 831-842. Society of Petroleum Engineers, Richardson, TX. SPE paper 17628.

Detailed discussion of high-angle, horizontal, and extended-reach well completions with emphasis on cementing.

313. Fontenot, John E. and M. Vikran Rao (March 4, 1988). MWD poised for future - 4. MWD aids vital drilling decisions. Oil & Gas Journal 86 (11), 60, 62-65.

MWD can help in making drilling decisions and improving drilling efficiency. MWD advantages include: reduced time for directional surveys; reduced number of trips; timely formation evaluation using gamma-ray and resistivity logs; improved pore-pressure determination. Other advantages are discussed in this article.

314. Fontenot, John E. and M. Vikran Rao (March 28, 1988). MWD poised for future - conclusion. Measurement-while-drilling essential to drilling. Oil & Gas Journal 86 (13), 52-55, 58.

Limitations of MWD are: size of borehole, hole diameters of 7.75 inches or larger; temperature, most MWD systems are restricted to less than 257 degrees Fahrenheit; data sampling rates vary but most require 8 to 60 seconds with some requiring up to 1,800 seconds (gravity, magnetic and temperature surveys). Drilling safety (fluid influx detection, mud density, etc) and drilling efficiency/trouble avoidance (mud viscosity, annular pressure, etc) data will be added to MWD in the future.

315. Forrest, J. K. (1990). Horizontal gravel packing studies in a full-scale model wellbore. Proceedings 1990 SPE Annual Technical Conference & Exhibition, Production Operations and Engineering volume, September 23-26, 1990, New Orleans, pp. 647-655. Society of Petroleum Engineering, Richardson, TX. SPE paper 20681.

Full-scale, 45- and 100-ft models were used to study gravel-packing on high-angle (greater than 60 degrees) and horizontal wells. Conventional viscosified carrier fluids gave incomplete annular packing. Pack efficiency can be improved by eliminating blank sections between screen joints. Low-viscosity carrier fluids may give better gravel packs in some conditions. One method that gave better results was a two-stage packing technique that used a viscous carrier and low-viscosity carrier in combination. Low-side perforation packed completely in all tests while top-side perforations rarely packed. Details of the experiments are given in the paper.

316. Fox, Chris (July, 1988). Moving towards tomorrow's 'well architects.' Offshore Engineer, 27

Interview with Jean Francois Giannesini, president of Horwell. Giannesini believes that within ten years vertical wells will be used only for exploration and appraisal. Production wells will be designed by "well architects" who will design a network of conduits using horizontal and vertical wells tailored to maximize recovery from reservoirs.

317. Fox, Chris (August, 1988). Downhole electromagnetic controls face testing time. Offshore Engineer, p. 101.

Baker Oil Tools develops electromagnetic control equipment to improve control of downhole completion equipment.

318. Fox, Chris (July, 1990). North Sea drills into the nineties with horizontal wells high on the agenda. Offshore Engineer, pp. 37-30.

A review of the 1990 horizontal drilling activity in the North Sea. Operations are summarized in a two-page table listing projects by operator.

319. Fraser, Lindsay (1990). Effective ways to clean and stabilize high-angle holes. Petroleum Engineer International 62 (11), 30, 32, 34-35.

Discussion of the problems of cuttings transport in high-angle and horizontal wells. Guidelines are given for drilling fluid design.

320. Freeborn, W. R., B. Russel, A. MacDonald (1989). South Jenner horizontal wells - a case study. Proceedings of Heavy Oil Special Interest Group of Petroleum Society of CIM, April 20, 1989, Calgary. The Petroleum Society of CIM, Montreal.

Not examined.

321. Freeborn, W. Randy, Boyd Russell and Alistair J. MacDonald (1990). South Jenner horizontal wells a water coning case study. Journal of Canadian Petroleum Technology 29 (3), 41-46.

Horizontal drilling was used in the South Jenner field to combat the effects of water coning. Two horizontal wells with displacements of 420m and 1042m were targeted to recover oil from a thin pay zone in a bottomwater reservoir. The operator, AEC Oil and Gas, found that drilling costs fell with experience (steep learning curve) and that careful consideration needs to be given to the economics of the project as production rates are significantly different for the two cases (vertical well vs horizontal well). Some of the recommendations of AEC in drilling horizontal wells are: (a) drill in areas with precise geologic control; (b) use steerable systems with known build rates; (c) use long radius of curvature; (d) avoid intermediate casing if possible; (e) production logs are not generally useful; (f) production testing is not required; (g) use bullet guns when low shot density is desired; and (h) reasonable estimates of recovery can be made using a modified Butler correlation.

322. Freisatz, Wayne B. (1990). Fracture-enhanced porosity and permeability trends in Bakken formation, Williston basin, western North Dakota. A.A.P.G. Bulletin 74 (8), 1324.

Abstract. The Bakken formation has low matrix porosity and permeability, high organic content, high thermal maturity, and high relative lateral homogeneity. Effective porosity and permeability in the Bakken are the results of natural fractures. LANDSAT derive lineament maps can show the orientation and extent of the fractures.

323.
Freisatz, Wayne B. (1990). Fracture-enhanced porosity and permeability trends in Bakken formation, Williston basin, western North Dakota. in Hansen, William B. (editor), Bakken Formation Short Course, p. 19. Montana Geological Society, Billings. Study Notes No. 1.

Abstract. Effective porosity and permeability of the Bakken are the result of natural fractures. Landsat data and large scale images show regional lineaments and fractures in the Williston basin.

324. Fritz, Mary (July, 1990). Something old, something new: Horizontal wells give Florence a future. AAPG Explorer 11 (7), 1, 12-13.

Florence field in Colorado, discovered in the 1860's, is the oldest continuously producing oil field in the United States. Recent horizontal wells, actually high-angle wells, have raised hopes for a revitalization of the field. Three wells have been drilled, one reaching an angle of 79 degrees, have intersected oil-filled fractures. Production has been rather low but encouraging.

325. Fritz, Mary (August, 1990). Promising play endures: Horizontal drilling brings home the Bakken. A.A.P.G. Explorer 11 (8), 1, 14-15.

Overview of the Bakken formation horizontal drilling play in the Williston basin.

326. Fritz, Mary (June, 1991). Horizontal drilling comes full circle. A.A.P.G. Explorer 12 (6), 1, 18.

Columbia Gas Development has successfully completed a horizontal well in the Cane Spring shale member of the Pennsylvanian Paradox formation. The Kane Spring Federal #21-1, located in Utah, tested 2,302 BOPD and 627 MCFGPD through a 14/54-inch choke. The Cane Spring reservoir is a naturally fractured reservoir. The well has a measured depth of 8,244 ft, TVD is 7,248 ft, and a horizontal length in the reservoir of 1,000 ft.

327. Fritz, Richard D. and Christopher L. Johnson (1991). Horizontal drilling potential in the Mid-Continent region. in Chuber, Stewart (editor), Abstracts and Short Papers of the South Texas Geological Society Austin Chalk Exploration Symposium, February 24-26, 1991, San Antonio. South Texas Geological Society, San Antonio.

Abstract. Four reservoir types have been identified as potential horizontal drilling targets. Type 1 has little or no matrix porosity with the effective porosity provided by natural fractures. Type 2 has fair to good matrix porosity and low permeability. Type 3 has fracture enhanced porosity of good matrix porosity with fair permeability and is frequently associated with paleokarst. Type 4 reservoirs have good porosity and permeability and have gas and/or water coning problems.

328. Frost, E., D. Stedman and W. H. Fertl (1982). Formation evaluation in the Texas Cretaceous chalk trend. World Oil 194 (7), 213-214, 217, 219, 222, 225, 228, 231, 233, 236. Abstract in The Engineering Index Annual (1982) 81, Part IV, 5195.

Review of formation evaluation methods that indicate the presence of open and potentially productive fractures in the Austin Chalk.

329. Fryt, Monte (1990). An evaluation of log responses in the Bakken shale. in Hansen, William B. (editor), Bakken Formation Short Course, pp. 19-20. Montana Geological Society, Billings. Study Notes No. 1. Abstract also printed in Hansen, William B. (editor) (1991), 1991 Guidebook to Geology and Horizontal Drilling of the Bakken Formation, p.113, Montana Geological Society, Billing, Montana.

Abstract. Useful maps of the Bakken can be made by mapping log parameters such as deep resistivity, bulk density, and gamma ray curves. Bulk density, for example, closely approximates the variation in total organic carbon in the Bakken shale members.

330. Fuh, G. F., D. L. Whitfill, and R. P. Schuh (1988). Use of borehole stability analysis for successful drilling of high-angle holes. Proceedings 1988 IADC/SPE Drilling Conference, February 28-March 2, 1988, Dallas, pp. 483- 491. Society of Petroleum Engineers, Richardson, TX. IADC/SPE paper 17238.

Removal of rock during drilling changes the stress in the near borehole rocks, and may lead to instability of the borehole itself. Borehole collapse may be reduced by using optimum mud weights to control hole stability. Pore pressure alone is not enough to predict hole stability. Lowest mud weight used should be the borehole collapse gradient computed from the local borehole stresses, pore pressure, and rock strength. The borehole collapse gradient decreases with increasing hole angle and may be greater than the pore pressure gradient. Compressive failures of borehole are progressive and originate from lack of hydrostatic support of and increased load on borehole walls as more rock is removed by drilling.

331. Fuh, G.-F. and P. K. Loose (1989). Horizontal wellbore stability for openhole completions. Proceedings 1989 SPE Annual Technical Conference & Exhibition, Production Operations and Engineering volume, October 8-11, 1989, San Antonio, pp. 155-164. Society of Petroleum Engineers, Richardson, TX. SPE paper 19717.

Horizontal wells may be useful for recompleting vertical wells. One recompletion in the North Sunshine field, Wyoming, entailed removal of the liner from a horizontal well. The well had been recompleted by drilling laterally 1,000 ft from the vertical wellbore and completing through 4.5 inch liner. Production was lower than expected so borehole analysis was done to determine if the borehole was stable enough to remain open without the liner. The analysis concluded that the liner could safely be removed. After removal the borehole performed as predicted and later a 300-ft section of the open hole was successfully perforated. Details of the borehole stability analysis are given in the article.

332. Fuh, Giin-Fa, E. G. Dew, C. A. Ramsey, and K. Collins (1990). Borehole stability analysis for the design of first horizontal well drilled in the U.K.'s southern 'V' fields. Proceedings 1990 SPE Annual Technical Conference & Exhibition, Drilling volume, September 23-26, 1990, New Orleans, pp. 31-42. Society of Petroleum Engineers, Richardson, TX. SPE paper 20408.

Borehole stability analysis was used to aid in planning the first horizontal well in the U.K. 'V' fields. The well was the first medium-radius well drilled in the U.K. and the fourth horizontal well drilled on the U.K. continental shelf. Details of the analysis are given in the article.

333. Fultz, Jim and Fred J. Pittard (1990). Bottomhole system works over, completes horizontal well. World Oil 210 (3), 48-50. Abstract in The Engineering Index Monthly (June, 1990), 28 (6), 671.

Slimdril International Inc has developed a system for horizontal well workovers which is described in the article.

324. Fultz, J. D. and F. J. Pittard (1990). SlimDril horizontal workover system. in ASME Petroleum Division Publication PD v. 27 Drilling Technology Symposium, January 14-18, 1990, New Orleans, pp. 181-184. ASME, New York. Presented at the Thirteenth Annual Energy-Sources Technology Conference and Exhibition. Abstract in The Engineering Index Monthly (June, 1990), 28 (6), 669.

Not examined.

335. Gallup, Andrew, B. L. Wilson and Robert Marshall (1990). ESP's placed in horizontal lateral increase production. Oil & Gas Journal 88 (25), 58-63.

Electric submersible pumps (ESP) are effective in lifting fluids in horizontal wells. ESPs may require special considerations for proper installation and operation. To maximize the potential of the ESP in horizontal wells the following practices should be followed: (a) drill the well as large as is economically possible; (b) drill the horizontal section with a continuous down-hill inclination to minimize gas locking; (c) design a straight tangent section in the build curve to provide an alternate location for the ESP; (d) install a reduced-diameter liner below the ESP to increase superficial velocity; (e) use only bag-type seal chambers; (f) centralizers and cable protection devices may be needed, they should be used with caution; (g) variable-speed drive should be considered.

336. Garber, Raymond A. and Paul M. Harris (1987). Similarity of Grayburg/San Andres dolomitic reservoirs - Eastern edge of Central Basin Platform, Permian basin. A.A.P.G. Bulletin 71 (2), 238.

Abstract. The dolomitic reservoirs of the Grayburg and San Andres formations of the eastern edge of the Central Basin Platform, Permian basin, have similar vertical sequences. This study uses data from McElroy field/Goldsmith field, and Waddell field to illustrate this sequence. The sequence has three parts: basal open-shelf; middle shallow shelf; and upper variable intertidal/supratidal deposits. Editor's Note: The San Andres and Grayburg formations are horizontal drilling targets.

337. Gatens, J. M., III, C. W. Harrison, C. W., III, D. E. Lancaster, F. K. Guidry (1990). In-situ stress tests and acoustic logs determine mechanical properties and stress profiles in the Devonian shales. SPE Formation Evaluation 5 (3), 248-254. SPE paper 18523 presented at the 1988 SPE Eastern Regional Meeting, Charleston, WV, November 1-4, 1988.

In-situ stress profiles are important for designing hydraulic fracture treatments. Vertical fracture height or growth is determined by the stress contrast between layers. In-situ stress tests were conducted in the Devonian shale of the Appalachian basin in two wells: Sterling Drilling & Production Inc.-Jarvis No. 1143, Calhoun County, West Virginia, and Ashland Exploration Inc.-E.J. Evans No. 91, Breathitt County, Kentucky. The shales evaluated have a low stress-gradient and few horizontal barriers to vertical fracture-height growth. Methods of measurement are discussed in the paper.

338. Gates, R. L. and G. Schwab (1984). Specialized drilling systems set new world records in high-angle drilling. Journal of Petroleum Technology 36 (2), 241-248. SPE paper 10421 presented at the 1982 Offshore South East Asia Conference, Singapore, February 9-12. Abstract in The Engineering Index Annual (1984), 83, Part IV, 6665.

Two high-angle wells (up to 72 degrees) were successfully completed in the Gippsland basin of Australia. The two wells are producing oil from a previously isolated fault block. The well was drilled with aluminum drillpipe to reduce weight and a "controlled activity" oil-continuous mud was used to reduce torque, drag, and wall-sticking tendencies.

339. Gautier, Donald L. and Dudley D. Rice (1982). Conventional and low-permeability reservoirs of shallow gas in the northern Great Plains. Journal of Petroleum Technology 34 (7), 1600-1608. SPE paper 9846 presented at the 1981 SPE/DOE Low-Permeability Symposium, Denver, May 27-29.

The northern Great Plains have significant shallow gas resources at depths of less than 4000 ft (1200 m). Much of this gas is in low-permeability stratigraphic traps - unconventional reservoirs. Low-permeability reservoirs are known in the Gammon Shale and Milk River Formation. Conventional reservoirs are known in the Gammon and the Eagle Sandstone. Editor's Note: The low-permeability reservoirs discussed in this article are potential targets for horizontal drilling.

340. Gavignet, A. A. and I. J. Sobey (1986). A model for the transport of cuttings in highly deviated wells. Paper presented at the 1986 SPE Annual Technical Conference & Exhibition, October 5-8, 1986, Las Vegas, Nevada. Society of Petroleum Engineers, Richardson, TX. SPE paper 15417. Also published in Journal of Petroleum Technology (September, 1989), 41 (9), 916-921.

A simple model is presented for prediction of cuttings- bed formation in high-angle and horizontal wellbores. Cuttings-bed formation always occurs when the flow is turbulent. Other important factors affecting bed formation are: drillpipe eccentricity; particle size; pipe size; and borehole size. The friction coefficient of the cuttings against the wall of the hole strongly influences bed formation.

341. Genele, F. J., K. L. Schultz, and G. S. McJannet (1991). Horizontal wells in a thick steeply dipping reservoir at NPR-1, Elk Hills, California. Pacific Oil World 83 (3), 35.

Abstract. Two horizontal wells have been drilled in the Elk Hills field, California. Both wells were completed in the Stevens sand. The horizontal wellbores cut across the steeply dipping (60 degrees) beds, where drainage is along the bedding-planes. Some details of the wells are given in the abstract.

342. George, Charles (1988). Cement sheath vital in horizontal well. The American Oil & Gas Reporter 31 (12), 30-32.

A good cement sheath around the casing is a vital factor is completing a horizontal well. Displacement of the drilling mud from the annulus ahead of the cement is the most important factor in establishing the cement sheath. Pipe movement, either reciprocal or rotational, is also important in cementing casing. These and other factors in cementing horizontal wells are briefly discussed in this article.

343. Giannesini, J. F. (January/February, 1984). Horizontal wells and today's technology. Petroasia.

Not examined.

344. Giannesini, J. F. and J. Bosio (1988). Horizontal wells cut offshore production costs. Paper presented at the 7th Annual SPE Offshore SE Asia Conference. Society of Petroleum Engineers, Richardson, TX. SPE paper 18172.

Not examined.

345. Giannesini, Jean-Francois (1989). Horizontal drilling is becoming commonplace: here's how its done. World Oil 208 (3), 35-40. Abstract in The Engineering Monthly Index (June, 1989), 27 (6), 419.

Horizontal wells have increased from a few wells in the early 1980's to over 100 drilled in 1988. A horizontal well is one in which a section of borehole is inclined more than 80 degrees. Horizontal wells may be applied to producing hydrocarbons from vertically fractured reservoirs, thin reservoirs, and tight formations. Horizontal sections of wells have reached over one mile in length. Further details of horizontal drilling techniques are given in the article.

346. Giannesini, Jean Francois (1989). Production technology for horizontal wells takes new direction. World Oil 208 (5), 46-48, 50.

Three methods of logging horizontal wells have been developed: drillpipe-conveyed method, coiled-tubing-conveyed method, and pump-down stinger method. The coiled-tubing-conveyed method appears to be the best method for logging horizontal wells. Two types of horizontal completions are: continuous entry flow completions and discreet entry flow completions. In the continuous entry flow completion the entire horizontal drainhole is open to flow. In the discreet entry flow method, liner or casing is set and fluid entry into the borehole is restricted to discreet intervals by perforating and selective cementation. In the discreet entry method, multiple-zone hydraulic fracturing may be done. Horizontal well design is a multi-disciplinary team effort by geologists, geophysicists, and drilling, reservoir, production, and mud engineers. Giannesini makes the point that "it is symptomatic to note that all operators that have dealt with horizontal wells have done it through specially set up task forces."

347. Gianzero, S., Roland Chemali, Sheymin Su, Lin Min, Yih-Yih, and Mackay Foster (1985). A new resistivity tool for measurement-while-drilling. Transactions SPWLA 26th Annual Logging Symposium, June 17-20, 1985, Dallas, Paper A. Society of Professional Well Log Analysts, Houston.

Discussion of Gearhart's focused resistivity (dual resistivity) tool for use with MWD.

348. Gianzero, S., R. Chemali, and S. M. Su (1989). Induction, resistivity, and MWD tools in horizontal wells. The Log Analyst 30 (2), 96. Abstract of paper presented at the SPWLA 30th Annual Logging Symposium, Denver, June 11-14, 1989.

Abstract. Focused resistivity devices like those used with MWD are more sensitive to conductive than resistive beds. Results of modeling of these tools are discussed.

349. Giger, F. M. (December, 1982). L'equipment de forages horizontaux. [Horizontal well equipment]. Petrole et Technique No. 294, 40-42. In French.

Indexed but not annotated.

350. Giger, F. (May/June, 1983). [Theoretical evaluation of the effect of water cresting on production by horizontal wells]. Revue de l'Institut Francaise du Petrole, p. 38. In French.

Not examined.

351. Giger, F. May-June, 1983). Reduction du nombre de puits par l'utilisation de forages horizontaux. Revue de l'Institut Francais du Petrole 38 (3), 351-360. In French with English abstract. Abstract in The Engineering Index Annual (1983) 85 Part IV, 4928.

Abstract. Analysis of the areal productivity of horizontal wells.

352. Giger, F., J. Comb, and L. H. Reiss (1983). L'interet du forage horizontal pour l'exploitation des gisements d'hydrocarbures [Advantages of horizontal drilling for production from hydrocarbon fields]. Revue d l'Institu Francais du Petrole 38 (3), 329-350. In French with English abstract. Abstract in The Engineering Index Annual (1983) 85 Part IV, 4928.

Abstract. Use of horizontal wells to reduce gas and water coning during multi-phase flow.

353. Giger, F. M. (1983). Evaluation theoretique de l'effet d'arret d'eau sur la production par puits horizontaux [Theoretical evaluating the effect of water cresting on production by horizontal wells]. Revue de l'Institut Francais du Petrole 38 (3), 361-370. In French with English abstract. Abstract in The Engineering Index Annual (1983) 85 Part IV, 4938.

Abstract. Determination of the critical flow rate and water cresting in horizontal wells.

354. Giger, Francois M. and Andre P. Jourdan (1984). The four horizontal wells producing oil in Western Europe. Paper presented at the Fifth Annual Advances in Petroleum Recovery & Upgrading Technology Conference, June 14-15, 1984, Calgary.

Four horizontal wells have been drilled by Elf Aquitaine in Europe. These wells are: two wells in Lacq field, France; one well in Castera Lou field, France; and one well in Rospo Mare field, Adriatic Sea, Italy. All of the wells are in fractured, dolomitized, or karstic carbonate reservoirs. Elf was able to drill, log, and core these horizontal wells at acceptable cost. The projects are described in the article.

355. Giger, F. M., L. H. Reiss, A. P. Jourdan (1984). The reservoir engineering aspects of horizontal drilling. Paper presented at the 59th Annual SPE Technical Conference & Exhibition, September 16-19, 1984, Houston. Society of Petroleum Engineers, Richardson, TX. SPE paper 13024.

This paper presents criteria for choosing horizontal wells over conventional wells. The recommendations are based on the experience of Elf Aquitaine in drilling four horizontal wells in France and Italy. Methods are presented to estimate probable productivity of horizontal wells. Reservoir anisotropy, reservoir thickness, and water or gas coning (critical rate) are important factors in choosing horizontal over vertical drilling. Horizontal drilling in favorable circumstances may improve productivity up to five times that of a vertical well except in the case of a naturally fractured reservoir where improvement may be greater. Case histories of four wells are given.

356. Giger, F. M. (1985). Horizontal wells production techniques in heterogeneous reservoirs. Proceedings SPE Fourth Middle Eastern Oil Show, March 11-14, 1985, Bahrain, pp. 239-246. Society of Petroleum Engineers, Richardson, TX. SPE paper 13710.

Horizontal wells, unlike vertical wells, provide information about the lateral heterogeneities of petroleum reservoirs. Horizontal wells in stratified reservoirs intersect one or at most a few productive horizons. Horizontal wells are, however, able to take advantage of reservoir inhomogeneities, either vertical or horizontal, by exposing a greater length of wellbore to the permeable zones within the reservoir than do vertical wells. Horizontal wells are more effective in thinner reservoirs than they are in thick reservoirs with good vertical permeability. In thin reservoirs, a horizontal well will have lateral flow lines with vertical flow only very near the wellbore. Details of flow models are given in the article.

357. Giger, F. M. (1986). Analytic 2-D models of water cresting before breakthrough for horizontal wells. Paper presented at the 61st Annual SPE Technical Conference and Exhibition, October 5-6, New Orleans. Society of Petroleum Engineers, Richardson, TX. SPE paper 15378. Also in SPE Reservoir Engineering (1989), 4 (4), 409-416. Abstract in The Engineering Index Monthly (April, 1990), 28 (4), 663.

A two-dimensional model is presented to evaluate the effect of water coning on horizontal well performance. Three cases are examined: lateral-edge drive; gas-cap drive; and bottomwater drive. This model allows determination of the best position of the horizontal wellbore with respect to production for the gas-cap drive with a non-active aquifer. The model shows that for the bottom-water-drive case the critical flow rate value decreases with cumulative production.

358. Giger, F. M. (1987). Low-permeability reservoirs development using horizontal wells. Proceedings SPE/DOE Low Permeability Reservoirs Symposium, May 18-19, 1987, Denver, pp. 173-184. Society of Petroleum Engineers, Richardson, TX. SPE/DOE paper 16406.

This paper addresses the following aspects of horizontal drilling: (1) criteria for comparison of horizontal and vertical well productivity; (2) horizontal well depletion profiles; and (3) aerial productivity index and replacement ratio equations. A method is presented for determining the optimum spacing of horizontal wells.

359. Gilhuus, T. (1988). A study of possible oil production from the thin oil zone of the Midgard field. Paper presented at the Recovery from Thin Oil Zones Conference, April, 1988, Stavanger.

Not examined.

360. Gillman, J. R. (1991). Discussion of productivity of a horizontal well. SPE Reservoir Engineering 6 (1), 147-148. SPE paper 21610.

Critical discussion of Babu and Odeh (1989), SPE Reservoir Engineering 4 (4), 417-421.

361. Goetz, Joseph F., Douglas D. Seiler, and Charles S. Edmiston (1991). Fracture description in the Austin Chalk using CAST images. in Chuber, Stewart (editor), Abstracts And Short Papers of the South Texas Geological Society Austin Chalk Exploration Symposium, February 24-26, 1991, San Antonio, pp. 51-63. South Texas Geological Society,

A discussion of the Circumferential Acoustic Scanning Tool (CAST) for detecting fractured intervals in boreholes with examples from Austin Chalk wells.

362. Goode, P. A. and R. K. M. Thambynayagan (1987). Pressure drawdown and buildup analysis of horizontal wells in anisotropic media. SPE Formation Evaluation 2 (4), 683-697. SPE paper 14250 presented at the 1985 Annual Technical Conference and Exhibition, Las Vegas, September 22-25. Abstract in The Engineering Index Annual (1988), 87, Part IV, 6,041.

Analytic solutions are presented for pressure response during pressure drawdown and buildup of a horizontal well. Equations are given that allow calculation of directional permeabilities, average pressures, and mechanical skin factors. The horizontal well may be approximated by replacing the strip source by an effective well radius of one-fourth of the strip length.

363. Goode, P. A. and D. J. Wilkinson (1989). Inflow performance of partially open horizontal wells. Proceedings 1989 SPE Eastern Regional Conference, October 24-27, 1989, Morgantown, pp. 309-320. Society of Petroleum Engineers, Richardson, TX. SPE paper 19341.

This paper presents a method for calculating the Productivity Index (PI) of a horizontal well that has several open intervals in the reservoir. An analytic formula is presented for the determination of inflow pressure of a horizontal well with several open intervals.

364. Gorody, Anthony W. October 1, 1984). TEDSI develops horizontal drilling technology. Oil & Gas Journal 82 (40), 118, 120, 125-126.

Texas Eastern Drilling Systems Inc. (TEDSI) has developed a system for drilling short-radius horizontal boreholes. Using this system several drainholes in one or more horizons may be drilled and completed. The TEDSI system has been successfully used in Grassy Trail Creek field, Utah, where multiple-zone completions have been made in the Triassic Moenkopi formation, a low-permeability, fractured reservoir. The system is described in detail in the article.

365. Gosar, A. J. (1960). Stratigraphic and structural traps in Big Piney-La Barge area, Wyoming. A.A.P.G. Bulletin 44 (6), 957.

Abstract. Brief discussion of the stratigraphy of the Big Piney and La Barge fields, Wyoming.

366. Gose, Matt (1991). Tectonic features of the northwestern Maverick basin. in Chuber, Stewart (editor), Abstracts and Short Papers of the South Texas Geological Society Austin Chalk Exploration Symposium, February 24-26, 1991, San Antonio, pp. 65-71. South Texas Geological Society, San Antonio.

The Maverick basin has oil production from most of the Cretaceous carbonate formations in the basin. Much of this production is from fractured reservoirs. The natural fractures can be related to regional tectonics of the Maverick basin as discussed in this paper.

367. Gosnold, W. D., Jr., R. D. Lefever, and J. J. Crashell (1989). Thermal history of the Bakken shale in Williston basin. A.A.P.G. Bulletin 73 (12), 1491.

Abstract. Discussion of the thermal history of the Bakken formation, Williston basin.

368. Gradeen, G. and B. Lowen (1982). Slant hole drilling for shallow gas. Journal of Canadian Petroleum Technology 21 (2), 40. CIM paper 82-33-44. Abstract of paper presented at the 33rd Annual Technical Meeting of The Petroleum Society of CIM, June 6-9, 1982, Calgary, Canada.

No information is given in the abstract.

369. Graham, Stephen A. and Greg Nazzal (March 18, 1991). Second lateral in horizontal well solves water problem. Oil & Gas Journal 89 (11), 111-114.

Cachara Oil & Gas Co. has completed a Pearsall field Austin Chalk well with two horizontal drainholes. The first drainhole was drilled and completed in the lower B2 and C horizons of the Austin Chalk. This lateral drainhole produced excessive water, so another lateral was drilled and completed in a higher zone, the B1 zone. The second drainhole initially produced 5,754 barrels of oil and 1,100 barrels of water in a twelve day test. During the period of operation to date the well has produced enough oil to cover the operating costs and royalties plus a modest profit. The water produced in these drainholes appears to be water lost during drilling of nearby wells rather than formation water. Details of the lateral drainhole drilling and completions are given in the article.

370. Greener, J. M., K. A. Sauvageau, and I. Pasternack (1988). Horizontal drilling of eolian Nugget sandstone in the Overthrust Belt. Proceedings of the 1988 SPE Rocky Mountain Regional Meeting, May 11-13, 1988, Casper, pp. 429-434. Society of Petroleum Engineers, Richardson, TX. SPE paper 17527.

Horizontal drilling was used to reduce water and gas coning in the Nugget sandstone reservoir of the Rykman Creek field, Wyoming. The horizontal well penetrated a greater interval of the reservoir than would a vertical well, allowing lower drawdown pressures, and increasing the drainage area. The horizontal well was a medium-radius well. Details of the project are given in the paper.

371. Greenip, John F., Jr (1989). How to design casing strings for horizontal wells. Petroleum Engineer International 61 (1), 34-38. Abstract in The Engineering Index Monthly (June,1990), 28 (6), 668.

Considerations for casing design of horizontal wells include: axial and torque loads can be estimated by analyzing the separate intervals; separate analysis is required for load magnitudes of pick up, set down, and neutral states; and development of relationships of parameters which facilitate casing string design. An example is given using a build rate of 20 degrees/foot. The torque required to rotate changes significantly with different build rates. A program for an HP-41 calculator may be obtained to do the necessary calculation for designing a casing string.

372. Gregory, Jenny (March, 1990). Taking a calculated risk on Crawford's complex sands. Offshore Engineer, pp. 36-38.

Hamilton Brothers Oil & Gas will continue to develop Crawford field, U.K., North Sea. Crawford field is a complex field with reservoirs in the Paleocene Andrew, Sele, and Tarbert formations and in Jurassic and Triassic sands. Horizontal wells will be used in the field.

373. Greif, M. A. and C. A. Koopersmith (1986). Petrophysical evaluation of thinly bedded reservoirs in high-angle/displacement development wells with the NL recorded lithology logging system. The Log Analyst XVII (5), 29-38.

High-angle wells were successfully logged in the South Timbalier 300-301 field, offshore Louisiana, using the Gearhart TOOLPUSHER drillpipe-conveyed logging system. Logs run include the NL Recorded Lithology Log (RLL) and MWD gamma-ray log. The tool measures electromagnetic wave resistivity (EWR) which gives better resistivity measurements in 1.5 ft thick beds and can resolve 8 inch thick beds.

374. Griffin, P. J. and P. N. Trofimenkoff (1984). Laboratory studies of the steam-assisted gravity drainage process. Paper presented at the Fifth Annual Advances in Petroleum Recovery & Upgrading Technology, June 14-15, 1984, Calgary.

Report of the simulations and modeling done for the steam-assisted heavy-oil pilot project at Cold Lake by Esso Resources Canada Ltd.

375. Grigorian, A. M. (1969). [Forage des couches productives par puits horizontaux et a completion multiple]. Nedra 190. French translation of the original Russian language article. IFP 127746, February, 1980.

Not examined.

376. Grossman, U. and C. Marx (1990). Bohrkleintraspot in geneigten und horizontalen Bohrungen [Transport of cuttings in deviated and horizontal wells]. Erdoel Erdgas Kohle 106 (2), 58-62. In German with English abstract. Abstract in The Engineering Index Monthly (August, 1990), 28 (8), 599.

Abstract. A laboratory simulator for cuttings transport was developed using Buckingham's [pi]-theorem. Discussion of the results of the experiments.

377. Grout, Marilyn A. and Earl R. Verbeek (1989). Prediction of fracture networks at depth in low-permeability reservoir rocks, Piceance and Washakie basin, western United States. A.A.P.G. Bulletin 73 (9), 1158.

Abstract. Prediction of subsurface fracture patterns at depth from surface data is complex and depends on the nature of the rocks. The DOE Multiwell Experiment site fractured-reservoir fracture network is very different from the surface fractures. Fractures in the C-a mine in the Piceance basin do correspond to surface fractures.

378. Guevara, Edgar H. and Noel Tyler (1991). The Spraberry Trend, Midland basin, Texas: Development, innovation, and reserve growth potential in a mature giant. A.A.P.G. Bulletin 75 (3), 586.

Abstract. The Spraberry Trend field, Texas, is a stratigraphic trap on a west-dipping monocline. The field is 60 miles long and 40 miles wide. The field was originally underpressured. Recovery efficiency of the solution-gas drive field is estimated to be 8%. The Spraberry Trend has been the site of several experimental projects that include horizontal drilling, imbibition flooding, and air-drilling. The field produces from naturally fractured Permian age Wolfcamp, Dean, and Spraberry formations. The reservoir rocks are very-fine grained sandstones and siltstones with porosities less than 10% and permeabilities that average less than 1 md. Editor's Note: The porosities and permeabilities of the Spraberry Trend sandstones and siltstones are comparable to the those of the Austin Chalk fractured reservoirs of the Pearsall and Giddings fields.

379. Gulf Coast Oil World (1990). Regional News: Smako claims horizontal record. Gulf Coast Oil World 8 (4), 22.

Smacko Ltd. has successfully drilled a horizontal well in the Smackover formation of Huxford field, Alabama. The well has a vertical depth of 14,672 ft and a measured depth of 14,936 ft. The well was kicked-off at 14,322 ft measured depth.

380. Gussis, G. L. (1985). Simulation of steam injection through horizontal wellbores for viscous oil recovery. in UNITAR 3rd International Conference on Heavy Crude and Tar Sands, Long Beach, California, v. 2, 699-725. Paper No. HGTF/CF.3/8.17.

Not examined.

381. Gust, D. A. (1983). Norman Wells expansion project directional drilling. Paper presented at the 34th Annual Technical Meeting of the Petroleum Society of CIM. Meeting Report: The Journal of Canadian Petroleum Technology, May 11-13, 1983, Banff. CIM, Montreal. CIM paper 83-34-02. Abstract in The Journal of Canadian Petroleum Technology (1983), 22 (2), 27.

Abstract. Brief description of Esso Resources Canada Ltd horizontal drilling project at Norman Wells field, Canada.

382. Gust, D. A. and R. R. MacDonald (1989). Rotation of a long liner in a shallow long-reach well. Journal of Petroleum Technology 41 (4), 401-404. SPE paper 17183. Also published in Proceedings 1988 IADC/SPE Drilling Conference, Dallas, TX, February 28-March 2, 1988, pp. 51-58.

Proper cementing of horizontal wells is complicated by the difficulty in cleaning the hole and in preventing free water cavities in the cement. Rotation of the liner during cementing was found to solve these problems in an 81 degree well in Norman Wells field, Canada. Drilling and completion programs are discussed in detail in this article.

383. Gust, Douglas (July 24, 1989). Horizontal drilling evolving from art to science. Oil & Gas Journal 87 (30), 43-46, 49-52. Abstract in The Engineering Monthly Index (September, 1989), 27 (9), 787.

Horizontal drilling has developed steadily since 1978 when Esso drilled horizontal wells at Cold Lake, and later at Norman Wells field. New developments in horizontal drilling included steerable systems, downhole motors, MWD, better directional surveys, improved drilling fluids, and better completion methods. Applications of horizontal drilling are: reducing gas and water coning; low-permeability reservoirs; fractured reservoirs; extended-reach wells; reducing the number of development wells needed; acceleration of production; gravity drainage; EOR applications; and connecting multiple-layered zones.

384. Haas, R. C. and Stokley, C. O. (1989). Drilling and completing a horizontal well in fractured carbonate. World Oil 209 (4), 39-40, 42, 44-45. Abstract in The Engineering Index Monthly (June, 1990), 28 (6), 671.

Proco, Inc. (Pinnacle Royalty & Operating Co.) drilled and completed an Austin Chalk horizontal well in the Pearsall field that produced 1,550 BOPD and 200 BWPD. The well was drilled on seismic targets. The horizontal section of the well was 1,250 ft long. Details of the project are given in the article. Mudlogging located a fractured zone not predicted from the seismic data.

385. Hagoort, J. (1978). Oil recovery by gravity drainage. Paper presented at the 53rd SPE Annual Fall Technical Conference and Exhibition, October 3, 1978, Houston. Society of Petroleum Engineers, Richardson, TX. SPE paper 7424.

Gravity drainage may result in recoveries of the oil- in-place of up to 87%. Gravity drainage in this case indicates that gas replaces the oil removed and that gravity dominates over gas displacement. Experiments to determine the effectiveness of gravity drainage of oil from reservoirs were conducted using a variety of rocks including Berea sandstone, Kurkan dolomite, Weeks Island sandstone, and Bentheim sandstone. The experiments show that high oil recovery may be obtained from a water-wet, connate, water-bearing reservoir, depending on the magnitude of gravitational forces relative to viscous forces, shape of oil relative permeability, and reservoir geometry and heterogeneity.

386. Hale, Brent W. (1986). Analyses of tight gas well production histories in the Rocky Mountains. SPE Production Engineering 1 (4), 310-322. SPE paper 11639 presented at the 1983 SPE/DOE Symposium on Low Permeability, Denver, March 14-16.

Analysis of gas production from tight gas-reservoirs in the San Juan basin and Green River basin, Rocky Mountain region. Included in the analysis are data from the Mancos formation and Mesaverde Group wells, both targets for horizontal drilling.

387. Hall, Dan (1987). Steerable system saves time in straight-hole drilling. Petroleum Engineer International 59 (9), 24-25, 28, 30.

Report on the use of steerable drilling systems in drilling straight holes. The examples given are wells in the Mancos Shale, San Juan basin, and Frontier formation, Wyoming. Both formations are now horizontal drilling targets. Steerable systems using downhole motors were found to be more economical than conventional drilling systems.

388. Hanley, Edward J. and Larry E. VanHorn (1982). Niobrara development program, Washington County, CO. Journal of Petroleum Technology 34 (3), 628-634. SPE paper 9850 presented at the SPE/DOE Low-Permeability Gas Symposium, May 27-29, Denver.

The Niobrara formation, a Cretaceous chalk, in the Denver basin produces gas from the Smoky Hill member, a moderately clean fine-grained carbonate rock. Porosity of the Smoky Hill ranges from 30% to 45% with permeabilities of 0.1 to 10 millidarcies. The reservoirs are underpressured with respect to the normal hydrostatic gradient. Niobrara wells are low-volume producing gas wells with production rates of 25 to 400 MCFGPD. Three-thousand ft deep Niobrara wells can be drilled and cased in 72 hours. Niobrara wells are characterized by a sharp decline in production rate in the first 6 to 12 months after start-up with stabilization at about 5% of original production. The expected producing life of these wells is 30 years and the expected recovery is 638 MMCFG/well. The normal stimulation method is foam fracturing.

389. Hanson, P., T. Trigg, G. Rachal, and M. Zamora (1990). Investigation of barite 'sag' in weighted drilling fluids in highly deviated wells. Proceedings 1990 SPE Annual Conference & Exhibition, September 23-26, 1990, New Orleans, pp. 223-230. Society of Petroleum Engineers, Richardson, TX. SPE paper 20423.

Mud weight may vary significantly after trips in highly deviated wells. This phenomenon is termed "sag," and may indicate that pack off or lost circulation is imminent. Four case histories are given for directional wells in the Gulf of Mexico (South Pelto, Green Canyon, Ewing Bank, and High Island).

390. Hansen, William B. (editor) (1990). Bakken Formation Short Course. Montana Geological Society, Billings. Study Notes No. 1.

Miscellaneous notes, list of participants of the short course, abstracts of papers presented, and data on horizontal wells drilled in the Bakken formation. Abstracts are separately listed in this bibliography.

391. Hansen, William B. and George I. W. Long (1990). Introduction to the Bakken formation, U.S. and Canada (or how far does the fairway fly?). in Hansen, William B. (editor), Bakken Formation Short Course, pp. 20-21. Montana Geological Society, Billings. Study Notes No. 1.

Abstract. Bakken production was established in the Antelope field two years after the first discovery of oil in the Williston basin. Bakken discoveries include Elkhorn Ranch field, North Dakota; Rocanville field, Saskatchewan; and Salt Lake field, Montana. Many wells drilled for targets deeper than the Bakken show evidence of naturally fractured reservoirs in the Bakken. Many of these reservoirs have suffered formation damage by injection of salt-based mud during drilling. An indication of formation damage is wide separation of Microspherically Focused Log and Deep Resistivity Laterolog in the Bakken shale interval. The best Bakken producer has produced over 300,000 barrels of oil and was still flowing in January, 1990. The middle member of the Bakken is now known to be a potential reservoir where porosities exceed 10%. The down-dip limit of Bakken production appears to coincide with the 50% water saturation line in the Bakken middle member.

392. Hansen, William B. and George I. W. Long (1990). Criteria for horizontal and vertical prospect generation in the Bakken formation, Williston basin, Montana and North Dakota. A.A.P.G. Bulletin 74 (8), 1325.

Abstract. Three criteria can be used to locate fractures and generate prospects in the Bakken formation. Drillstem tests in vertical wells exhibit rapid pressure build up when the wellbore is in a fractured reservoir. Wide separation between the Dual Laterolog-Microspherically Focused log resistivity curves indicates a fractured zone. Production information can be used to locate fractured zones.

393. Harding, T. G. and S. M. Farouq Ali (1984). Mine-assisted heavy oil recovery technology. Proceedings 1984 SPE California Regional Meeting, April 11-13, 1984, Long Beach, pp. 555-564. Society of Petroleum Engineers, Richardson, TX. SPE paper 12787.

Review of mine-assisted heavy-oil recovery methods including drainhole drilling with steamflood from underground mine tunnels. Emphasis is on Canadian heavy-oil sands in Alberta (Suncor mine; Saline Creek tunnel) and Soviet mine-assisted recovery projects (Yarega field; Baku-area fields).

394. Hardman, P. (1989). Beckingham 36 horizontal well. SPE Drilling Engineering 4 (1), 17-23. SPE paper 15895. Also in Proceeding 1986 SPE European Petroleum Conference, London, October 20-22, pp. 429-428. Abstract in The Engineering Monthly Index (June, 1989), 27 (6), 420.

The first horizontal well in the U.K. was drilled by BP Petroleum Development Ltd. in the Beckingham field in 1985. The target was a thin (49.2 ft - 16.4 ft) sandstone in the Eagle sand. The well (BK 36) was drilled to a vertical depth of 3,369.4 ft (1,027 m) with a measured depth of 5,134.4 ft

(1,565 m). The horizontal displacement was 2,811.6 ft (857 m) with a total horizontal length in the reservoir of 941.6 ft (287 m). The well was put on production test in February, 1986.

395. Harrell, J. W., R. T. Dickinson, and W. W. Wiley, III (1988). The application of a modular measurement while drilling system for directional control and formation evaluation in horizontal wells. Proceedings SPE International Meeting on Petroleum Engineering, November 1-4, 1988, Tianjin, China, pp. 641-652. Society of Petroleum Engineers, Richardson, TX. SPE paper 17612.

Measurement-while-drilling systems can now be designed for specific drilling conditions. Modules that can be used in such systems include: directional tools; pressure sensors; and gamma-ray logs (standard or focused). Formation evaluation tools, such as resistivity tools, may also be used with MWD. This article addresses the use of these tools and the conditions such as hole curvature the limit tool compatibility.

396. Hart, Don (1990). Regional News: D-J spurs massive drilling program; New firms fatten Niobrara ranks. Western Oil World 47 (10), 7-11.

HS Resources Inc. will acquire Amoco's Codell-Niobrara rights on 73,400 acres in Colorado. Elk Exploration Inc., a wholly-owned subsidiary of HS Resources, will begin drilling in October, 1990. A summary of Niobrara horizontal drilling is also given. Horizontal wells will be drilled in the Sand Wash basin and Denver-Julesburg basin.

397. Hart, K. W. and L. V. Jankowski (1984). The application of slant hole drilling in development of shallow heavy oil deposits. Journal of Canadian Petroleum Technology 23 (1), 51-55.

Slant hole drilling at the Fort Kent Thermal project proved economically and technically efficient in developing production in the field. As of May, 1983, 100 wells had been successfully drilled. The Fort Kent reservoir is at a depth of 350 m. Slant wells were used to develop a larger area of the field from a pad. The 100 slant wells are expected to increase production in the "huff and puff" project from 180 BOPD to 650 BOPD. Details of the project are discussed in the paper.

398. Hartley, A. G. (1977). The Nido reef oil discovery in the Philippines, its significance. Proceedings of the First ASEAN Council on Petroleum Conference, Jakarta, pp. 263-277.

Not examined.

399. Hartley, R. and M. G. R. Bosma (1985). Fracturing in chalk completions. Journal of Petroleum Technology 37 (1). 73-80.

Report of the results of laboratory experiments in acid-fracturing chalk reservoirs. Some of the conclusions reached are: oil-based fluids must be used in chalks with high oil saturation; fracture propagation in chalks appears to be radial; proppants are required in "soft" chalks; high drawdowns may cause pore collapse. See also: Whitson, C. H. (1985), Discussion of fracturing in chalk completions, Journal of Petroleum Technology 37 (9), 1498; and Hartley and Bosma (1985), Author;s reply to discussion of fracturing in chalk completions, Journal of Petroleum Technology 37 (9), 1498.

400. Hartley, R. and M. G. R. Bosnma (1985). Author's reply to discussion of fracturing in chalk completions. Journal of Petroleum Technology 37 (9), 1498. See also: Hartley and Bosma (1985), Journal of Petroleum Technology 37 (1), 73-79.

Reply to "Discussion of fracturing in chalk completions:, C. H. Whitson, Journal of Petroleum Technology 37 (9), 1498.

401. Harvey, A. Herbert (1988). How well logs detect fractures. Part 1 - Techniques used in identification. Petroleum Engineer International 60 (7), 45-46,51.

The detection of open fractures with well logs, in wells with a uniform bore, is straightforward in homogeneous formations. Sealed fractures may not be detected. Sonic logs respond to matrix porosity, while density logs respond to total porosity. The difference between the two logs should be due to secondary porosity (fractures, vugs, etc). Neutron logs also respond to total porosity, and may be used instead of sonic logs to determine secondary porosity. Cementation factor may also be used to detect fractures, if the in-situ value can be determined. Cross-plots of various log parameters may be used in fracture detection when developed for specific reservoirs. Resistivity logs such as Microlaterolog or Microspherically-Focused Log (both Schlumberger trademarks) can sometimes indicate fractured zones.

402. Harvey, A. Herbert (1988). How well logs detect fractures. Part 2 - Logging tools. Petroleum Engineer International 60 (8), 40, 43-45.

Several types of logging tools are especially suited to fracture detection. The borehole televiewer is an acoustic tool the produces an "image" of the inside of a borehole. The Formation Microscanner measures electrical conductivity and provides a high-resolution "image" of microresistivity variations of the wellbore surface. This tool can detect fractures at least as small as 1 mm, and can distinguish between open and closed fractures. Full wave-form sonic logs measure the travel time of shear and compressional waves in the logged formations. Signal strength can detect the probable presence or absence of fractures. Gamma-ray logs can show the presence of fractures if radioactive materials have been deposited in the fractures by circulation of formation water. Caliper logs, particularly four-arm dipmeters, may also be used to detect fractures. Circulation of relatively cool drilling mud can cool fractured zones more than unfractured zones allowing a temperature log to detect fractures. Other logs discussed in this paper are density logs, photoelectric density logs, cross-hole tomography, and circumferential microsonic logs.

403. Harvey, Floyd (November 5, 1990). Horizontal wells - 4. Fluid program built around hole cleaning, protecting formation. Oil & Gas Journal 88 (45), 37-41.

Drilling fluids used in horizontal wells must be chosen for their hole-cleaning ability as well as for formation damage control. Mud systems for horizontal wells are divided into five categories: oil-based, polymer, gel/water, clear water/sweeps, and sized salt. The role of drilling fluids in the horizontal well is addressed in this article.

404. Hasan, A Rashid and C. Shah Kabir (1988). Predicting multi- phase flow behavior in a deviated well. SPE Production Engineering 3 (4), 474-482. SPE paper 15449 presented at the SPE Annual Technical Conference and Exhibition, New Orleans, October 5-8, 1988.

This paper examines the flow behavior of two-phase mixtures in deviated wells. A model has been developed to predict flow regime and pressure gradient in a deviated wellbore. In this model four flow regimes are recognized: bubbly, slug, churn, and annular. Experimental work determined the bubbly/slug transition, and void fraction and pressure drop in bubbly and slug flow regimes. Appropriate modifications are proposed to handle bubbly and slug flow in annular channels. It was assumed that transition criteria for churn and annular flow remain unchanged in deviated wells.

405. Hassen, B. R. and A. J. MacDonald (1990). Field comparison of medium- and long-radius horizontal wells in the same reservoir. Proceedings of the 1990 IADC/SPE Drilling Conference, February 27-March 2, 1990, Houston, pp. 637-652. Society of Petroleum Engineers, Richardson, TX. IADC/SPE paper 19986.

Two horizontal wells were drilled by AEC Oil & Gas Co. in the heavy oil South Jenner reservoir in Alberta, Canada. One well was a medium-radius well, the other was a long-radius well. They were drilled to minimize water coning by placing the horizontal wellbores as far above the oil/water contact as possible. Both wells were successful with the medium-radius well producing 126 BOPD, water free, for 2.5 months and the long-radius well producing 252 BOPD, water free, for 4.5 months. Both wells are being produced in a controlled fashion to minimize water coning. The ultimate recovery seems to be the same as for a vertical well of the same spacing, however productivity is higher in the horizontal wells. The operator has chosen to use long-radius wells for further development. Details of drilling and completing both wells are given in the article.

406. Hatten, J. (1986). Lateral Drilling - Ein Kormplettierungsverfahren. [Lateral drilling - A completion method]. Erdol Erdgas Kohle 102 (7/8), 338-341. In German with an English abstract.

Abstract. A description of the TEDSI horizontal drilling system developed by Texas Eastern Drilling Systems operated by Eastman Whipstock Inc.

407. Haugen, Sigurd A., Oystein Lund, Leif Hoyland (1988). Statfjord field: Development strategy and reservoir management. Journal of Petroleum Technology 40 (7), 863-873. SPE paper 16961 presented at the 1987 SPE Annual Technical Conference & Exhibition, September 27-30, 1988, Dallas. Society of Petroleum Engineers, Richardson, TX.

Although horizontal drilling in Statfjord field is not addressed in this article, it is good review of the reservoir geology, development, and EOR projects in Statfjord field.

408. Haymond, Doug (April, 1991). The Austin Chalk - An overview. Houston Geological Society Bulletin 33 (8), 27-33, 32, 34.

Overview of Austin Chalk geology, exploration methods, drilling history, and horizontal drilling.

409. Heacock, D. W. (1986). An on-site method for optimizing oil recovery during drilling of deviated and off-target wells at Norman Wells. Journal of Canadian Petroleum Technology 25 (3), 60-65. CIM paper 84-35-101. Abstract in The Journal of Canadian Petroleum Technology (1984), 23 (2), 38.

A method was developed to properly place directionally drilled wells in a five-spot waterflood pattern in the Norman Wells field, Canada. Off target results can be costly in an anisotropically fractured reservoir such as the one at Norman Wells. Wells need to be oriented parallel to or perpendicular to the prevailing fractures in the reservoir. A TI-59 program was developed to evaluate the off target penalties for up to six reservoir zones and to provide summations in after tax present-value dollars.

410. Heaton, Mark and Wayne Rotel (1990). Medium-radius lateral drilling applications in Canada. Journal of Canadian Petroleum Technology 29(3), 36-40.

Medium-radius horizontal wells are those with a build-up rate (BUR) with an arc of 90 m - 95 m. Fixed angle-build motors were used to drill three medium-radius horizontal wells. These fixed angle-build motors can not be rotated but they have the advantage of predictable build rates. Constant control of the build rate can be achieved by using a steerable motor. The fixed angle-build motor was used to build the angle to 90 degrees and then was replace by the steerable motor to drill the straight, horizontal portion of the hole. All three cases described in this paper were successfully drilled. Current medium-radius technology allows flexibility in the BUR of horizontal wells drilled to different depths.

411. Henmphill, Terry (November 26, 1990). Tests determine oil-mud properties to watch in high-angle wells. Oil & Gas Journal 88 (48), 64-70.

Baroid Drilling Fluids Inc. conducted a battery of tests with oil-based mud to determine the best drilling fluids to use in cleaning high-angle boreholes. Conclusions reached in this study include:

(a) cuttings accumulation is dramatically affected by small changes in pump rate (45 degrees requires the lowest pump rate); (b) at high angles, increases in viscosity increase the critical flow rates; and (c) critical flow rates are higher for base oils than for oil-mud systems. Details of the tests and the results are given in this article.

412. Herring, T. R. (1989). Simultaneous gas and water coning in horizontal wells - A simulation study. M.S. Thesis. Rogaland University Centre, Stavanger, Norway.

Not examined.

413. Hester, Timothy C. and James W. Schmoker (1991). Regional geologic characteristics relevant to horizontal drilling, Woodford shale, Anadarko basin, Oklahoma. A.A.P.G. Bulletin 75 (6), 1128.

Abstract. The Woodford shale of the Anadarko basin has several characteristics similar to the Bakken formation of the Williston basin. These characteristics include: division into three members, two depocenters divided by a paleotopographic high, and thermal-maturity trends shaped by post-depositional burial history. No horizontal drilling recommendations are given.

414.
Heuillon, B. (1972). Oil fields of Neocomian of Paris basin, France. in King, Robert E. (editor), Stratigraphic Oil and Gas Fields - Classification, Exploration Methods, and Case Histories, A.A.P.G. Memoir 16, 599-609. A.A.P.G., Tulsa.

Discussion of the geology of the Paris basin oil fields, including Chateaurenard.

415. Hester, Timothy C., James W. Schmoker, and Howard L. Sahl (December 3, 1990). Woodford shale in the Anadarko basin: Could it be another 'Bakken type' horizontal target? Oil & Gas Journal 88 (49), 73-78. This article has been published as U.S. Geological Survey Bulletin 1866-D, 'Log-derived Regional Source-Rock Characteristics of the Woodford Shale, Anadarko basin, Oklahoma.'

The Woodford shale, Anadarko basin, is an organic-rich Devonian-Mississippian shale similar to the Bakken Formation of the Williston basin, New Albany shale of the Illinois basin, Antrim shale of the Michigan basin and the Exshaw formation of the Alberta basin. These shales are, when thermally mature, good source rocks. The Woodford shale is thought to be a major source of Anadarko basin hydrocarbons. The geological setting, depositional patterns, stratigraphy, and organic carbon content of the Woodford are discussed in the article. Editor's Note: The Woodford is now considered to be a good target for horizontal drilling. A useful reference list is included in the article.

416. Higgins, R. V. (May, 1953). Factors that influence gravity drainage. Petroleum Engineering 25 (5), B83-B84, B86, B90-B92.

Report of experiments in oil recovery by gravity drainage from unconsolidated sand. Oil recovery by gravity drainage from unconsolidated sand is reduced by the presence of interstitial water, finer grain-size of the sand, and higher viscosity of the oil. Thermal recovery of oil is less effective with light-grade crude oils.

417. Hinds, Gregory S. (1990). Interpretation of well log response in the Austin Chalk. A.A.P.G. Bulletin 74 (5), 675.

Abstract. Three distinct zones of organic maturity have been identified in the Austin Chalk. These zones are depth related: immature zone (0-6000 ft), intermediate (6000-7000 ft), and deep (7000-

9500 ft). The deep zone is the potential reservoir and migration zone, and is marked by an increase in resistivity, gamma-ray, and density porosity with decreasing neutron porosity.

418. Hinrichs, Paul D., F. Jerry Luccia, and Robert L. Mathis (1987). Permeability distribution and reservoir continuity in Permian San Andres shelf carbonates, Guadalupe Mountains, New Mexico. A.A.P.G. Bulletin 71 (2), 239.

Abstract. Outcrop studies of the San Andres formation, Permian basin, show that reservoir permeability is continuous over a 40 acre unit but the permeability path is tortuous. These results suggest that pay may be bypassed in re-entrants and low-permeability pockets.

419. Hobbs, G. Warfield, IV (1990). Powder River Basin: Drilling sideways in Thompson Creek. Western Oil World 47 (8), 24-25.

Karlton Terry Oil Co. is drilling horizontal wells in the Thompson Creek field, Wyoming. The wells will produce from the Muddy formation. Details of the project and participants are given in the article.

420. Hobbs, G. Warfield and Robert O. Winkler (1990). Coalbed methane shows its potential. Petroleum Engineer International 62 (6), 26-33.

Review of coalbed methane economics, coalbed methane in the Black Warrior basin, and coalbed methane in the San Juan basin.

421. Hocker, Morgan T. (1989). Characterization and probable origin of fractures in Florence oil field, Fremont County, Colorado. A.A.P.G. Bulletin 73 (9), 1160.

Abstract. Florence field has produced from the fractured Pierre shale since 1876. Fracture orientations are sub-parallel, northwest-southeast and northeast-southwest joint sets. Florence field has produced over 15 million barrels of oil. Editor's Note: Horizontal wells are now being used to further develop the Florence field.

422. Holbert, Don R. (February 9, 1981). Drainhole drilling - 1. Variables in kickoff remain first consideration during drainhole drilling. Oil & Gas Journal 79 (6), 121-122, 124, 127-128.

Discussion of kicking-off of a horizontal drainhole and correction of directional errors.

423. Holbert, Don R. (February 16, 1981). Drainhole drilling - conclusion. Special practices allow high-angle drilling. Oil & Gas Journal 79 (7), 91-94, 99.

Use of wiggly stabilizers and improved drilling techniques allow drilling of drainholes of up to 1,200 ft.

424. Holbert, D. R (1981). New interest in drainhole drilling revives technology. World Oil 192 (4), 57-58, 61-62, 64, 66, 70, 72.

Drainholes are horizontal wellbores drilled from vertical wellbores into a reservoir to improve drainage. Drainholes are usually drilled with a radius of curvature of 12 to 40 ft until the desired angle is reached and then a larger radius is used. Drainhole drilling began in the 1950's as an attempt to improve productivity, but proved to be too expensive and not competitive with fracture treatments of the time. The rising price of oil and improvements in technology have caused renewed interest in drainhole drilling. This article reviews the technology of drainhole drilling as of 1981. Editor's Note: This article is a useful review of the technology, but recent developments have rendered much of the information out of date. An example is the belief that because of torque and drag drainholes could be drilled only 600 to 1,200 ft, whereas now they are routinely drilled to over 10,000 ft horizontally.

425. Holbert, Don R. (August 29, 1983). Directional theory - conclusion. Tangential attitude theory validated. Oil & Gas Journal 81 (35), 50-53. Abstract in The Engineering Index Annual (1984), 83, Part IV, 6683.

Presentation of field results of drilling the curved portion (angle-build section) of a directional well using tangential attitude theory.

426. Holbert, Don (June 3, 1985). Conventional tubulars can cut cost of drilling horizontal holes. Oil & Gas Journal 83 (22), 68-73. Abstract in The Engineering Index Annual (1985), 84, Part IV, 6187.

Horizontal wells with curvatures of 0.25 to 0.40 degrees/ft can be economically drilled, cased, and produced with API tubular goods. Shorter-radius holes can be drilled if clean steels, polished or treated surfaces, and special field handling are specified.

427. Holditch, Stephen A. and David E. Lancaster (August 9, 1982). Economics of Austin Chalk production. Oil & Gas Journal 80 (32), 183-189.

The Austin Chalk in Texas is a low-permeability reservoir which produces oil and gas from fractures in a narrow trend parallel to the Gulf coast. The Austin Chalk is at the base of the "Resource Triangle", a low-grade, huge-volume reservoir. Ultimate recovery from a chalk well is about 60,000 barrels of oil. Better wells can be drilled if they intersect several natural fracture- systems. Most Austin Chalk wells produce less than 20,000 barrels except in "sweet spots", such as the Pearsall and Giddings fields. Economics of the Austin Chalk wells are tied to well costs and the price of oil, as average recoveries are low. The entire Austin Chalk trend is seen as one field. A useful historical review of Austin Chalk exploration is given. Editor's Note: This article, in my opinion, seriously misjudges the Austin Chalk geology and reservoir properties. The entire trend is not productive, the "sweet spots" can be predicted by application of geological analysis, and there are discrete fields in the Austin Chalk trend.

428. Holifield, Ray H. and Bill Rehm (1989). Recompleting by horizontal drilling pays off. World Oil 208 (3), 42-43, 50. Abstract in The Engineering Monthly Index (June, 1989), 27 (6), 419.

Twenty Austin Chalk wells in the Giddings field, Texas, have been recompleted by drilling horizontal drainholes from pre-existing vertical wells. The new horizontal intervals range in length from 300 to 1,200 ft. Payout of these recompletions occurs in 3 to 24 months. The technique used is slim-hole, medium-radius drilling. Experience has lowered costs by 75%. Total return on investment is 2 to 4 times. Details of the project are given in the article.

429. Holmes, Allen B. (1987). New generation of MWD systems show promise. Petroleum Engineer International 59 (5), 36, 39-44.

Discussion of a vortex-controlled mud pulsing system for MWD data transmission. This device can produce pulses at rates up to 26 Hz up a 10,000-ft pipe.

430. Hordern, M. J. (1988). Changing development patterns in the Gippsland basin, offshore, southeast Australia. Proceedings International Meeting on Petroleum Engineering, November 1-4, 1988, Tianjin, China, pp. 903- 911. Society of Petroleum Engineers, Richardson, TX. SPE paper 17635.

Development of fields in the Gippsland basin, offshore Australia, has been in three phases: (1) low-angle (vertical or "normally" deviated) wells drilled crestally to drain good reservoirs; (2) high-angle and deep wells drilled to drain more difficult target reservoirs; and (3) small fields developed using mini-platforms, sub-sea satellites and single-leg structures.

431. Hovland, S., C. Jones, and T. Whittle (1990). Planning implementation and analysis of the first Troll horizontal well test. SPE European Petroleum Engineering Conference, October 22-24, 1990, The Hague, pp. 197-208. Society of Petroleum Engineers, Richardson, TX. SPE paper 20963.

The first horizontal well has been successfully drilled and completed in the Sognefjord sandstone, Troll field, Norwegian North Sea. The well was drilled 502 meters horizontally in the reservoir, four meters above the oil/water contact. The productivity index of the horizontal well was forty times that of vertical wells in Troll field. Details of the drilling, completing, and testing of the well are given in the article.

432. Howell, E. P., L. J. Smith, C. G. Blount (1989). Coiled-tubing logging system. SPE Formation Evaluation 3 (1), 37-39. SPE paper 15489. Abstract in The Engineering Index Annual (1988), 87, Part IV, 6011.

Coiled-tubing-conveyed logging is advantageous in high-angle or horizontal boreholes, high-temperature holes, and unstable holes when circulation needs to continue during logging operations.

433. Hower, T. L. (1990). Using depletion ratios to select and evaluate drilling locations in naturally fractured reservoirs. Proceedings 1990 SPE Annual Technical Conference & Exhibition, Reservoir Engineering volume, September 23-26, 1990, New Orleans, pp. 429-437. Society of Petroleum Engineers, Richardson, TX. SPE paper 20758.

Mapping depletion ratios of wells can highlight the most productive zones in a trend or reservoir. This type of analysis is particularly useful in fractured reservoirs. This technique was used in the Pictured Cliffs sandstones of the San Juan basin. The Pictured Cliffs sandstones have four types of secondary porosity classified as poor dissolution porosity, good dissolution porosity, very good dissolution porosity, and fracture porosity. The producing sands in each well were assigned to one of the porosity types by cross plotting net pay vs production. The wells are plotted on a map using color codes to identify each reservoir type. The plots show linear trends which can be projected into un-drilled areas to predict high probability drilling targets. Selection of drilling targets can be further enhanced by mapping depletion ratios, and concentrating on the indicated areas of maximum productivity. A twenty-well drilling program was completed using this method with a 90% success rate, higher than that predicted by the model (70% success rate).

434. Howes, John (August, 1988). Rospo Mare: setting the pace for the world. The Oilman, p. 71.

Elf Aquitaine is developing the Rospo Mare field, Italy, using horizontal wells. The field produced from a karstic limestone reservoir. The reservoir has 1.8% porosity and a permeability of up to 1,500 darcies. Some details of the project are given.

435. Howes, John (August, 1988). Coping with coring in an 89 degree well. The Oilman, p. 72.

British Petroleum has cored its first horizontal well, the Beckingham 36. The well was cored with a 30-ft core barrel. Details of the operation and equipment are given in the article.

436. Howes, John (August, 1988). Horizontal drilling from a vertical bore. The Oilman, p.74.

Petrolphysics Inc. has developed a waterjet drilling method for drilling ultra-short-radius horizontal drainholes. The system is known as the Radial Production System and is briefly described in the article.

437. Howes, John (August, 1988). Cement and casing versus openhole approach. The Oilman, p. 80.

Brief discussion of casing and cementing of horizontal wellbores.

438. Howes, John (August, 1988). Reservoir modelling key to horizontal ells. The Oilman, p. 81.

Brief interview with William Ahlefeldt of International Petroleum Engineering Consultants about horizontal drilling and reservoir modelling.

439. Howes, John (April, 1988). Horizontal and extended reach drilling come of age. The Oilman, p. 15.

Brief descriptions of horizontal and extended-reach drilling.

440. Howes, John (April, 1988). The right angle for oil companies. The Oilman, pp. 16-17.

A review of horizontal and extended-reach drilling projects worldwide. The projects are indexed.

441. Howes, John (April, 1988). Horizontal endeavors. The Oilman, p. 18.

Maersk has embarked on drilling horizontal wells in Dan field, Danish North Sea, to improve productivity of the field. Tests and modeling indicate that horizontal wells may have three to four times the productivity of vertical wells in Dan field. Some details of the project are given in the article.

442. Howes, John (April, !988). Lateral thinking. The Oilman, p. 20.

Elf Aquitaine and IFP have set up Horwell, a new company, to promote horizontal drilling.

443. Howes, John (April, 1988). The drilling program. The Oilman, p. 22.

Brief discussion of the problems of drilling extended-reach wells.

444. Howes, John (April, 1988). Research on reach. The Oilman, p. 24.

Conoco UK has proposed drilling an extended-reach well, with a record horizontal displacement of six miles, in the central North Sea. The previous record horizontal displacement well was drilled in the Bass Straits, Australia, by Esso Australia. The current (1988) North Sea record extended-reach well was drilled by Marathon in Brae field.

445. Howes, John (April, 1988). A cheaper way to drill. The Oilman, p. 26.

British Gas used a slant rig to drill high-angle, extended-reach wells in Morcambe Bay field, U.K., Irish Sea. Morcambe Bay field produces gas from the Triassic Keuper sandstone. The extended-reach wells have allowed development of the field from three platforms instead of the seven platforms that conventional drilling would have required.

446. Howes, John (April, 1988). The problems with logging horizontal wells. The Oilman, pp. 30-32.

A method for logging horizontal wells has been developed by Elf Aquitaine and Institut Francais du Petrole. The system is called SIMPHOR. SIMPHOR has been used successfully in Lacq field, France. SIMPHOR is a drillpipe-conveyed logging system. This type of system is used by several well logging companies, including Western Atlas, Schlumberger, and Gearhart. Another system for logging horizontal wells uses coiled-tubing-conveyed logging tools. The CT-conveyed system is used by Preussag AG and has been successfully used to log wells in the Berkhoepen field, Germany, the Paris basin, Prudhoe Bay field, Alaska, and the Lacq field, France.

447. Howes, John (April, 1988). The three degrees. The Oilman, pp. 30, 32.

Brief description of short-radius, medium-radius, and long-radius horizontal drilling.

448. Hsiao, C. (1988). A Study of horizontal-wellbore failure. SPE Production Engineering 3 (4), 498-494. SPE paper 16927 presented at the SPE 1987 Annual Technical Conference and Exhibition, Dallas, September 27-30.

Maximum-normal-stress theory and Drucker-Prager failure theory are used to develop a model of failure of horizontal wellbores. Breakdown pressure and borehole pressure required to prevent borehole compressive failure have been correlated with borehole orientation.

449. Hsu, Kai, Alain Brie and Richard A. Plumb (1987). A new method for fracture identification using array sonic tools. Journal of Petroleum Technology 39 (8), 677-683.

Development of array-sonic logging tools has made it possible to detect fractures by recording and analyzing an array of waveforms. Low-frequency Stonely waves can be used to detect open fractures. A case example is given and compared with full-core data and borehole televiewer data.

450. Huang, W. S. and M. A. Hight (1989). Evaluation of steamflood processes with horizontal wells. SPE Reservoir Engineering 4 (1), 69-76. SPE paper 14130 presented at the 1986 SPE International Meeting on Petroleum Engineering, March 17-20, Beijing.

A three-dimensional, numerical model, using data from a California heavy oil project, was used to test the use of horizontal wells to reduce steam override in a mature steamflood project. The simulation indicates that horizontal wells are effective in recovering oil from blind spots in a mature steamflood project, and that oil recovery will be faster than when only vertical wells are used. Higher injection and production throughputs may be obtained in horizontal wells as new, better technology improves the lifting of hot fluids in large amounts. Horizontal wells can improve efficiency of the steamflood if used from the beginning of the project. A combination of vertical and horizontal wells can reduce well density by 50% over that of the inverted 13-spot vertical-well array.

451. Hubbard, David and Nolan O. Pierson (1986). New frac treatment boosts Austin Chalk productivity. World Oil 203 (3), 31-32, 34.

The authors propose a "dendritic" hydraulic fracturing method for increasing production in depleted Austin Chalk wells.

452. Hughes, B. D. and T. L. Logan (1990). How to design a coalbed methane well. Petroleum Engineer International 62 (5), 16-20.

Description of coalbed methane production, well design, and coalbed properties.

453. Hurley, Neil F. and Ron Budros (1991). Albion-Scipio and Stoney Point fields, Michigan basin, U.S.A. A.A.P.G. Bulletin 75 (3), 599.

Abstract. Lateral drainhole drilling has been used with limited success in the Albion-Scipio field, Michigan basin. The Albion-Scipio and Stoney Point fields produce from dolomitized Trenton and Black River Groups carbonates of Ordovician age. The fields are very heterogeneous with vugs, fractures, and caverns providing most of the deliverability.

454. Huygen, H. H. A. and J. B. Black (1982). Steaming through horizontal wells and fractures - scaled model tests. Paper presented at the 2nd European EOR Symposium, November 8-10, 1982, Paris.

Not examined.

455. Hyde, G. G. and K. H. Smith (1984). Drilling and completions activities for Cold Lake Production Project. Paper presented at the 35th Annual Technical Meeting of the Petroleum Society of CIM, June 10-13, 1984, Calgary. The Petroleum Society of CIM, Montreal. CIM paper no. 84-35-82. Abstract in The Journal of Canadian Petroleum Technology (1984) 23 (2), 38.

Abstract. Report of plans to drill 220 directionally drilled steam injection wells in the Cold Lake heavy oil field, Alberta, Canada, by Esso Resources Canada Ltd.

456. Hyland, C. (1984). Drainhole drilling — An old idea whose time has come. Paper presented at the 1984 SPE California Regional Meeting, March 25-26, 1984, Long Beach. Society of Petroleum Engineers, Richardson, TX. SPE paper 12792. Not printed in the Proceedings volume.

Horizontal drainholes are a mechanism for increasing the length of exposure of the borehole to the reservoir. This article gives a brief history and discussion of the applications of horizontal drainhole drilling.

457. Illich, Harold A. (1991). Geologic aspects of horizontal drilling in self-sourcing reservoirs. A.A.P.G. Bulletin 75 (3), 599.

Abstract. Most of the current horizontal drilling targets are in self-sourcing formations such as the Austin Chalk, Bakken shale, Mesaverde Group, and Niobrara formations. Reservoirs in self-sourcing formations reduce the importance of migration and accumulation of hydrocarbons.

458. Inglis, T. A. (1987). Chapter 11. Highly Deviated and horizontal wells. Directional Drilling, pp. 196-211. Graham & Trotman, London. Petroleum Engineering and Development Studies v. 2.

General discussion of high-angle and horizontal wells. Subjects include "applications of horizontal wells," "operational problems related to horizontal wells," and "drilling practices for horizontal wells."

459. Islam, M. R. and A. E. George (1989). Sand control in horizontal wells in heavy oil reservoirs. Proceedings 1989 SPE California Regional Meeting, April 5-7, 1989, Bakersfield, pp. 437-452. Society of Petroleum Engineers, Richardson, TX. Also in Journal of Petroleum Technology (1991) 43 (7), 844-863. SPE paper 18789.

This paper gives the results of sand-control research in three Canadian heavy oil reservoir sands: Tangleflags, Pikes Peak, and Aberfeldy.

460. Islam, M. R. and A. Chakma (1990). Comprehensive physical and numerical modeling of a horizontal well. Proceedings 1990 SPE Annual Technical Conference & Exhibition, Production Operations and Engineering volume, September 23-26, 1990, New Orleans, pp. 111-123. Society of Petroleum Engineers, Richardson, TX. SPE paper 20627.

This paper presents physical and numerical models of the horizontal wellbore and the behavior of the reservoir in the presence of horizontal wellbores. A physical model (1 meter long tube) was used to observe the fluid flow into perforations, and the pressure drop along the tube when multiphase fluids were used. The presence of perforations within the wellbore changes the flow considerably from that in an unperforated horizontal pipe (pipeline). Details of the numerical model and experimental results are given in the paper.

461. Islam, M. R. and A. Chakma (1990). A new recovery technique for heavy oil reservoirs with bottomwater. Proceedings SPE/DOE Seventh Symposium on Enhanced Oil Recovery, April 22-25, 1990, Tulsa, pp. 823-832. Society of Petroleum Engineers, Richardson, TX. SPE/DOE paper 20258.

Scaled-model experiments were conducted to test the feasibility of using horizontal wells in conjunction with electromagnetic heating to recover heavy oil from reservoirs with bottomwater. The model considered variables such as oil-to-water zone thickness ratios, oil viscosity, horizontal well placement, inert gas injection rate, and production well temperature. Using this method recovery rates as high as 77% was obtained.

462. Isvan, A., I. D. Wheatley, and P. M. Barry (1986). High angle development wells in the northwest Java Sea Bima field. Paper presented at the 15th Annual Indonesian Petroleum Association Convention, October 1-15, 1986, Jakarta.

Not examined.

463. Iyoho, A. W. (1980). Drilled-cuttings transport by non-Newtonian drilling fluids through inclined, eccentric annuli. Ph. D. Dissertation. University of Tulsa, Tulsa.

Not examined.

464. Jain, S. and A. Khosla (1985). Predicting steam recovery of Athabasca oil through horizontal wells. Journal of Canadian Petroleum Technology 24 (2), 26. CIM paper 85-36-28. Paper presented at the Joint Annual Meeting of The Petroleum Society of CIM and the Canadian Society of Petroleum Geologists, June 2-5, 1985, Edmonton.

Abstract. Discussion of strategies used in steamflood recovery of oil from the Athabasca heavy oil sands.

465. Jamison, D. and W. Clements (1990). A new test method to characterize setting/sag tendencies in extended reach drilling. in ASME Drilling Technology Symposium, 1990, 109-113. Paper PD-27.

Not examined.

466. Jan, Yih-Min and John W. Harrell (1987). MWD directional focused gamma ray - A new tool for formation evaluation and drilling control in horizontal wells. in 1987 SPWLA Annual Logging Symposium, June 29-July 2, 1987, London. Reprinted in SPE Reprint Series No. 30, Directional Drilling (1990), pp. 189-197.

A Gearhart Industries MWD gamma-ray tool with focussed detection of gamma-rays has been developed and was used by Arco Oil & Gas Co. to log horizontal wells. The design and use of the tool is described in the paper. The focused gamma-ray MWD tool enables bed boundaries to be detected with better resolution. The MWD tool also allows logging of the hole before washouts can occur in later operations such as fishing.

467. Jennings, Arnold H. (1991). Williston basin oil exploration: Past, present, and future. A.A.P.G. Bulletin 75 (6), 1129.

Abstract. History of exploration in the Williston basin with mention of horizontal drilling.

468. Johancsik, C. A., D. B. Friesen and R. Dawson (1984). Torque and drag in directional wells - Prediction and measurement. Journal of Petroleum Technology 36 (7), 987-992. SPE paper 11380 presented at the 1983 IADC/SPE Drilling Conference, New Orleans, February 20-23, 1983.

A computer model and a calibrating model were used to predict torque and drag forces in directional wells. Simple sliding friction between the drillstring and borehole wall is the primary cause of drillstring torque and drag. Sliding friction coefficients, in seawater-based mud, lie between 0.25 and 0.40.

469. Johancsick, C. A., A. E. Lerand, B. G. Petovello, D. A. Gust, and B. W. Smith (1985). Application of measurement while drilling in a shallow, highly deviated drilling program. Journal of Canadian Petroleum Technology 24 (5), 37-42.

MWD was chosen as the primary survey system used in directional drilling in Norman Wells field, Northwest Territories, Canada, because other systems did not provide the needed accuracy. The cost penalty of missing the drilling target was greater than the additional cost of the more expensive MWD system. In addition to MWD, survey systems considered in this paper are: magnetic single shot; electronic single shot; single shot gyroscope; and wireline steering tool. The merits of all five systems are compared.

470. Johnson, Bill and Eric Hajas (March 11, 1991). Horizontal wells drilled successfully under turnkey contract. Oil & Gas Journal 89 (10), 68, 70.

Turnkey drilling contracts are now being used in horizontal drilling. Turnkey contracts were used by Union Pacific Resources Co. for two wells in North Dakota, with Forwest Inc. the contractor. The target formation was the Bakken. Details of the wells are given in the article.

471. Johnson, Rod (1991). Pay zones: Pay zones for horizontal exploration, Wilson county, Texas. Popular Horizontal 1 (1), 14.

Brief discussion of the "pay zones" of the Austin Chalk in Wilson County, Texas.

472. Johnson, Sandra (1990). Drill bits test Sand Wash's horizontal potential. Western Oil World 47 (12), 27-28, 31.

Review of horizontal drilling projects in Sand Wash basin, western Colorado. The play in Sand Wash basin is the fractured Niobrara formation and Cozzette formation (Mesaverde group).

473. Johnson, Sandra (1990). Bakken ignites Williston mini-boom. Western Oil World 47 (5), 37, 39-41, 43-44.

Review of the horizontal drilling activity in the Bakken formation, Williston basin, North Dakota. Editor's Note: Individual operators and fields are indexed.

474. Joly, E., A. M. Dormigny, G. L. Catala, and F. P. Pincon (1985). A new production logging technique for horizontal wells. Paper presented at the 1985 SPE Annual Technical Conference and Exhibition, September 22-25, 1985, Las Vegas. Society of Petroleum Engineers, Richardson, TX. SPE paper 14463. Also in SPE Production Engineering 3 (3), 328-332.

Production logging of horizontal wells was first done in Rospo Mare field, Italy, a karstic, heavy-oil field. Production logging was done using a pumped-down stinger attached to the tools and a locomotive which seals the tubing. After the tool string has been lowered by gravity as far as possible, fluid is pumped to force the locomotive and tool string into the horizontal portion of the borehole to the desired depth. Logging may be done as the tool-stinger is extended into the hole or as it is withdrawn by pulling on the cable. Results of logging at Lacq field, France, and Rospo Mare, Italy, are discussed.

475. Joly, E. L., A. N. Dormigny, G. M. Catala, F. P. Pincon, and A. J. Louis (1988). New production logging technique for horizontal wells. SPE Production Engineering 3 (3), 328-338. SPE paper 15469 presented at the 1986 SPE Technical Conference and Exhibition, New Orleans, October 5-8.

Production logs may be run in a horizontal well by pumping the logging tools, attached to a rigid stinger, down a service tubing. This technique has been used successfully at Lacq field, France, and Rospo Mare field, Italy. Case histories are given.

476. Jones, A. H. (1986). Horizontal well stability, Deep Coal Seam Project. in "Development and Evaluation of Technology for Methane Production from a Deep Coal Seam in the Piceance Basin." Gas Research Institute Quarterly Report, June 1, 1986-August 31, 1986.

Not examined.

477. Jones, Austin (1991). Mud motors: Tripsaver D101. Popular Horizontal 1 (1), 18, 20-21. Mud motors are used extensively in drilling horizontal wells in the Austin Chalk. Many of these wells are drilled under-balanced with fresh water or brine fluids which do not provide good lubrication for mud motors. In many cases mud motors are operated outside of their design parameters. Tripsaver D101, an extreme pressure lubricant, can provide better lubrication of mud motors and prolong their operating life. Tripsaver D101 is non-fluorescing, non-toxic, and environmentally safe. Tests indicate that Tripsaver D101 may prolong mud motor rotor life by 50-130% and that it extended the life of bearings, as well.

478. Jones, S. E. (1986). The effect of horizontal wellbore injection versus point-source injection on the oil recovery by CO2. M.S. Thesis. University of Wyoming, Laramie.

Not examined.

479. Jones, Warren (October 22. 1990). Horizontal wells - 3. Unusual stresses require attention to bit selection. Oil & Gas Journal 88 (43), 81-85.

Bit selection for horizontal wells can affect well economics significantly. This article discusses the parameters that need to be considered in planning the bit program for a horizontal well. No rules or analysis are given for bit programs.

480. Jordan, R. O. (1985). Casing running considerations in highly deviated wellbores. Journal of Canadian Petroleum Technology 24 (5), 43-48.

Difficulty in running casing in highly-deviated wells at Norman Wells field were determined to be the result of inadequate hole cleaning. At deviations of greater than 10 degrees from vertical, cuttings preferentially settle on the low side of the wellbore, eventually forming a "cuttings bed". The solution to the casing running problem at Norman Wells was an increase in circulation rates and annular velocities, and avoidance of correction runs along hold portions of the wellbore.

481. Jorgensen, Lars Nydahl (1991). Dan field - Danish North Sea. A.A.P.G. Bulletin 75 (3), 604.

Abstract. The Dan field, Danish North Sea, produces from a low-permeability, unfractured chalk of Upper Cretaceous-Lower Paleocene age. In 1987, the first horizontal well was drilled in Dan field. This well was the first horizontal well to be hydraulically fractured. Massively fractured and propped horizontal wells have more than doubled the production per well than conventional wells in Dan field. Geological input was necessary in the well planning phase of operations.

482. Joshi, S. D. and C. B. Threlkeld (1984). Laboratory studies of thermally aided gravity drainage using horizontal well. Paper presented at the Fifth Annual Advances in Petroleum Recovery and Upgrading Technology Conference, June 14-15, 1984, Calgary.

Results of experiments on thermally-aided gravity drainage using horizontal wells. The experiments modeled three well-configurations using a steel box, with a transparent window, packed with glass beads to simulate a reservoir. The three well-configurations were: horizontal injection and production pair; vertical steam injection well and horizontal production well; and single vertical well used for both injection and production. Experiments were done using both light-oil and heavy-oil plus water as reservoir fluids. The experiments showed that steam-assisted gravity drainage using the dual horizontal well pair was the most efficient method for recovering oil, and had the highest production rate. The effect of fractures was also modeled, and indicated that the presence of vertical fractures aids the gravity drainage.

483. Joshi, S. D. (1986). A laboratory study of thermal oil recovery using horizontal wells. Proceeding of the Fifth SPE/DOE Symposium on Enhanced Oil Recovery, April 20-23, 1986, Tulsa, pp. 457-467. SPE/DOE paper 14916.

Steam-assisted gravity drainage of heavy-oils using horizontal wells is an efficient recovery method. Experiments were conducted using 17-gravity oil in a sand-packed box and oil/steam ratios of about 0.7.

484. Joshi, S. D. (1986). A review of horizontal well technology. 2nd Western Research Institute/U.S. DOE Tar Sand Symposium, July, 1986, Jackson, Wyoming, Paper 3-1.

This article is an early version of the In-Situ paper (next reference) published in 1987.

485. Joshi, S. D. (1987). A review of thermal recovery using horizontal wells. In-Situ 11 (2-3), 211-259. Abstract in The Engineering Index Annual (1988), 87, Part IV, 614-615.

This paper reviews the state-of-the-art for using horizontal wells in the thermal recovery of heavy-oil and tar sands. Reservoir engineering, drilling and completion technology, and horizontal and drainhole drilling are discussed. Both steanflood and gravity drainage of heavy- oil and tar sands seem to give better recovery with horizontal wells than with vertical wells. Data is given for many horizontal drilling projects, worldwide, including horizontal steamflood projects in heavy oil and tar sands.

486. Joshi, S. D. (1987). A review of horizontal well and drainhole technology. Proceedings 1987 SPE Annual Technical Conference & Exhibition, Reservoir Engineering volume, September 27-30, 1987, Dallas, pp. 613-628. Society of Petroleum Engineers, Richardson, TX. SPE paper 16868. Also in Proceedings 1988 Rocky Mountain Regional Meeting, May 11-13, 1988, Casper, Wyoming, pp. 89-108.

A review of the state of the art of horizontal and drainhole drilling.

487. Joshi, S. D. (1988). Production forecasting methods for horizontal wells. Proceedings International Meeting on Petroleum Engineering, November 1-4, 1988, Tianjin, China, pp. 303-321. Society of Petroleum Engineers, Richardson, TX. SPE paper 17580.

This paper considers horizontal well productivity in cases of single-phase flow in homogeneous, naturally fractured, bottomwater drive, and solution-gas drive reservoirs. Horizontal wells are most effective in thin reservoirs, reservoirs with high vertical-permeability, and reservoirs with water coning problems. Methods for forecasting production in these cases are given in the paper.

488. Joshi, S. D. (1988). Augmentation of well productivity with slant and horizontal wells. Journal of Petroleum Technology 40 (6), 729-739. SPE paper 15375. Also in Proceedings 1986 SPE Annual Technical Conference & Exhibition, New Orleans, October 5-8, 1986. Abstract in The Engineering Index Annual (1988), 87, Part IV, 6040.

Horizontal wells are suitable for reservoirs that are thin, have high vertical-permeability, or have gas- or water-coning problems. Slant-wells perform better than horizontal wells in thick reservoirs. An equation and its derivation are given to predict the productivity of horizontal wells.

489. Joshi, S. D. (December, 1989). Proper completion critical for horizontal wells. American Oil and Gas Reporter, pp. 11-15.

Brief review of completion and stimulation methods for horizontal wells.

490. Joshi, S. and Clay Durham (September 21, 1990). Geologic & Reservoir Considerations in Horizontal Drilling. Houston Geological Society Short Course Notes. Houston Geological Society, Houston.

Notes for a short course on horizontal drilling and the Austin Chalk formation. Joshi reviews the technology of horizontal drilling, and its applications, using examples from the Austin Chalk, Bakken shale, Lacq field, Rospo Mare field, Zuidwal field, Helder field, and others. Data on many horizontal wells are listed. Included is a reprint of 'Proper completion critical for horizontal wells' by S. Joshi, American Oil and Gas Reporter, January, 1990. Durham reviewed the geology and stratigraphy of the Austin Chalk.

491. Joshi, Sada D. and Susan L. Lacy (1991). International trends for future horizontal drilling. A.A.P.G. Bulletin 75 (3), 604.

Abstract. Little information given in the abstract.

492. Joshi, Sada D. (1991). Horizontal Well Technology. PennWell Publishing Co., Tulsa Oklahoma. 535 pages. ISBN 0-87814-350-5.

Comprehensive coverage of all aspects of horizontal drilling, with many case histories of fields and reservoir types.

493. Jourdan, Andre P. and Guy A. Baron (1981). Elf drills 1,000+ ft horizontally. Petroleum Engineer International 53 (11), 51-52, 56, 58.

Elf Aquitaine and IFP have successfully completed a second horizontal well in Lacq field. This well was extended to 1,124 ft (370m) horizontally, five cores were cut, logs were run, and a 7-inch slotted liner run to TD. The hole was kicked-off at 279 ft (85m) with a bent sub (2 degrees, 5 minutes). Inclination was increased until it was 23-degrees where a constant angle and azimuth was maintained until 1,184 ft (361m) was reached. At that point surface casing was run.

494. Jourdan, A. and C. Wittrisch (February, 1982). French drill horizontal drainhole using MWD. Drilling Contractor.

Not examined.

495. Jourdan, A., G. Baron, and F. Goger (1982). Horizontal drilling technology. Paper presented at the 3rd Congresso Latino-Americano de Perforacao, September, 1982, Rio de Janeiro.

Not examined.

496. Jourdan, Andre, Louis Reiss, and Guy Alexandre Baron (April 15, 1982). One, two, and three... horizontal wells in France and Italy. Petrole Informations No. 1567, pp. 17-18, 23.

Brief description of the three wells drilled by Elf Aquitaine in Lacq field, France, and Rospo Mare field, Italy, and the logging methods used (SIMPHOR).

497. Jourdan, A. (December, 1982). Le forage de puits horizontaux. Revue de l'AFTP Petrole et Techniques" No. 294, pp. 37-39. In French.

Article is indexed but not annotated.

498. Jourdan, A. P. and Guy A. Baron (1984). Horizontal well proves productivity advantages. Petroleum Engineer International 56 (12), 23-25. Abstract in The Engineering Index Annual (1985), 84, Part IV, 6190.

Elf Aquitaine and Institut Francais du Petrole (IFP) completed the fourth, and final, horizontal well in their horizontal drilling research and development project. The well, drilled in the Castera

Lou field, southwest France, is the Castera Lou 110 (CLU 110). The Castera Lou field, with seven producing wells, produces oil with a high water-oil ratio from the Mano dolomite. The CLU 110 horizontal well was drilled to test the Garlin dolomite well above the oil-water contact. The CLU 110 was drilled in 117 days to a final drilled depth of 12,025 ft with 4,106 ft of horizontal displacement, of which 1,112 ft were in the reservoir.

499. Jourdan, A. and G. Baron (1985). Horizontal drilling technology for enhanced oil recovery. in DeBouw, R., E. Millich, J. P. Joulia, and D. Van Asselt (editors), New Technologies for the Exploration and Production of Oil amd Gas Resources, v. 1, pp. 187-195. Graham & Trotman, London. ISBN 0 86010 675 6.

Between 1979 and 1983, Elf Aquitaine drilled three onshore and one offshore horizontal wells. Horizontal hole segments totaling 1,500 meters were drilled, with 1,000 meters of that being in the reservoirs. Two of the wells were in Lacq field, France, one in Castera Lou field, France, and one in Rospo Mare field, offshore Italy. The project proved that economic horizontal drilling could be done to improve production of oil, that horizontal wells can be logged, and that selective completions can be made in horizontal drainholes.

500. Jourdan, A. P., L. H. Reiss, and E. Joly (1986). Horizontal wells: The worldwide state-of-the-art as of 1986. Paper presented at the SPE International Meeting on Petroleum Engineering, March 17-20, 1986, Beijing. Society of Petroleum Engineers, Richardson, TX. SPE paper 14067.

Not examined.

501. Jourdan, A. P., P. Armessen and P.Rousselet (1988). Horizontal Well Operations - 4. Elf has set up rules for horizontal drilling. Oil and Gas Journal 86 (22), 33-40. Abstract in The Engineering Index Annual (1988), 87, Part III, 6000.

Horizontal drilling encompasses several different techniques, including drilling horizontally from a mine shaft. Horizontal drilling in oil and gas exploration and production is normally done by first drilling a vertical or inclined borehole and then deviating it to the desired angle. Horizontal drilling may classified by the turning radius of the curved part of the borehole: ultra-short- radius (30 cm or less), short-radius (6 to 20 m), medium- and long-radius (20 m). Elf Aquitaine has developed an efficient method for horizontal drilling. The rules developed by Elf are: (1) radius of curvature should not be less than 180 m; (2) kick-off point must be located as deep as possible and the curvature built gradually ("J" curve); (3) when a large off-set is required, the extended-reach profile should be used; (4) a straight section of borehole should be drilled just above the target reservoir to allow identification of geological markers and to act as a safety margin; (5) the geometry of the rig, and the horizontal section of the hole should be a straight-line to minimize azimuth corrections; (6) a progressive curvature profile appears to be the ideal drilling path. Drilling techniques followed by Elf are: (1) keep the drillstring rotating to overcome friction and limit sticking; (2) restrict motors with bent subs to essential passes; (3) use rotary drilling to build angle after kick off; (4) use stiff turbodrilling assembly for highly inclined rectilinear phases; (5) take care to avoid steering errors, excessive bends, and major corrections during the vertical drilling and buildup phase; (6) simplify the BHA and limit the number of trips; (7) favor one-piece, long-life bits in the horizontal phase to reduce the number of trips. The motor-driven stabilizer is designed for horizontal drilling and will drill a highly stable trajectory with excellent penetration rates. A hydraulically-actuated bent-sub (Telepilote) allows the angle of the sub to be changed during drilling, without a trip. Cores can be cut using a downhole motor with a stabilized core barrel. An inverted drillstring can improve drilling efficiency in horizontal wells. Steerable motors are unrivaled for correcting hole trajectory without interrupting drilling. MWD is the most satisfactory method of conducting surveys.

502. Jourdan, A. P. and C. Mariotti (1989). How to build and hold a 90 [degree] angle hole. Proceedings 1989 SPE/IADC Drilling Conference, February 28-March 3, 1989, New Orleans, pp. 737-748. Society of Petroleum Engineers, Richardson, TX. SPE/IADC paper 18707. Abstract in The Engineering Monthly Index (November, 1989), 27 (11), 761-762.

This paper gives some guidelines for drilling horizontal wells. Examples are given from five wells drilled in Rospo Mare field, Italy. A horizontal well program must consider the reservoir geology, and reservoir engineering, as well as the drilling engineering aspects of the project.

503. Journal of Canadian Petroleum Technology (1981). Slant-hole drilling and production planned by Suncor. Journal of Canadian Petroleum Technology 20 (2), 38.

Suncor Inc. proposes to drill slantholes in its $88- million Fort Kent in-situ heavy-oil project. The use of slanthole drilling will allow several wells to be drilled from a small surface area, avoiding disturbance of large areas of land.

504. Journal of Petroleum Technology (1981). Technical advances broaden the use of highly deviated and horizontal drilling methods. Journal of Petroleum Technology 33 (2), 283-285.

Highly deviated and horizontal wells are used for achieving optimum production in certain circumstances. Directional drilling developed from its early beginnings in the 1920's with primitive and inaccurate directional survey methods to fairly accurate methods used in the 1950's. A patent for drilling a horizontal wellbore from a vertical well was sought in 1919, and was followed in the 1920's by four more patents. In 1929, Robert E. Lee of Coleman, Texas, drilled a well near Texon, Texas, the included two horizontal or lateral drainholes from the vertical wellbore. The two lateral holes extended 23 to 24 ft into the reservoir and increased production from 1.5 to 60 BOPD from the St. Andrews Lime. ARCO has used lateral drainholes in the Empire Abo Unit of New Mexico. An early example of horizontal drilling for oil "mining" was used in 1941 at McConnelsville, Ohio, where 1.5 miles of lateral drainholes were drilled, 100 ft of casing set in each lateral and "shot" with nitroglycerine. Horizontal drilling was used in the Wilmington field, Long Beach, California, in the 1950's, to correct problems caused by subsidence. Mobil Research & Development Corp. is working on methods to allow drilling out to 40,000 ft from the surface location.

505. Journal of Petroleum Technology (1990). Technology Digest: Oryx reports higher returns from horizontal wells in Austin Chalk. Journal of Petroleum Technology 42 (1), 112-113.

Oryx Energy Co. has cumulative production from seven Austin Chalk horizontal wells in the Pearsall field, Texas, of 800,000 barrels of oil. Current production is 3,000 BOPD. One well has produced 206,000 barrels of oil since May, 1988. Field production is now 16,000 BOPD, an increase of 100 fold over pre-horizontal drilling. Since November, 1989, when temporary field rules for horizontal wells became effective, 92 rigs were operating in the Austin Chalk as opposed to 19 before the new rules.

506. Jurgens, Rainer, Ron Bitto, Bruce Henderson, Cameron White, and Gus Mullins (1991). Horizontal drilling and completions: A review of available technology. Part 1 - Short- and medium-radius horizontal drilling. Petroleum Engineer International 63 (2),14-15, 17-18, 21.

Review of current short- and medium-radius horizontal drilling and completion technology with two case histories. Case history No. 1 is a re-entry in a 45-ft thick porous limestone between layers of dolomite at a depth of 4,500 ft TVD. Case history No. 2 is a new well in an 8-ft thick fractured shale zone with hard limestone above and dolomite below at a depth of 10,500 ft TVD.

507. Jurgens, Rainer, Ron Bitto, Bruce Henderson, Cameron White, and Gus Mullins (1991). Horizontal drilling and completions: A review of available technology. Part 2 - Medium- and long-radius horizontal drilling. Petroleum Engineer International 63 (3), 32-34, 36-37.

Review of horizontal drilling and completion technology for medium- and long-radius wells with two case histories. Case history No. 1 is redrilling an 8.5-inch medium-radius well from 9 5/8-inch casing. Case history No. 2 is drilling an extended-reach well from an offshore platform with a target of gas-bearing chalk below shale. The article concludes with predictions about horizontal drilling in the future.

508. Jurgens, R. and V. Kruger (1991). Neue Entwicklungen und Felderfahrungen auf dem Gebiet steuerbarer Motorsysteme fur Tiefbohrungen auf Erdol und Erdgas [New developments of steerable motorsystems for drilling oil and gas]. Erdol Erdgas Kohle 106 (4), 163-167. In German with English abstract.

Abstract. Review of drainhole drilling using steerable downhole motors.

509. Kairon, Saddon and D. R. Schroeter (1986). High-angle shallow- depth drilling in the South China Sea. Paper presented at the Sixth SPE Offshore SE Asia Conference, January 28-31, 1986, Singapore. Society of Petroleum Engineers, Richardson, TX. SPE paper 14615.

High-angle (70-75 degrees) wells were used to develop the Semangkok field, Malay basin, South China Sea. The Miocene reservoir is at relatively shallow depths (1050- 1380 meters TVDss). Horizontal displacements of the wells range from 2,000 meters to 3,000 meters. The reservoir covers an area 7 kilometers by 3.7 kilometers. Use of high-angle, extended-reach wells allows the field to be developed with only two platforms. Details of the project are given in the paper.

510. Karcher, B. J., F. M. Giger, and J. Combe (1986). Some practical formulas to predict horizontal well behavior. Paper presented at the 1986 SPE Annual Technical Conference & Exhibition, October 5-8, 1986, New Orleans. Society of Petroleum Engineers, Richardson, TX. SPE paper 15430.

Formulas are given for predicting the productivity index for horizontal wells in both homogeneous and anisotropic reservoirs. These formulas indicate that, in homogeneous media, horizontal wells should have 2 to 5 times the productivity of a vertical well of the same length. In the case of low vertical-permeability, a slanted-hole may have higher productivity than an equivalent horizontal well. Horizontal wells that cross natural fractures may have up to ten times the productivity of a vertical well. Coning may be reduced by using horizontal drainholes; critical rates may be twice as high as in vertical wells and recoveries may be three times as high in horizontal wells.

511. Karlsson, H., R. Cobbley, and G. E. Jaques (1989). New developments in short-, medium-, and long-radius lateral drilling. Proceedings 1989 SPE/IADC Drilling Conference, February 28-March 3, 1989, New Orleans, pp. 725-736. Society of Petroleum Engineers, Richardson, TX. SPE/IADC paper 18706. Abstract in The Engineering Monthly Index (November, 1989), 27 (11), 761.

Lateral drilling allows large portions of the wellbore to remain in the zone of interest; exposing a greater area of pay zone. Medium-radius wells are defined here as having a radius-of-curvature of 285 ft to 700 ft, corresponding to a build rate of 8 to 20 degrees per ft drilled. Two motor designs are used in drilling medium-radius wells — one to build angle and one to hold angle. Short-radius wells have a build rate of 1.5 to 3 degrees per ft drilled. Technology used to drill horizontal wells now includes steerable motors, measurement-while-drilling (MWD systems), and navigation drilling.

512. Karlsson, Haraldur and Ron Bitto (1989). Worldwide experience shows horizontal well success. World Oil 208 (3), 51-54, 56. Abstract in The Engineering Monthly Index (June, 1989), 27 (6), 419.

This article reviews worldwide horizontal drilling and briefly reviews the techniques used.

513. Keelean, L., S. A. Harris, and Nick Petronio (1989). Short radius drilling technology utilizing mobile service/workover rig: Part I. Well Servicing 29 (4), 20, 23-24. This article along with Part 2, are reprints of a paper presented at the SPE/IADC Drilling Conference, New Orleans, February 28-March 3, 1989.

Trendwell Oil Corp. has drilled five lateral drainholes in the naturally fractured Antrim shale reservoir in Otsego County, Michigan. Short-radius drainholes were drilled. The details of the project, including hardware, is discussed in the article.

514. Keelean, L., S. S. Harris, and Nick Petronio (1989). Short radius drilling technology utilizing mobile service/workover rig: Part II. Well Servicing 29 (5), 9-11. This article and Part 1 are reprints of the paper presented at the IADC/SPE Drilling Conference, New Orleans, February 28-March 3, 1989.

This article describes the five Antrim shale lateral drainholes drilled by Trendwell Oil Corp. in Michigan. See part one of the article for a discussion of the technology involved in the project.

515. Keelean, L., S. S. Harris, N. Petronio (1989). Short radius drilling technology utilizing mobile service/workover rig. Proceedings 1989 SPE/IADC Drilling Conference, February 28-March 3, 1989, New Orleans, pp. 765-772. Society of Petroleum Engineers, Richardson, TX. SPE/IADC paper 18711. This article was reprinted in Well Servicing (July/August, 1989), Part 1, and (August/September, 1989), Part 2. Abstract in The Engineering Monthly Index (November, 1989), 27 (11), 762.

A multi-well, short-radius horizontal drilling project was conducted in northern Michigan, with the Antrim shale as the target. The wells were planned to kick-off at the top of the Antrim "Hot Zone" at about 1,400 ft depth. The lateral drainholes were to be kept within the upper 60-70 ft of the Antrim. Horizontal displacement was planned to be 300-400 ft. The wells were drilled with Eastman Christensen's short-radius tools. A Cabot-Franks 300 series self-propelled service/workover rig was used for the project. The choice of a service/workover rig reduced costs by 20%-25%. As the drilling crew became more familiar with the lateral drilling operations, the drilling hours per ft were reduced from 0.3 to 0.17 hours/ft for the five well program.

516. Keller, S. R., R. J. Crook, R. C. Haut and D. S. Kulakofsky (1987). Deviated-wellbore cementing: Part 1 -Problems. Journal of Petroleum Technology 39 (8), 955-960. SPE paper 11979 presented at the 1983 SPE Annual Technical Conference and Exhibition, San Francisco, October 5-8.

In deviated wells, settling of solids from the drilling fluid to the low-side of the borehole may adversely affect the displacement of mud during cementing. If the solids settle, they may form a continuous permanent mud-channel in the cemented annulus. Excess water in the cement slurry can cause a water channel on the high side of the annulus.

517. Kempthorne, R. H. and J. P. R. Irish (1981). Norman Wells - A new look at one of Canada's largest oil fields. Journal of Petroleum Technology 33 (6), 985-991. SPE paper 9477 presented at the 58th Annual SPE Technical Conference and Exhibition, Dallas, September 21-24, 1980.

Norman Wells field, Northwest Territories, Canada, was discovered in 1920, but was not developed until the 1940's, because of its remote location. Production peaked at 4,400 BOPD in 1944. After World War II, Norman Wells continued to operate, and now [1981] produces 3,000 BOPD. By the end of 1980, only 4% of the estimated 630 million barrels of original oil in place had been produced. In 1979, an additional development phase was started, resulting in seven highly deviated and one horizontal well being drilled for water injection and production. Norman Wells produces from the Devonian Kee Scarp formation, a fractured, limestone reef reservoir. The trap is stratigraphic and has no gas cap. Porosity ranges up to 24% but permeability is less than 10 md. Horizontal permeability is 2 to 4 times that of vertical permeability. The fractures forming the reservoir are vertical (or near vertical), and have a mean orientation of N 26 E with a standard deviation of 21 degrees. Geological and engineering studies indicate that 43% of the oil-in-place can be recovered by pattern waterflooding, with a peak production of 25,000 BOPD. Editor's Note: Norman Wells field has been the site of several horizontal wells which have been used to reduce water coning and increase the ultimate recovery of the field.

518. Kerr, Callin Joe (1988). PDC drill bit design and field application evolution. Journal of Petroleum Technology 40 (3), 327-332.

Descriptions of PDC bits and their characteristics.

519. Kerr, D. and K. Lesley (1988). Mechanical aspects of medium radius well design. Proceedings 1988 International Meeting on Petroleum Engineering, November 1-4, 1988, Tianjin, China, pp. 719-726. Society of Petroleum Engineering, Richardson, TX. SPE paper 17618.

Techniques are presented for designing a bottomhole assembly for drilling a medium-radius horizontal well.

520. Kerr, Denny (1988). Medium-radius holes practical with new drilling technology. The American Oil & Gas Reporter 31 (12), 27-30.

General discussion of the technology and methods of drilling medium-radius horizontal wells.

521. Kerr, D. and A. Odell December, 1988). Here's how to plan a horizontal well. Drilling Contractor, pp. 23-26.

Not examined.

522. Kerr, Denny (1990). How to drill a smooth medium-radius well. World Oil 210 (3), 46-47. Abstract in The Engineering Index Monthly (June, 1990), 28 (6), 671.

Medium-radius wells are prone to sticking problems caused by frequent changes in tool-face orientation to correct the tendency of the medium-radius assembly to displace the bit from the proposed azimuth. The problem may avoided if the tool-face heading is kept constant during kickoff. This article discusses, in general terms, the sticking problem and its cure.

523. Kerr, Denny (1991). Designing tangent sections for medium-radius horizontal wells. World Oil 212 (3), 45-47.

Discussion of efficient design of medium-radius horizontal wells.

524. Kholsa, A. and G. Cordell (1985). Potential of horizontal well processes. Paper presented at the AOSTRA Sixth Annual Advances in Petroleum Recovery and Upgrading Technology Conference, June 6-7, 1985, Edmonton.

Not examined.

525. Kientz, J. (1990). Horizontal Drilling - Zeitsparender und umweltschonender Dukerbau. Erdol Erdgas Kohle 106 (11), 432-433. In German.

Indexed but not annotated.

526. King, G. E. (1989). Perforating the horizontal well. Journal of Petroleum Technology 41 (7), 671-672. SPE paper 19888. Abstract in The Engineering Monthly Index (November, 1980), 27 (11), 773.

Perforating horizontal wells is complicated by the decentralization of the perforating gun by gravity and the difficulty of positioning the gun in holes of greater than 55 degrees of deviation. Decentralization changes the hole size and depth of penetration of the perforations. Tubing-conveyed perforating guns may be used to properly place perforations in horizontal wells.

527. King, G. R. and T. Ertekin (1988). Comparative evaluation of vertical and horizontal drainage wells for the degasification of coal seams. SPE Reservoir Engineering 3 (2), 720-734. SPE paper 13091 presented at the 1984 Annual SPE Technical Conference and Exhibition, September 16-29, Houston.

Coal-seam degasification has been modeled using a two-phase, single well, numerical simulator applicable to either vertical wells or horizontal wells drilled from the bottom of a shaft. Two cases, one vertical well and one horizontal well, were modeled. Some conclusions reached were: (1) methane production from coal seams can be divided into three distinct time-periods (a) early-time period when production is dominated by depletion of natural fractures, (b) second flow period production is defined by two dimensionless factors and the dimensionless adsorption isotherm; and (c) late flow period dominated by production from micropores; (2) the length of horizontal boreholes is more important than the number of them.

528. King, R. L. and J. H. Stiles (1970). A reservoir study of the Hawkins Woodbine field. Paper presented at the SPE 45th Annual Fall Meeting, October 4-7, 1970, Houston. Society of Petroleum Engineers, Richardson, TX. SPE paper 2972.

Hawkins field, Texas, produces oil from the Woodbine formation. The field has a water drive and a significant gas cap. Recovery of oil was less than 50% when water breakthrough occurred. The field was unitized and converted to gas cap drive by injecting gas to maintain pressure and using gravity drainage. The oil recovery improved to 80% after unitization. Details of the project are given in the article.

529. King, William (1990). Selecting bits for extended-reach and horizontal wells. World Oil 210 (4), 53, 55, 57, 59-60.

Bit selection for drilling extended-reach and horizontal wells is very important because of the additional problems caused by building angle in these wells. Sideloading causes bit wear, as well as reducing the life of the downhole motors normally used in these wells. This article discusses in general terms bit selection for horizontal and extended-reach wells.

530. Kittridge, Mark G., Larry W. Lake, F. Jerry Lucia, and Graham E. Fogg (1990). Outcrop/subsurface comparisons of heterogeneity in the San Andres formation. SPE Formation Evaluation 5 (3), 233- 240. SPE paper 19596 presented at the 1989 SPE Annual Technical Conference & Exhibition, San Antonio, October 8-11, 1989.

Data taken from the San Andres formation outcrop and from the San Andres in the Wasson field, Texas, is compared to determine the reservoir permeability heterogeneity. Permeability determined from outcrop samples varied from 0.1 md. to 2,000 md. Mean permeability is significantly higher in the outcrop than in the subsurface. Substantial variability in permeability was noted within individual units. The outcrop of the San Andres is a good analog of the Wasson field. Editor's Note: No permeability data for the Wasson field are given in the conclusions. From Table 2 in the text, it appears that the Wasson field permeability range is 62 to 271 md.

531. Knebel, Robert M. (1991). Stockdale (Austin Chalk) field, Wilson County, Texas. in Chuber, Stewart (editor), Abstracts and Short Papers of the South Texas Geological Society Austin Chalk Exploration Symposium, February 24-26, 1991, San Antonio, 73-74. South Texas Geological Society, San Antonio.

Abstract. The Stockdale field produces oil from a naturally fractured zone, in the Austin Chalk, associated with a down-to-the-north normal fault. This type of target is seen as a potential horizontal drilling target in Wilson County, Texas, and analogous areas. This paper was given as a poster session.

532. Knight, O. M. (1983). Toolpusher (TM) logging and aid to advanced directional drilling programs. Paper presented at the First Northern European Drilling Conference, October 24, 1983.

Not examined.

533. Kodl, Edward J. and J. C. Eacman (1990). The floating oil pool - A unique trapping mechanism at Kern River field. A.A.P.G. Bulletin 74 (5), 696.

Abstract. Discussion of the theory that the Kern River field reservoir is a pool of oil floating on top of the ground water, with no conventional trapping mechanism. Gravity or barometric pressure is the updip "seal".

534. Kossack, C. A., J. Kleppe, T. Aasen (1987). Oil production from the Troll field: A comparison of horizontal and vertical wells. Proceedings 1987 SPE Annual Technical Conference & Exhibition, Reservoir Engineering volume, September 27-30, 1987, Dallas, pp. 629-641. Society of Petroleum Engineers, Richardson, TX. SPE paper 16869.

Vertical wells in the Troll field, offshore Norway, typically cone gas and/or water within 2-3 days of start of production. The field contains 5.7 billion barrels of oil-in-place, with an oil column of 0-92 ft thick. It has a gas cap of 60 Tcf and is underlain by an aquifer. Numerical simulations of the field show that horizontal wells may more efficiently recover the oil and gas than will vertical wells. Wells with a 1500 ft horizontal section in the reservoir produce as well as two vertical wells and wells with a 2000 ft horizontal section will out perform three vertical wells. Production rates are higher for horizontal wells, but ultimate recovery may be less because of interference between wells. Optimal configuration of horizontal wells is an economic question determined by drilling costs and oil prices.

535. Kovalenko, K. I. (1981). [Promising production plans for exploratory oil and gas drilling using horizontal wells]. Tr. Vses. Neft. Naukno-Issled. Geologorazved. Inst., 92-99. In Russian.

Not examined.

536. Kuchuk, F. J., P. A. Goode, D. J. Wilkinson, R. K. M. Thambynayagam (1988). Pressure transient behavior of horizontal wells with and without gas caps or aquifer. Proceedings 1988 SPE California Regional Meeting, March 23-25, 1988, Long Beach, pp. 141-154. Society of Petroleum Engineers, Richardson, TX. SPE paper 17413. Also printed in SPE Formation Evaluation (1991), 6 (1), 86-94.

New analytical solutions are presented for pressure transient well testing to determine formation parameters using Laplace transforms for horizontal wells bounded by two horizontal boundary planes above and below the wellbore. The top and bottom boundaries are impermeable, and there is no gas cap or aquifer present in these calculations. Also presented is the solution for a case with one of the boundary planes having a constant pressure i.e. gas cap or active aquifer.

537. Kuchuk, F. J., P. A. Goode, B. W. Brice, D. W. Sherrard R. K. M. Thambynayagan (1988). Pressure transient analysis and inflow performance for horizontal wells. Proceedings 1988 SPE Annual Technical Conference & Exhibition, Reservoir Engineering volume, October 2-5, 1988, Houston, pp. 383-398. Society of Petroleum Engineers, Richardson, TX. SPE paper 18300.

A method of pressure transient analysis is developed and then applied to a horizontal well in the Prudhoe Bay field. Because of the complex geometry of horizontal wells, short time pressure tests do not allow determining of system parameters. Drawdown and buildup tests must be combined with downhole flow rate measurements to be properly interpreted. In addition to an interpretive methodology, analytical formulas are given for inflow performance of horizontal wells, both with and without the presence of gas cap and aquifer.

538. Kuchuk, Fikri J., Peter A. Goode, Bradley W. Brice, David W. Sherrard, and R. K. Thambynayagan (1990). Pressure-transient analysis for horizontal wells. Journal of Petroleum Technology 42 (8), 974-979, 1028-1031. SPE paper 18300 presented at the 1988 SPE Annual Technical Conference and Exhibition, Houston, October 2-5.

Well tests of horizontal wells are difficult because of the lack of radial symmetry, 3D nature of flow geometry, wellbore storage effects, and zonal variations of vertical permeability. Proper testing of a horizontal well requires drawdown and buildup tests with acquisition of the downhole flow rate with pressure to estimate the reservoir parameters accurately. Details of testing of a real well at Prudhoe Bay and a synthetic well are given.

539. Kuich, Nanette M. (1989). Seismic fracture identification and horizontal drilling: Keys to optimizing productivity in fractured reservoir, Giddings field, Texas. A.A.P.G. Bulletin 73 (9), 1186.

Abstract. Seismic-data interpretation can identify fracture swarms in the Austin Chalk of the Giddings field. Horizontal drilling allows wellbores to intersect multiple fracture swarms spaced 100 to 300 feet apart.

540. Kuich, Nanette (1990). Seismic fracture identification and horizontal drilling; keys to optimizing productivity in a fractured reservoir, Giddings field, Texas. Bulletin of the South Texas Geological Society 30 (6), 28-39. Reprinted from Transactions of the Gulf Coast Association of Geological Societies XXXIX, pp. 153-158. Abstract in A.A.P.G. Bulletin (1989), 73 (9), 1186.

Fracture swarms in the Austin Chalk of Giddings field, Texas, can be identified on seismic and seismic attributes data. The fracture swarms in the Austin Chalk appear to be separated, horizontally, by 100 to 300 ft of impermeable chalk. Production in the field is obtained by intersecting the fracture swarms with a wellbore. Horizontal wells may intersect multiple fracture swarms. Details of the seismic analysis and the local geology are given in the article.

541. Kume, Jack (1963). The Bakken and Englewood Formations of North Dakota and Northwestern South Dakota. North Dakota Geological Survey, Grand Forks. Bulletin 39.

The Englewood formation of Mississippian age is composed of three units in the outcrop in the northern Black Hills: a lower shale, a middle argillaceous and shaly limestone, and an upper dolomitic limestone. Ten miles north of the outcrop, the Englewood has a lower unit of calcareous siltstone, a middle unit of silty limestone, and an upper unit of calcareous siltstone. The Bakken does not outcrop and also has three units: a lower black shale, a middle sandstone, silty limestone or limestone, and an upper black shale unit. The Bakken thins toward the rim of the Williston basin and pinches out. This publication describes both formations in detail and gives measured sections, well sample descriptions, and has an extensive bibliography.

542. Kuuskraa, V. A., J. P. Brashear, L. E. Elkins, and F. Morra, Jr. (1981). Gas recovery from tight formations - A function of technology and economics. Journal of Petroleum Technology 33 (8), 1545-1556. SPE paper 7916.

Natural gas production from tight gas reservoirs is now 5% of the total U.S. natural gas production. The total tight gas resource in the U.S. is estimated to be 400 Tcf in place. In 1981, the threshold price for economic production of tight gas is $4.50/Mcf. For tight gas to become a larger part of the U.S. energy picture, either gas prices must increase or technology must lower the cost of finding and production. Detailed economic analysis of tight gas in the U.S. is given in this article.

543. Kuuskraa, V. A., D. E. Wicks, W. K. Sawyer, and P. R. Esposito (1982). Production potential from Devonian gas shales: Effective production strategies. Proceedings 57th Annual SPE Technical Conference and Exhibition, September 26-29, New Orleans, pp. 169-174. Also SPE Eastern Regional Meeting, Champion, Pennsylvania, November 9-11, 1983.

The Devonian shales of Ohio contain 6.2 to 22.5 Tcf of gas that is recoverable with current (1983) technology. The Devonian shales produce from naturally fractured, anisotropic reservoirs. Artificial stimulation such as hydraulic fracturing is necessary for optimum productivity. Reduction of well spacing from the traditional 120 acres to 80 acres will increase the recoverable reserves substantially. Editor's Note: Horizontal drilling was not considered in this study. Horizontal drilling may improve productivity and recovery of gas from the Devonian shales.

544. Kwakwa, K. A., A. S. Batchelor, and R. C. Clark (1991). Assessment of the mechanical stability of high-angle wells in Block 22/11, Nelson field discovery. SPE Drilling Engineering 6 (1), 25-30. SPE paper 19240.

A model is presented for assessing the mechanical stability of high-angle and horizontal wellbores in a tri- axial stress field. Borehole instability in horizontal wells can affect the project economics and therefore the ability to develop marginal reservoirs or the efficient development of fields. The model discussed uses Nelson field, U.K., as its example.

545. Kydlund, T., P. M. Haugen, G. Bousquet, and S. O. Havig (1988). Application of unconventional techniques in constructing an integrated reservoir simulation model for Troll field. SPE Reservoir Engineering 3 (3), 967-976. SPE paper 15623 presented at the 1986 SPE Annual Technical Conference & Exhibition, New Orleans, October 5-8, 1986.

Discussion of the model and simulation of the Troll field reservoirs. The Troll field is a giant gas field producing from Jurassic formations in the Norwegian North Sea. Editor's Note: The Troll field was developed, in part, with horizontal wells. Horizontal wells are used to recover oil from a thin oil zone in the field, and to reduce gas and water coning.

546. Labastie, A. and A. Vio (1981). Chateaurenard polymer pilot field test. Paper presented at the European Symposium on Enhanced Recovery, Bournemouth, England.

Not examined.

547. Laferriere, Alan P. and Donald E. Hattin (1989). Use of rhythmic bedding patterns for locating structural features, Niobrara formation, United States Western Interior. A.A.P.G. Bulletin 73 (5), 630-640.

Geological study of the Niobrara formation in Colorado, Kansas, and New Mexico. Emphasis is on Milanovitch-type bedding cycles and regional tectonics.

548. Lancaster, James R. and Asher Atkinson (199). Drilling parameters are utilized to compute fracture porosity and in-situ permeability in a horizontally drilled Bakken well. A.A.P.G. Bulletin 74 (8), 1332-1333.

Abstract. Drilling porosity and in-situ permeability of the Bakken formation can be calculated using drilling parameters, drilling fluid properties, and textural properties. Drilling porosity minus sonic porosity gives the fracture porosity.

549. Landrum, B. L. and P. B. Crawford (February, 1955). Effect of drainhole drilling on production capacity. Journal of Petroleum Technology 7 (1), 45-47. Also in Transactions, A.I.M.E. 204, pp. 271-273.

Electrical model studies indicate that if the drainhole length is near 0.2 times the drainage radius, the production capacity may be 1.8, 2.8, and 2.33 for the cases of 1, 2, and 4 drainholes. The details of the model and the analysis are given in the article.

550. Lang, William J. and Marion B. Jett (September 24, 1990). Horizontal wells-1. High expectations for horizontal drilling becoming reality. Oil & Gas Journal 88 (39), 70, 72, 74, 76, 79.

Horizontal drilling is becoming increasingly important as technology improves. Horizontal wells were drilled in the U.S.S.R. in the 1950's, when 43 wells were drilled. The number of horizontal wells drilled per year is now approaching 1,000 worldwide. Most companies involved in horizontal drilling have formed specialized horizontal drilling groups. There are several reason to drill horizontal wells, including: connecting vertical fractures; producing from thin reservoirs; decreasing gas and water coning; better sand control; and increasing injectivity of fluids. Completion methods are steadily improving, and now include zone isolation, liners, cementing improvements, multiple zone fracing, better gravel packing, etc. North America leads in horizontal drilling because of the activity in the Austin Chalk of Texas. Almost 10% of the land rigs operating in the U.S. are in the Austin Chalk trend. Some of the reports of Austin Chalk production have been overblown. Seventy-five percent of the horizontal Austin wells have initial production of 500 BOPD or less. Twelve percent of the Austin horizontal wells have initial production of over 1,000 BOPD. The Bakken formation of the Williston basin has the next highest level of activity with an estimated 150 horizontal wells in 1990. Other horizontal drilling targets in the U.S. are western coalbed methane and Niobrara formation. Canada leads in applying horizontal drilling to steam injection and gravity drainholes in heavy oil and tar sands. In Europe, horizontal drilling has been concentrated in the North Sea where more than 32 wells have been drilled. Horizontal drilling is increasing in the Far East, Latin America, Middle East, and Africa. As experience is gained in horizontal drilling, costs are decreasing. In Canada, one operator announced that horizontal wells cost 1.5-4 times more than vertical wells while producing 2.5 to 5 times as much oil.

551. Latos, Jeff and Dale Chenery (1988). Logging with a coiled tubing system. Journal of Canadian Petroleum Technology 27 (2), 81-84.

Coiled tubing systems are excellent for logging and completing highly deviated (horizontal and high-angle) wells because they can place tools accurately in such boreholes. Systems are described, using 25.4 mm coiled tubing, for perforating and logging highly deviated boreholes,

552. Lattimore, G. M., R. S. Carden, and T. Fisher (1987). Grand Canyon directional drilling and waterline project. Paper presented at the 1987 SPE/IADC Drilling Conference, March 15-18, 1987, New Orleans. Society of Petroleum Engineers, Richardson, TX. SPE paper 16169.

A high-angle (71 degrees) borehole was drilled from the South Rim of the Grand Canyon (Grand Canyon National Park, Arizona) to a breakout point in the Muav limestone above Indian Gardens. The well was part of a project to supply water to Indian Gardens campground in the National Park. The well was drilled according to National Park Service guidelines, which included well-site location 400-500 ft from the canyon rim, sufficient diameter to allow installation of an 8 inch waterline, and easily accessible exit point. The 5,075 ft (measured depth) well was drilled in 49 days. Problems encountered include fractured zones, solution cavities, hole cleaning, and plugback and sidetrack resulting from fishing operations. Air/mist was used as a circulating medium because of the extensive fractures and well location near the canyon rim. The well is considered a valuable example of drilling high-angle wells with air in low-pressure reservoirs.

553. Laubach, S. E. and E. W. Collins (1991). Fracture geometry and fracture intensity associated with faults and folds in the Austin Chalk and other Cretaceous limestones of Central Texas. in Chuber, Stewart (editor), Abstracts and Short Papers of the South Texas Geological Society Austin Chalk Exploration Symposium, February 24-26, 1991, San Antonio, pp. 77-80. South Texas Geological Society, San Antonio.

Natural fractures occur in the Austin Chalk and other Cretaceous carbonates of Texas in association with faults and folds. Fracture intensity and geometry varies with rock type, bed thickness, and structural position. The results are given of a field investigation near Waxahachie and Richland Creek, Texas.

554. Layne, Abbe W. and H. J. Siriwardane (1988). Insights into hydraulic fracturing of a horizontal well in a naturally fractured formation. Proceedings of the SPE Annual Technical Conference and Exhibition, October 2-5, 1988, Houston, pp. 725-736. Society of Petroleum Engineers, Richardson, TX. SPE paper 18255. Abstract in The Engineering Monthly Index (November, 1989), 27 (11), 723.

The U.S. Department of Energy funded a horizontal drilling experiment in Wayne County, West Virginia, to study the effects of horizontal wells on productivity from naturally fractured reservoirs in Devonian shale. This experiment included hydraulic fracturing of the target formation. Hydraulic fracturing of a naturally fractured reservoir may not create new fractures because of the existence of natural, multiple, low resistance paths. In this experiment, using tracers and tracer logs, it was observed that the fracture fluids used entered the natural fractures and returned to the wellbore along natural intersecting fractures. Details of this experiment are given in the article.

555. Lea, James F. and Herald W. Winkler (May, 1991). What's happening in artificial lift. World Oil 212 (5), 73-74, 76, 78, 80.

Electric submersible pumps are being designed for use in horizontal wells.

556. Leabo, Jim (1991). Safety check points: Safety in horizontal drilling. Popular Horizontal 1 (2), 22-23.

Horizontal drilling in the Austin Chalk has been remarkably safe to date. Only one Austin Chalk blowout has resulted in fire, fatal injuries, and destruction of the rig. Other blowouts have been less severe, although several fatalities have occurred.

557. Lebihan, M. L. and A. G. Bourgeois (1985). Production test on Rospo Mare heavy oil reservoir (Italy). in DeBouw, R., E. Millich, J. P. Joulia, and D. Van Asselt (editors), *New Technologies for the Exploration and Exploitation of Oil and Gas Resources,* v. 2, pp. 820-826. Graham & Trotman, London. ISBN 0 86010 676 4.

The Rospo Mare field reservoir is a karstic limestone with no matrix porosity. The danger to the reservoir is early breakthrough of water, because of the small difference between the densities of the oil and water, and the high contrast between the mobilities of oil and water. Production tests were carried out on one vertical and one deviated well in the field to determine the dynamic characteristics of the reservoir.

558. Leibman, Paul R. (1990). Bakken shale typifies horizontal drilling success. World Oil 211 (6), 53-58.

The Bakken shale horizontal drilling play has become a significant investment opportunity since Meridian Oil started the play in 1987. The success rate for vertical Bakken wells in the Williston basin was 20-25% with expected ultimate reserves of 136,000 barrels of oil. Meridian Oil identified the Bakken as a potentially economic target for horizontal drilling. The Bakken target is usually 6-9 ft thick at depths over 10,000 ft, and requires drilling a 2,000 ft lateral wellbore into the Bakken. Meridian's first Bakken horizontal well was in Elkhorn Ranch field, North Dakota. This well was drilled horizontally 2,603 ft horizontally at a depth of 10,418 ft, and was completed with an IP of 258 BOPD of oil and 299 MCFGPD. This first well has produced 240,000 barrels of oil, through April, 1990, and was still flowing at a rate 171 barrels per day. Since this first discovery, the Bakken play has increased to 50% of the drilling activity in North Dakota, with an average of 13 rigs working per month. Meridian increased its drilling activity in the Bakken until, in 1990, forty wells were planned in a development project. Drilling experience indicates that individual wells may be quite variable and success depends on picking the "sweet spots." An example economic analysis, given in the article, shows a rate of return of 62% after-tax discounted cash flow and reaches payout in 1.54 years. This result is based on a well cost of $1.1 million. In the case of 25% after-tax return, the well cost must not exceed $1.5 million. The article describes, in some detail, the experience of Meridian and other operators in the Bakken play, and provides data on economics and reserves.

559. Lefever, Julie A. (1990). Horizontal drilling the Bakken formation, Williston basin: A new approach. A.A.P.G. Bulletin 74 (5), 702-703.

Abstract. Horizontal drilling can increase the probability of success and improve the economics of drilling in the Bakken formation of the Williston basin. The Bakken produces from naturally fractured reservoirs. The Bakken matrix is a poor reservoir, with only 5% effective porosity, and permeability ranging from 0.1 to 0.2 md. Production from horizontal wells is generally 2 to 4 times that of vertical wells, with payout in 1.5 years instead of 3.5 years.

560. Lefever, Julie A. (1990). No title given. in Hansen, William B. (editor), Bakken Formation Short Course, pp. 21-22. Montana Geological Society, Billings. Study Notes 1.

Abstract. Bakken production was established in 1953, in Antelope field, by Stanolind Oil & Gas Corp. By 1977, the Bakken was producing from only three fields, Antelope, Buckhorn, and Elkhorn Ranch. In 1987, Meridian drilled and completed the first horizontal well in the Bakken. To date (September, 1989) the Bakken has produced 19,623,036 barrels of oil from 50 fields, 35 of them active. Horizontal drilling has increased Bakken production. Horizontal wells now average 151 barrels of oil/day and account for 40% of the Bakken oil produced in the first nine months of 1989.

561. Legris, B. and G. Nazzal (1989). Specifics of horizontal drilling in the Zuidwal gas field. Paper presented at the 1989 SPE European Offshore Conference: Offshore Europe 89, Aberdeen.

Not examined.

562. Lien, S. C., R. Seines, S. O. Havig, T. Kyland (1990). Experience from the ongoing long-term test with a horizontal well in the Troll oil zone. Proceedings 1990 SPE Annual Technical Conference & Exhibition, Reservoir Engineering volume, September 23-26, 1990, New Orleans, pp. 621-632. Society of Petroleum Engineers, Richardson, TX. SPE paper 20715.

The results of the first five months of a twelve month test of a horizontal well in the Troll field, Norway, shows that a thin (22 meter thick) oil column exists in the gas reservoir. The well was drilled 500 meters horizontally and is placed 4 meters from the oil/water contact and 12 meters from the oil/gas contact. Oil production is more than 4 times that of a vertical well in the same area.

563. LeMay, William J. (1972). Empire Abo field, New Mexico. in King, Robert E. (editor), *Stratigraphic Oil and Gas Fields - Classification, Exploration Methods, and Case Histories.* A. A. P. G. Memoir 16, 472-480. A.A.P.G., Tulsa.

The Empire Abo field, New Mexico, reservoir is the Permian Abo barrier reef. This stratigraphic trap is estimated to contain 250,000,000 barrels of recoverable oil. A description of the geology, stratigraphy, and history of the field is given in the paper.

564. Lennon, Rickye (1991). Meeting the needs. Horizontal drilling: New information technologies needed to match the drilling technologies. Popular Horizontal 1 (1), 22-23.

Agency Information Consultants Inc. (AIC) was formed to provide comprehensive data to operators of horizontal wells. AIC provides five services: (1) tracking horizontal drilling permits and completions on a daily basis; (2) computer maps showing horizontal well locations and cumulative production of horizontal and straight wells; (3) production histories and decline curves for leases in given fields; (4) complete records of horizontal wells including directional surveys and production records; and (5) customized prospect maps and data packages.

565. Leraad, A. E., B. G. Petovello, and C. A. Johansik (1984). Application of measurement while drilling in a shallow, highly deviated drilling program. Paper presented at the 35th Annual Technical Meeting of The Petroleum Society of CIM: Recovery and Beyond, June 10-13, 1984, Calgary. CIM, Montreal. CIM paper 84-35-116. Abstract in The Journal of Canadian Petroleum Technology (1984), 23 (2), 38.

Abstract. Measurement-while-drilling was chosen for directional surveys in Norman Wells Expansion Project, Canada.

566. Lesage, M., I. G. Falconer and C. J. Wick (1988). Evaluating drilling practice in deviated wells with torque and weight data. SPE Drilling Engineering 3 (3), 248-252. SPE paper 16114 presented at the SPE/IADC Drilling Conference, March 15-18, 1988, New Orleans.

Friction coefficients can be estimated on a foot-by-foot basis during drilling of deviated wells using MWD and surface values of weight-on-bit and torque. The rotating-friction factor is used to evaluate the transfer of torque between the surface and the MWD tool. Sliding- friction factor (drag) is used to evaluate the transfer of weight-on-bit between the surface and the MWD tool. Dimentionless torque monitors the drillstring below the MWD tool.

567. Lessi, J. and A. Spreux (1988). Completion of horizontal drainholes. Proceedings 3rd SPE International Meeting on Petroleum Engineering, November 1-4, 1988, Tianjin, China, pp. 209-218. Society of Petroleum Engineers, Richardson, TX. SPE paper 17572.

Horizontal wells may be expected to achieve two results: (1) increase productivity; and (2) reduce water and/or gas coning. This article addresses these expectations and the completion methods that may be used to achieve them. Subjects considered are: cementing, formation packers, logging, and production evaluation.

568. Levinson, L. M., O. V. Garris, M. M. Khakimov, Ya. S. Kotskulich, and K. I. Vdovin (1982). [Development of wells at fields of highly viscous and non-flowing oils of the Tatar ASSR]. Tekhnol. Buren. Neft. Gazov. Skoazhin, 98-102. In Russian.

Not examined.

569. Li, T. K., V. Chandelle and J. Brych (1986). Lateral drilling: A new application shows promise. World Oil 202 (6), 68-71.

A horizontal drainhole has been successfully drilled in a thin coal seam in Thulin, Belgium, by Institution pour le Developpment de la Gazeification Souterraine (IDGS), a joint German/Belgian government project. The short-radius horizontal well was drilled from an existing vertical well.

570. Lichtenberger, G. J. (1990). Pressure buildup test results from horizontal wells in the Pearsall field of the Austin Chalk. Proceedings 1990 SPE Annual Technical Conference & Exhibition, Formation Evaluation and Reservoir Geology volume, September 23-26, 1990, New Orleans, pp. 835-850. Society of Petroleum Engineers, Richardson, TX. SPE paper 20609.

Transient pressure tests of horizontal wells, in the Pearsall field Austin Chalk reservoir, indicate predominantly linear flow, preceded by short period of radial flow. Indicated permeabilities due to fracture porosity range from 0.5 to 10 md. Based on radial flow permeability estimates, net thickness drained by horizontal wells is in the range 100 to 350 ft. The Austin Chalk displays multiple porosity behavior.

571. Littleton, Jeff H. (1985). Sohio studies extended-reach drilling for Prudhoe Bay. Petroleum Engineer International 57 (11), 28, 32, 34.

Sohio Petroleum Co. plans to use extended-reach (horizontal) wells to enhance recovery in the Prudhoe Bay field of Alaska's North Slope. A test well was drilled in the Olmos sand of McMullen County, Texas, to test extended-reach drilling techniques. The well targeted a small oil zone, of marginal quality, at 10,300 ft. The well was drilled at 25 degrees inclination to 10,957 ft, logged and cored, and plugged back. The horizontal portion of the hole was kicked-off, and an angle of 83 degrees was built. The top of the Olmos was hit at 10,300 ft TVD. Log and cuttings analysis indicated that a small vertical fault was crossed, so angle was built to 96.7 degrees and the Olmos target was reentered from below. On completion, the well had 1,908 ft of borehole in excess of 83 degrees. The horizontal hole was logged by pushing the tools with drill pipe. The horizontal section was drilled with a downhole motor. Sohio plans to use this technique and technology to enhance recovery from areas of Prudhoe Bay that are not economically viable using conventional vertical wells.

572. Littleton, J. (1986). Standard Oil applies extended-reach drilling to Prudhoe Bay. Petroleum Engineer International 58 (4), 43-44, 46.

Discussion of the first horizontal well (JX-2) drilled at Prudhoe Bay by Sohio.

573. Locke, H. A., J. B. Johnson, and D. C. Jewkes (1988). Extended reach appraisal well: A case history. Proceedings International Meeting on Petroleum Engineering, November 1-4, 1988, Tianjin, China, pp. 123- 132. Society of Petroleum Engineers, Richardson, TX. SPE paper 17564.

An extended-reach, high-angle appraisal well was drilled in Liuhua field, offshore People's Republic of China. The field produces from a highly-permeable carbonate reservoir at sub-sea depth of 4,000 ft (water depth is 1,000 ft). The reservoir has an aerial extent of 9 miles by 4 miles, has a water drive, and contains heavy oil. A high-angle appraisal well was drilled because of economics. The field can economically support a maximum of two platforms. Extended-reach wells are required to adequately develop the field from two platforms. The maximum angle required for these wells is 75 degrees with a maximum horizontal displacement of 10,000 ft. Details of the project are given in the paper.

574. Logan, Richard (1990). Application of MWD propagation resistivity and neutron tool. A.A.P.G. Bulletin 74 (5), 707.

Abstract. When MWD is used for formation evaluation, it should be borne in mind that the measurements are being made minutes after drilling, while formation invasion may still be occurring. This factor may produce a log response different from that of a common wireline log.

575. Logan, T. L. (1987). Application of medium radius horizontal drainhole drilling technology for underground coal gasification, Hanna basin, Wyoming. Paper presented at the 13th Annual Underground Coal Gasification Symposium, August 24-26, 1987, Laramie, Wyoming.

Not examined.

576. Logan, T. L., J. J. Schwoebel, and D. M. Horner (1987). Application of horizontal drainhole drilling technology for coalbed methane recovery. Proceedings SPE/DOE Low Permeability Reservoirs Symposium, May 18-19, 1987, Denver, pp. 195-206. Society of Petroleum Engineers, Richardson, TX. SPE/DOE paper 16409.

The Gas Research Institute (GRI) has funded projects to test the methane-recovery potential of horizontal wells drilled in coal seams. One project is the Soldier Canyon coal mine project, in which horizontal boreholes are drilled into the coal seams of the Soldier Canyon mine, Utah. The pilot project was conducted by Resources Enterprises, Inc. The project resulted in forty in-seam horizontal and slant wells with a total of 65,000 linear ft of borehole. Gas production has reached 200-250 MCFGPD. Another GRI project is the Deep Coal Seam Project, conducted at the Red Mountain site, Piceance basin. This project involves drilling horizontal wells from surface locations into the Cameo "D" coal seam of the Mesaverde Group. The average depth of the coal seams is 5500 ft. The Cameo "D" seam is 16-20 ft thick at the Red Mountain site. Short-radius horizontal drilling was used to drill two drainholes 241 ft in to the coal seam. Details of these projects are discussed in the paper.

577. Logan, T. L. (1988). Horizontal drainhole drilling techniques used for coal seam resource exploitation. Proceedings 63rd SPE Annual Technical Conference & Exhibition, October 2-5, 1988, Houston, pp. 711-723. Society of Petroleum Engineers, Richardson, TX. SPE paper 18254.

This article describes the short-radius and medium-radius horizontal drilling techniques used in methane drainhole-drilling in coal seams.

578. Long, John M. (1991). Old geology, new technology; horizontal drilling in Pearsall field. in Chuber, Stewart (editor), Abstracts and Short Papers of the South Texas Geological Society Austin Chalk Exploration Symposium, February 24-26, 1991, San Antonio, pp. 81-88. South Texas Geological Society, San Antonio.

Horizontal drilling has rejuvenated the 56-year old Pearsall field, Texas. Three hundred horizontal wells have been drilled in the Pearsall field Austin Chalk reservoir. The average IP of these wells is 500 BOPD. The wells average 200 BOPD production after six months. Successful wells should average about 150,000 barrels of oil cumulative production.

579. Longoria, Jose F. (1991). Microfacies analysis as a litho-correlation tool applied to the Austin/Anacacho problem: Surface and subsurface of Bexar County. in Chuber, Stewart (editor),

Abstracts and Short Papers of the South Texas Geological Society Austin Chalk Exploration Symposium, February 24-26, 1991, San Antonio, pp. 89-99. South Texas Geological Society, San Antonio.

Microfacies analysis offers a useful tool to distinguish the Austin Chalk from the overlying Anacacho formation. This paper offers an analysis of both the Austin and Anacacho in the Bexar County, Texas, area.

580. Lorenz, C. (January, 1987). Reservoir sedimentology of Mesaverde rocks at the Multiwell Experiment Site and east central Piceance basin. Sandia National Laboratories, Albuquerque. Report SAND 87-0040.

Not Examined.

581. Lorenz, John C., Sharon J. Finley and David I. Norman (1988). Tectonism, subsidence, and fracturing of Mesaverde reservoirs in the Piceance basin, Colorado. A.A.P.G. Bulletin 72 (2), 215.

Abstract. Most fracturing in the Piceance basin occurred during the Laramide orogeny. All fractures in the Mesaverde formation, at the Department of Energy Multiwell Experiment site, in the Piceance basin are oriented west-northwest.

582. Lorenz, John C., Lawrence W. Teufel and Norman R. Warpinski (1989). Mechanism for formation of regional fractures at depth in flat-lying reservoirs. A.A.P.G. Bulletin 73 (9), 1165.

Abstract. Regional fractures in flat-lying strata form slowly at depth, in extension and parallel to regional horizontal tectonic compression. This fracturing is not caused by high pore pressure, which results in hydraulic fracturing. The fractures form at flaws in the rock and propagate quasi-statistically normal to the least compressive stress.

583. Lorenz, John C. and Sharon J. Finley (1989). Differences in fracture characteristics and related production: Mesaverde formation, northwestern Colorado. SPE Formation Evaluation 4 (1), 11-16. SPE paper 16809 presented at the 1987 Annual SPE Technical Conference and Exhibition, Dallas, September 27-30, 1987.

A core taken from a well in the U.S. DOE Multiwell Experiment site, Colorado, shows that fracture characteristics in the Mesaverde formation change at the boundaries of depositional environments. The Mesaverde is a low-permeability sandstone that produces natural gas from naturally fractured reservoirs. Although a unidirectional, sub-parallel, regional fracture system occurs throughout the Mesaverde formation, the spacing, vertical and lateral distribution, and degree of mineralization of fractures are controlled by reservoir heterogeneity. Editor's Note: The Mesaverde formation is the target of horizontal drilling experiments to try to increase productivity of this low-permeability fractured reservoir. The conclusions of this report should be considered when other tight, naturally fractured reservoirs are being considered for drilling targets, i.e. Austin Chalk.

584. Lorenz, J. C., N. R. Warpinski, P. T. Branagan, and A. R. Sattler (1989). Fracture characteristics and reservoir behavior of stress-sensitive fracture systems in flat-lying lenticular formations. Journal of Petroleum Technology 41 (6), 615-622.

Regional fracture systems may control permeability in flat-lying reservoirs. These fractures occur in a continuum of sizes, occur in sub-parallel, en echelon patterns, and have few high-angle, orthogonal fractures because they may be caused by high pore pressures and relatively low tectonic stresses. Vertical continuity of the regional fractures is limited by inhomogeneities in the lithology. These types of fractures have been documented in cores from the Mesaverde formation in Rulison field, Piceance Creek basin, Colorado. These fractures form highly anisotropic reservoirs. Editor's

Note: Rulison field, Colorado, is the location of the U.S. Department of Energy Multiwell Experiment Site where horizontal drilling has been used to study gas production from tight gas sands.

585. Lorenz, John C. (1991). Fracture spacing in slant/horizontal core, Mesaverde formation, CO: Comparison with outcrop and vertical-core data. A.A.P.G. Bulletin 75 (3), 623.

Abstract. Cores were taken from the Department of Energy Slant Hole Completion Test Well in Colorado. The cores are from the flat-lying Mesaverde formation. The cores show that fracture spacing averages about 3 feet and is not proportional to bed thickness. Fracture spacing in flat-lying subsurface reservoirs may be closer than is commonly acknowledged. Additional details are given in the abstract.

586. Lorenz, John C. (1991). Making the most of the geometric relationships between deviated wellbores and regional natural fractures. A.A.P.G. Bulletin 75 (6), 1131.

Abstract. A horizontal wellbore can be made to intersect 87% of the natural fractures in a reservoir, if the fracture orientations are known. If the fracture orientations are not known then a randomly drilled horizontal wellbore has a 67% chance of intersecting 50+% of the fractures.

587. Loxam, D. C. (1982). Texaco Canada completes horizontal drilling program. Petroleum Engineer International 54 (11), 40, 42-44, 48, 50, 52.

Texaco successfully drilled three horizontal well in the Athabasca oil sands near Fort McMurray, Alberta, Canada. Three wells extended 984 feet (300m) from the surface location. All three were drilled from the same pad. Directional drilling was done with Navidrill or Dyanadrill mudmotors.

588. Lyle, Don (1989). Bakken jousters make horizontal tilt. Western Oil World 46 (5), 14-17.

A review of horizontal drilling in the Bakken formation, Williston basin, with emphasis on the activity of Meridian Oil Inc., Samuel Gary Jr. & Associates Inc., BWAB Inc., Canadian Hunter Exploration Ltd., and Petroleum Experience Inc. The horizontal Bakken play was opened by Meridian, which remains the leader in the play. All of the cited companies have assembled very large acreage positions individually or in partnerships. Horizontal wells in the Bakken are expected to produce up to three times as much oil as one vertical well, and will be as productive as two vertical wells.

589. Lyle, Don (1990). Austin Chalk: Oryx recharges chalk's batteries; racing to the 1-mi horizontal mark; Enron triples Pearsall line; chalk players close in on Gonzales [County]; Fleet launches horizontal attack. Gulf Coast Oil World 8 (3), 30-33, 35-39.

Five very short articles on horizontal drilling activity in the Austin Chalk from Pearsall to the Giddings field. The many operators and service companies mentioned in the article are indexed.

590. Lyle, Don (1990). Horizontal players test Saratoga Chalk. Gulf Coast Oil World 8 (5), 33-35.

Review of horizontal drilling projects in the Cretaceous Saratoga Chalk of Sabine Parish, Louisiana. Horizontal drilling has been tried in the Fort Jessup field, Louisiana. Several companies including Oryx, Woolf & Magee, Dynamic Exploration, Sonat Exploration, Wainoco Oil & Gas, JWR Exploration, and Sun Limited Partnership have leased acreage in the play.

591. Lyle, Don (1990). Horizontal drilling: Solid production boosts Giddings horizontal action. Gulf Coast Oil World 8 (5), 24-25, 27-28, 30-32.

Review of horizontal drilling in Austin Chalk Giddings field. Major horizontal drilling operators are Union Pacific Resources Co., United Oil & Minerals Co., Total Minatome, Clayton W. Williams Jr., and Texaco USA. Horizontal wells in Giddings field have lower IPs than Austin Chalk wells in

Pearsall field, generally in the range of 250-500 BOPD. Payout of these wells occurs in about one year.

592. Lyle, Don (1990). Horizontal well hits oil zone in Anacacho. Gulf Coast Oil World 8 (6), 30.

Two horizontal wells have been successfully completed in the Anacacho lime in Bexar County, Texas. Dorris Petroleum Inc. has completed a 127 BOPD well in Somerset field, and Anacacho Oil Co. has completed a commercial well in Fairfield field.

593. Lyle, Don (1990). Caldwell citizens enjoy horizontal drilling boom. Gulf Coast Oil World 8 (6), 31.

Caldwell, the county seat of Burleson County, Texas, has benefitted from the horizontal drilling "boom" in the Giddings field. 1990 oil and gas sales are up $59 million from 1989. Through September 17, 1990, 144 oil and gas permits have been issued for Burleson county. Real estate prices have also risen in 1990 to $1500/acre from $1000/acre.

594. Lyle, Don (1990). Firms attach MWD talent and skill to wireline tools for more efficient horizontal well completions. Gulf Coast Oil World 8 (6), 32.

Many old vertical wells in the Giddings field, Texas, have a small wellbore, 4 inches or less. Ensco Technologies has developed a method for using wireline logging tools in the hole while drilling as a substitute for small borehole MWD tools.

595. Lyle, Don (1990). Trivia quiz on Texas horizontal holes challenges talents of technology buffs. Gulf Coast Oil World 8 (6), 32.

A ten-question quiz on horizontal drilling in Texas.

596. Lyle, Don (1990). West Texans make tough, tight turns. Southwest Oil World 37 (7), 19-20.

Review of horizontal drilling activity in the Permian basin. Projects by several companies including Oryx, Arco, Southland Oil Co., Coastal Oil & Gas, Perker & Parsley Petroleum Co., Mobil, and Meridian are mentioned.

597. Lyle, Don (1990). Williston settles into comfortable pace. Western Oil World 47 (10), 22, 24-27.

Horizontal drilling activity has declined in the Williston basin. Nine rigs were operating in the Williston basin in the third quarter of 1990, as opposed to thirteen rigs in the first half of the year. Attempts to expand the Bakken fairway (100 miles by 30 miles) in North Dakota have failed. Recoverable reserves of oil from horizontal Bakken wells is estimated at 200,000 barrels of oil per well, vertical wells recover about 100,000 barrels of oil per well. Payout for horizontal wells is six months to two years. A brief discussion of drilling activity in the Bakken trend is given in the article.

598. Lyle, Don (1991). Eastern Colorado: Sideways wells hint at climbing profits. Western Oil World 48 (2), 17, 19, 20-22, 24-25.

Three very short articles mainly about horizontal drilling projects in Colorado. The projects include horizontal wells in Silo field with AFE well costs, a Precambrian horizontal well drilled by Nerdhlic Co., and various Niobrara formation wells.

599. Lyman, T. J. and A. W. Riddell (1984). Heavy oil mining technology and economic analysis. Proceedings 1984 SPE California Regional Meeting, April 11-13, 1984, Long Beach, pp. 565-574. Society of Petroleum Engineers, Richardson, TX. SPE paper 12788.

Heavy-oil mining is a method that promises to recover a substantial percentage of the heavy oil reserves of the United States. Heavy-oil mining may be done either by strip mining or by underground methods. Underground methods using steam-assisted gravity drainage has been used in the Northeast McKittrick field, California. This method involves tunneling under the reservoir, drilling drainholes upward into the reservoir and then parallel to the base of the reservoir, injecting steam, and allowing the oil to drain into production drainholes. Also discussed are surface and underground extractive methods, and economics of the various methods.

600. MacDonald, Ron R. (1987). Drilling the Cold Lake horizontal well Pilot No. 2. SPE Drilling Engineering 2 (3), 193-198. SPE paper 14428 presented at the 1985 SPE Annual Technical Conference and Exhibition, Las Vegas, September 22-25.

Esso Resources Canada Ltd. drilled the 43X well, Cold Lake field, Alberta, Canada, to test the feasibility of a long horizontal well, to evaluate downhole pump alternatives and performance, and to test the feasibility of multiple injector (steam injection) wells. Two problems encountered were: hole drag and wellbore position accuracy. Drag was higher than predicted in this well. The blocks were used to assist the drillpipe and heavyweight pipe into the hole until there was sufficient weight above the drill string. Running liner required two stands of collars immediately above the liner, followed by six stands of heavyweight drillpipe, followed by collars again. This sequence of heavyweight pipe and collars was repeated three times, and was limited only by the supply (330 meters) of collars on site. BORSURE, a commercially available computer program, was used with the MWD instrument to calculate the borehole position. BORSURE accounts for drillstring magnetization, bias, and scale error. A through-the-drillpipe gamma-ray log was used to check the vertical position of the borehole, a method effective because of the good stratigraphic control available in Cold Lake field. After the horizontal well was drilled, a seismic proximity locator method was used to check the location of the horizontal hole. A vertical observation well was drilled 20 meters from the estimated location of the horizontal wellbore. Fifteen hydrophones were suspended in the vertical well and primacord was detonated in the horizontal wellbore. When the horizontal well was thus located, vertical injection wells could be drilled to achieve the desired injector-well spacing.

601. MacDonald, C. J. (1988). Horizontal wells: An application in the North Sea. Proceedings SPE European Petroleum Conference, October 17-19, 1988, London, pp. 85-95. Society of Petroleum Engineers, Richardson, TX. SPE paper 18342. Abstract in The Engineering Index Monthly (February, 1990), 28 (2), 665.

Cyrus field, North Sea, produces from the thin submarine fan sands of the Paleocene Andrew formation. Modeling techniques are given for horizontal wells in Cyrus field. Conclusions reached concerning modeling of horizontal wells include: (a) a method for modeling horizontal wells using a conventional black oil simulator is given; (b) choice of grid block dimensions used is critical; (c) a method for avoiding numerical problems associated with grid refinement is given. Conclusions concerning the reservoir simulation include: (1) a 550 meter horizontal well will recover at least 1.5 times as much oil as a vertical well in Cyrus field; (2) the horizontal well will produce at a higher rate than the vertical well; (3) at Cyrus field, horizontal wells will delay, but not prevent, water coning; (4) the greater the length of the horizontal well in the reservoir the greater the ultimate recovery; (5) the ratio of vertical to horizontal permeability is important with regard to water coning; (6) water breakthrough can be delayed by increasing the standoff of the horizontal section above the oil-water contact.

602. MacEwen, H. (1988). Coiled-tubing conveyed logging systems. Proceedings European Petroleum Conference, October 17-19, 1988, London, pp. 167-171. Society of Petroleum Engineers, richardson, TX. SPE paper 18350. Abstract in The Engineering Index Monthly (February, 1990), 28 (2), 635.

Coiled-tubing-conveyed logging and completion systems are viable alternatives for use in highly-deviated and horizontal wells. A brief description of coiled-tubing-conveyed systems is given in this paper.

603. Mahoney, B. J.(October, 1988). Horizontal drilling use is on the rise: Why and how. World Oil 208 (4), 45-47, 49-50, 53-54, 57.

A general discussion of horizontal drilling.

604. Mainland, Glenn G. Hing Y. Lo (1984). Technological basis for commercial in-situ recovery of Cold Lake bitumen. Proceedings of the Eleventh World Petroleum Congress, London, pp. 235-242. John Wiley & Sons Ltd., London.

Discussion of the geology and steamflood/steam stimulation pilot tests conducted at the Cold Lake, Alberta, Canada, bitumen recovery project by Esso Resources Canada Limited.

605. Mall, T. and R. Fincher (June 9, 1986). Michigan operator salvages well using lateral drilling. Oil & Gas Journal 84 (23), 33-38. Abstract in The Engineering Index Annual (1986), 85, Part IV, 6522.

A horizontal completion was successful in a Niagaran reef in the Michigan basin. The original vertical well had a gas show in a dolomitic zone but could not be commercially completed. A horizontal drainhole drilled 251 ft into the reef tested 613.3 BOPD and 607.4 MCFGPD. Details of the project are given in this paper.

606. Malone, Mark F. and Alan Hippman (October 16, 1989). Short- radius horizontal well fails to improve production. Oil & Gas Journal 87 (42), 41-45. Abstract in The Engineering Index Monthly (January, 1990) 28 (1), 662. Article is based on a paper presented at the CADE/CAODC Spring Drilling Conference, April 26-28, 1989, Calgary.

A short-radius horizontal well, drilled in Drumheller field, Canada, was unsuccessful in improving production. The well was drilled with no unusual problems, except higher than expected costs caused by high bit wear. The use of short-radius drilling limited the information collected about the well because of the inability to log the borehole. MWD and logging were unavailable, so it could not be determined whether or not the target zone was drilled. The zone in which the well was completed was apparently a heterogeneous reservoir, based on cuttings analysis.

607. Manger, Katherine C., Stephen J. P. Oliver, and Richard J. Scheper (1991). The Antrim shale: Structural and stratigraphic influences on gas production. A.A.P.G. Bulletin 75 (3), 629.

Abstract. The productivity of the Antrim shale of the Michigan basin is influenced by both structural and stratigraphic factors. The present producing zones are in the lower part of the Antrim and have four distinct lithologies: black shale, green shale, carbonate, and gray shale. Detailed geological mapping shows allows prediction of production trends.

608. Manning, Kendall (1989). Planning, the critical factor in horizontal drilling. Proceedings Thirty-Sixth Annual Meeting Southwestern Petroleum Short Course, April 19- 20, 1989, Lubbock, Texas, pp. 59-68. Southwestern Petroleum Short Course Association Inc., Lubbock.

Pre-drilling planning for horizontal wells is critical because of the need to respond promptly to changes in formation characteristics with respect to the wellbore i.e. the need to keep the wellbore within the reservoir. This article addresses the case of the medium-radius horizontal well and gives two case histories.

609. Mari, J. L. (1989). VSP processing in horizontal well. in 51st EAEG Meeting, Berlin. Institut Francais du Petrole Internal Report No. 37085.

Not examined.

610. Mari, J. L., C. Wittrisch, R. Goepfer, and A. M. Spreux (1989). Vertical seismic profile in horizontal wells. Proceedings 1989 SPE Annual Technical Conference & Exhibition, Formation Evaluation and Reservoir Geology volume, March 8-11, 1989, San Antonio, pp. 579-590. Society of Petroleum Engineers. Richardson, TX. SPE paper 19856. Abstract in The Engineering Index Monthly (January, 1990) 28 (1), 663.

Three vertical seismic profiles (VSP) were made from the Lacq 91 horizontal well, Lacq Superieur field, France. The horizontal section of the well is in a 50-meter thick reservoir. The VSP's were acquired using three vibroseis trucks, a three-component downhole tool, including hydrophone, and an inclinometer. Sixty-nine different downhole tool positions were recorded, spaced irregularly vertically and with horizontal departures between 84 and 575 meters from the wellhead. The tool was pipe-conveyed in the horizontal section of the well. The VSP confirmed the geological and reservoir studies. The VSP was demonstrated to be a valid method for use with horizontal wells. Two anomalies detected by the VSP were a poorly cemented zone, detected by the geophone, and the casing perforations, detected by the hydrophone.

611. Maricotti, C. (February, 1988). Horizontal wells turn a profit for Elf. Noroil, 32-33.

Not examined.

612. Markle, R. D. (1987). Drilling engineering considerations in designing a shallow, horizontal well at Norman Wells, N.W.T., Canada. Proceedings 1987 SPE/IADC Drilling Conference, March 15-18, 1987, New Orleans, pp. 873-880. Society of Petroleum Engineers, Richardson, TX. SPE/IADC paper 16148.

Engineering problems encountered in drilling long-reach horizontal wells in Norman Wells field, Canada, were: managing torque and drag; drilling and cementing high quality wellbores; and obtaining geological information to better characterize the reservoir. Horizontal well K-50X was drilled in 38.9 days, 6 days longer than planned, with a horizontal segment of 1,223 meters and a TVD of 1,560 meters. Hole cleaning was a major problem during drilling. Intermediate casing depth was 938 meters. Engineering details are given in the text. A suite of openhole logs was run, drill-string conveyed.

613. Marott, A. (April 3, 1989). Slant rigs reduce time, cost of deviated wells. Oil & Gas Journal 87 (14), 52-54. Abstract in The Engineering Monthly Index (September, 1989), 27 (9), 787.

Slanthole drilling was used to develop the heavy oil fields of the Bolivar Coast area, Venezuela, to minimize the impact on the agricultural and urban environment. Several slanthole wells can be drilled from a common location so as to reach reservoirs beneath croplands and towns. In 1985, slant rigs were used in the Zuata field of the Orinoco heavy oil trend for this purpose. The oil in Zuata field is 8-14 API gravity, and is produced at depths between 1,200 and 4,000 ft. The trap is stratigraphic with monoclinal dips of 2-5 degrees. Pay zones are 125-154 ft thick with porosity of 37% and permeability of 1-4 darcies. Initial production was from a gas solution drive, with later steam injection. Details of the well spacing and drilling engineering are given in the article.

614. Mariotti, Christian and Evelyne Kou (1988). Elf improves horizontal drilling at Rospo Mare. Petroleum Engineer International 60 (8), 30, 32, 35. Abstract in The Engineering Index Annual (1988), 87, Part IV, 6004.

A remote-controlled bent sub was used to drill five horizontal wells in Rospo Mare field in the Adriatic Sea, Italy. The remote-controlled bent sub, called T-3000 Telepilote, allows the sub angle to be changed from the drill floor by temporarily increasing the flow rate of the drilling fluid. This system eliminates round trips to modify the bottomhole assembly.

615. Markle, R. D. and G. Lee (1985). Fatigue analysis of Norman Wells drill strings. Journal of Canadian Petroleum Technology 24 (2), 24. CIM paper 85-36-57. Paper presented at the Joint Annual Meeting of The Petroleum Society of CIM and the Canadian Society of Petroleum Geologists, June 2-5, 1985, Edmonton.

Abstract. Drill string problems in high-angle wells in Norman Wells field, N.W.T., Canada, have been less than expected.

616. Marsh, Jack (1988). Discussion of evaluation of directional survey errors at Prudhoe Bay. SPE Drilling Engineering 3 (1), 103. SPE paper 17467.

Critical discussion of Dubrule and Nelson "Evaluation of directional survey errors at Prudhoe Bay," SPE Drilling Engineering (1987) 2 (3), 257-267.

617. Marshall, Joseph W. (1952). Spraberry reservoir of west Texas. A.A.P.G. Bulletin 36 (11), 2189-2191.

The Permian Spraberry formation of West Texas is a thick (up to 950 ft) sequence of low porosity (10%) siltstone and shales. Production from the Spraberry is dependent on fracture porosity. Details of the general geology are discussed in the article.

618. Martin, M., C. Georges, P. Bisson, and O. Konirsch (1987). Transport of cuttings in directional wells. Proceedings 1987 IADC/SPE Drilling Conference, March 15-18, 1987, New Orleans, pp. 293-303. Society of Petroleum Engineers, Richardson, TX. SPE/IADC paper 16083.

This is a report of the results of experiments using tubes and annular space models to study the transport of cuttings in deviated wells. Parameters measured in these experiments include: hole and pipe diameters, inclination of wellbore, flow rates, drilling-fluid properties, and cuttings density.

619. Harbaugh, (1991). Exploitation of oil in a volcanic cone by horizontal drilling in the Elaine field, South Texas. A.A.P.G. Bulletin 75 (3), 630.

Abstract. Elaine field is located in the Uvalde Volcanic Field, South Texas. Production is from the interface between the underlying Austin Chalk and volcanic flows and ejecta. The first horizontal well drilled in Elaine field, by Horizontal Drilling and Production, Inc., had an initial potential of 1,609 BOPD. The horizontal section of the well extends 1,500 ft, but is not within the reservoir, continuously.

620. Martinson, R. (1989). Gas cresting towards a horizontal well - A simulation study. M.S. Thesis. Rogaland University Centre, Stavanger, Norway.

Not examined.

621. Matson, Ron and Rod Bennett (December 17, 1990). Horizontal wells - 7. Cementing horizontal holes becoming more common. Oil & Gas Journal 88 (51), 40-46.

Cementing is becoming more important in completing horizontal wells. Cementing of horizontal wells has been difficult because of the problems of controlling free water, of hole cleaning, pipe centralization, and isolating the borehole with a cement hydraulic seat. Cementing of horizontal wells should be considered and planned in advance of drilling. Computer simulations often are used to plan for a variety of possible well conditions. A good hydraulic seal can best be obtained if the wellbore is smooth and uniform, free of cuttings, and the hole is well conditioned (circulate two to three times with a targeted annular volume of 95% of the annular volume). The pipe must be properly centralized. Pipe movement is necessary to improve mud displacement. Proper cementing spacers are essential to a good cement job. Slurry design should have the objectives of controlling free water, fluid loss, gel-strength development, and gas invasion. Liner cementing is discussed in detail in this paper.

622. Mauduit, Dominique (1989). Determining the productivity of horizontal completions. World Oil 209 (6), 55-57, 61. Abstract in The Engineering Index Monthly (April, 1990), 28 (4), 620.

This article is a critical response to an article by Sparlin and Hagen in World Oil (November, 1988), pp. 55-57, "Controlling sand in a horizontal completion." Mauduit believes that the effect of skin damage productivity of horizontal wells is less than that shown by Sparlin and Hagen. Supporting arguments are given in the article.

623. McCabe, Charles (1990). Horizontal well opens up Troll oil development. Ocean Industry 25 (8), 19-24.

The oil zone in Troll field, Norwegian North Sea, is too thin for producing oil with a vertical well without gas or water coning to occur. A horizontal well has been proposed to eliminate the coning problem. Details of the project are given in the article.

624. McCabe, Charles (1990). Using MWD in short-radius horizontal wells. Ocean Industry 25 (8), 36-37.

Halliburton Geodata and Preussag Erdol und Erdgas have formed a partnership to market short-radius MWD directional drilling services. Short-radius horizontal wells have been drilled using this system in Vega field, Italy, and Reitbrook field, Germany. The Reitbrook set a world record for short-radius wells with a horizontal length of 1,230 ft. A brief description of the system is given.

625. McCaslin, John C. (July 14, 1986). New play set for oldest pool in the West. Oil & Gas Journal 84 (28), 111.

Sun Exploration & Production Co. plans new wildcat drilling in the Florence-Canon City field area, Colorado. The Florence-Canon City field is the oldest oil field in the West. The Niobrara formation is the target of the new drilling.

626. McCaslin, John (editor) (1989). International Petroleum Encyclopedia '89, pp. 22, 30-31, 171-172. Pennwell Publishing Co., Tulsa.

Brief reports of horizontal drilling in Australia, the U.K. North Sea, and Prudhoe Bay field.

627. McCaslin, John C. (June 12, 1989). Operators busy in western Williston basin. Oil & Gas Journal 87 (22), 52.

The first horizontal well completion in Montana was made, by Meridian Oil Inc., in the Bakken formation. The discovery is in the Rawhide prospect in North Fox Creek field area. Non-horizontal drilling activity is also reported.

628. McCaslin, John (1990). Don't overlook the Bakken. Oil & Gas Journal 88 (20), 66.

Meridian Oil Co. has established a new play in the Mississippian Bakken formation, North Dakota, by using horizontal drilling. Meridian had 100% success in its first 12 wells in the Bakken. The Bakken underlies most of the Williston basin. It was known as a locally important hydrocarbon producer since the 1950's. The "horizontal play" in the Bakken is in naturally fractured reservoirs.

629. McGhee, Michael D. (1991). Reservoir performance of the Austin Chalk formation, Giddings field area, Burleson, Lee, & Fayette counties, Texas. in Chuber, Stewart (editor), Abstracts and Short Papers of the South Texas Geological Society Austin Chalk Exploration Symposium, February 24-26, 1991, San Antonio, pp. 101-102. South Texas Geological Society, San Antonio.

Abstract. Production performance of the Austin Chalk in Giddings field, Texas, is dependent on the degree of natural fracturing at the drill-site. Vertical well performance should be an indicator of potential horizontal well performance. An integrated geological/well performance study is recommended for selection of drilling locations in this area.

630. McIntosh, R. A., K. N. Beckie, W. R. Freeborn, G. A. Robb, and K. J. Young (1990). Utilization of current geological, geophysical, and engineering technologies to increase oil recovery in the Suffield J pool, Alberta, Canada. A.A.P.G. Bulletin 74 (5), 715.

Abstract. Integrated geological and three-dimensional seismic interpretations have aided in proper placement of horizontal wells in Suffield field, Alberta, Canada. The field is in an esturine sand shoal deposit and is 5 miles long and 0.5 miles wide. The horizontal wells are placed along the top of the porous zone; designed to miss the shale filled channels.

631. McKissen, Paul (1984). Mud logging as a means of evaluating an elusive of subtle reservoir. Drilling 48 (2), 43, 63-65.

A mud logger can locate the potential productive zone in your well by detecting gas from the zone. He is also effectively "your well-site geologist" who provides sample descriptions. This information can result in unexpected producing zones.

632. McKissen, Paul (1985). Mud logging II. Preventing disaster. Drilling 46 (4), 50-51, 53.

Mud loggers can warn of impending problems and prevent disasters. The mud logger should detect gas in the mud, hydrogen sulfide, and loss of circulation. If the information is properly communicated to the driller disasters can be prevented.

633. McKown, G. K. (1989). Drillstring design optimization for high-angle wells. Proceedings 1989 SPE/IADC Drilling Conference, February 28-March 3, 1989, New Orleans, pp. 275- 282. Society of Petroleum Engineers, Richardson, TX. SPE/IADC paper 18650. Abstract in The Engineering Index Monthly (November, 1989), 27 (11), 762.

The drillstring has three basic functions: to transmit and support axial loads; to transmit and support torsional loads; and to transmit hydraulics. All three functions are more critical in high-angle and horizontal wells, where axial and torsional loads are greater, and hydraulics more critical, due to the use of downhole motors. Planning of the drillstring before the well is drilled must be done with the well objectives in mind. Limitations of the drillstring components, and of the drilling rig, must also be taken into consideration. Drilling conditions and wear as drilling progresses may reduce the efficiency and safety of the drillstring components. A suggested system for drillstring design is given in the article.

634. McLennan, J. D., J.C. Roegiers, and M. Economides (1989). Extended-reach and horizontal wells. in Economides, M. J. and K. G. Noble (editors), *Horizontal Drilling*. Prentice Hall, New York.

Not examined.

635. McLeod, H. O., Jr. and H. R. Crawford (1982). Gravel packing for high rate completions. Paper presented at the 57th Annual SPE Technical Conference and Exhibition, September 26- 29, New Orleans. Society of Petroleum Engineers, Richardson, TX. SPE paper 11008.

Gravel pack production efficiency can be adversely affected by improper perforation conditioning and formation damage. These problems can be reduced by employing proper procedures as explained in this article. The procedures described are for wells in the Gulf of Mexico.

636. McMillian, William H. (1981). Planning the directional well - A calculation method. Journal of Petroleum Technology 33 (6), 952-962.

Equations are given for the calculation of the three-dimensional path of directionally drilled wells from kick-off point to target.

637. McNally, Rich (1990). Horizontal drilling finding a niche. Petroleum Engineer International 62 (9), 38, 41.

Horizontal drilling is being used in the North Sea by Conoco in the U.K. and Dutch sectors. Medium-radius wells are being drilled in North Valiant field where the eolian Rotliegendes sandstone is the reservoir. The well, with a 2,000 ft horizontal section in the reservoir, has a maximum production rate five times higher than other wells in the field. Conoco has drilled a horizontal well in Kotter field, Dutch sector, with the goal of reducing water cut from the Vlieland sandstone reservoir. The Vlieland well has a 620 ft horizontal section and has doubled production from the field. Maersk Oil and Gas has drilled a horizontal well in Dan field. The well has a 3,000 ft horizontal section. Statoil, in the Norwegian sector, has drilled a Statfjord field extended-reach well with a world record 16,410 ft horizontal displacement. Norske Hydro plans a horizontal well for Oseberg field. This well will have a 1,600 ft horizontal liner in the oil zone. Norske Hydro has completed a horizontal well in Troll field.

638. McNally, Rich (1990). Kern County producers pairing costs and emissions. Petroleum Engineer International 62 (10), 16-18, 29.

Discussion of environmental restricting and their result on economics and operations in Kern County, California field. Elk Hills field horizontal drilling is mentioned.

639. McNeely, Branch M., Jr. (1989). Buoyed-aluminum proposed to reduce friction in high-angle well bores. Offshore 49 (2), 38-39, 42-43.

When drilling horizontal wells, aluminum drillpipe substitution for steel pipe below the curved-section of wellbore is suggested. In the horizontal section of a wellbore, the drillpipe does not contribute to the weight-on-bit of the drilling system. The drillpipe does contribute to friction. Details of the proposal are given in the article.

640. McNicoll, Rusty (1989). Drilling Shell's Cormorant Alpha 13 extended reach well: Case history of a world record. Drilling 50 (4), 23-25.

Shell E & P has drilled a well in Cormorant field with a horizontal displacement of 15,618 ft (4,760m) and a measured depth of 19,742 ft (6,017m). The well is an extended-reach rather than a horizontal well.

641. Mead, Homer N. (1988). Using finite-system buildup analysis to investigate fractured, vugular, stimulated, and horizontal wells. Journal of Petroleum Technology 40 (10), 1361-1371. SPE paper 16379 presented at the 1987 SPE California Regional Meeting, Ventura, April 8-10. See also Azeri, Mehdi (1991) Journal of Petroleum Technology 43 (1), 103-104, and Mead, H. N. (1991) Journal of Petroleum Technology 43 (1), 104-105, for critical discussion and reply.

Final shut-in static reservoir pressure may be accurately determined using the rectangular hyperbolic method (RHM). This method also calculates slope, skin, percentage of open reservoir thickness contributing to flow, effective drainage radius, and permeability. It appears to give reasonable answers in horizontal well buildups.

642. Mead, H. N. (1991). Author's reply to Discussion of Using finite-system buildup analysis to investigate fractured, vugular, stimulated, and horizontal wells. Journal of Petroleum Technology 43 (1), 104-105. SPE paper 22060. See also Azeri, Mehdi (1991) and Mead, H. N. (1988).

Reply to critical comments by Azeri.

643. Meissner, Fred F. (1976). Abnormal electric resistivity and fluid pressure in Bakken formation, Williston basin, and its relation to petroleum generation, migration, and accumulation. A.A.P.G. Bulletin 60 (8), 1403-1404.

Abstract. The Bakken shale shows an increase in resistivity from 2-15 ohms to very high resistivities (near infinity) at subsurface temperatures of 165 degrees. This change in resistivity may indicate replacement of normal pore water by hydrocarbons generated in place. These zones in the Bakken are overpressured and oil is the only producible fluid in place. The overpressuring caused by generation of oil may be the source of the fracture porosity in the Bakken.

644. Meissner, Fred F. (1984). Petroleum geology of the Bakken formation Williston Basin, North Dakota and Montana. in Demaison, Gerard and Roelef J. Murris (editors), *Petroleum Geochemistry and Basin Evaluation,* Memoir 35, pp. 159-180. A.A.P.G., Tulsa.

The Bakken formation is a thin shale-siltstone sequence of Upper Devonian-Lower Mississippian age. The shales are good source rocks and are believed to be the source of oil occurring in overlying formations. This paper reviews the geology, stratigraphy, petrophysical properties, and reservoir characteristics of the Bakken. The Bakken is overpressured in some areas. Overpressuring may be caused by hydrocarbon generation, which raised the pore-pressure enough to fracture the rock in-situ.

645. Meissner, Fred F. (1990). A geological/mechanical basis for creating fracture reservoirs in the Bakken. in Hansen, William B. (editor), Bakken Formation Short Course, p. 22. Montana Geological Society, Billings. Study Notes No. 1. Abstract also printed in Hansen, William B. (editor) (1991), 1991 Guidebook to Geology and Horizontal Drilling of the Bakken Formation, p. 165, Montana Geological Society, Billings, Montana.

Abstract. Naturally-fractured reservoirs in the Bakken formation are controlled by three parameters: (1) rock properties; (2) pore-fluid overpressures; and (3) critical amount of differential rock matrix stress. Hydrocarbon generation in the Bakken created overpressures because of volume changes. Bakken fluid pressure gradients indicate that the fracture gradients of confining rocks have been exceeded. Fracture spacings may be controlled by bed thickness and matrix permeability.

646. Meissner, Fred F. (1990). A geological/mechanical basis for creating fractured reservoirs in the Bakken. A.A.P.G. Bulletin 74 (8), 1336.

Abstract. Fractured reservoirs may be created in the Bakken formation by three basic and co-dependent factors: (1) rock properties; (2) critical amount of pore-fluid overpressure; (3) critical amount of differential rock matrix stress.

647. Mercer, J. C., H. R. Pratt, III, and A. B. Yost, II (1988). Infill drilling using horizontal wells: A field development strategy for tight fractured formations. Proceedings Gas Technology Symposium: Gas Supply - A Critical Balance of Politics, the Market Place, and Technology, June 13-15, 1988, Dallas, pp. 225-232. Society of Petroleum Engineers, Richardson, TX. SPE paper 17727. Abstract in The Engineering Monthly Index (July, 1989), 27 (7), 550.

Simulation of sixteen vertical gas wells, for a ten year period, was compared with simulated production from horizontal wells for a similar period. The simulation showed that as few as four unstimulated gas wells, extended 2,000 ft horizontally, can produce 1.5 times as much gas as sixteen vertical wells. Further, two hydraulically fractured horizontal wells may produce as much gas as four unstimulated horizontal wells. The simulations were done using the Appalachian basin Devonian shales as a model.

648. Merkulov, V. P. (1958). [Le debit des puits devies et horizontaux] [The flow of slanted and horizontal wells]. Neft. Khoz. 6, 51-56. In Russian.

Not examined.

649. Meszaros, G., A. Chakma, and M. R. Islam (1990). Scaled model studies and numerical simulation of inert gas injection with horizontal-wells. Proceedings 1990 SPE Annual Technical Conference & Exhibition, Enhanced Oil Recovery/General Petroleum Engineering volume, September 23-26, 1990, New Orleans, pp. 585-598. Society of Petroleum Engineers, Richardson, TX. SPE paper 20529.

Successful gas injection as an EOR technique requires gravity stabilization to avoid early gas breakthrough. In the absence of natural dip, horizontal wells may be used to achieve the required stability. A horizontal injection well can be placed at the top of the reservoir while producing through another horizontal well placed at the bottom of the reservoir. Partially scaled 2-dimensional model studies were made that indicate that up to 70% of the oil-in-place may be recovered from a heavy-oil reservoir by injecting an inert gas into the top of the reservoir. These experimental results and the model are discussed in this paper.

650. Michigan Oil & Gas News (August 12, 1988). Trendwell's Antrim shale laterals make up to 110 Mcfgd gas in Otsego. Michigan Oil & Gas News 24 (33), 1, 27.

Not examined.

651. Mitchell, Gary C., Fred E. Rugg and John Byers (September 25, 1989). East Central Utah - 1. The Moenkopi: Horizontal drilling objective in east central Utah. Oil & Gas Journal 87 (39), 120-124.

Short-radius horizontal drainholes drilled in the Moenkopi formation in Grassy Trail Creek field, Utah, improved production significantly. Vertical wells produced 18 to 105 BOPD using conventional drilling and completion methods. Horizontal wells in the same field produced 87 to 583 BOPD. Moenkopi fields produce from naturally fractured reservoirs.

652. Montgomery, S. L. (1989). Williston Basin Bakken Shale: Learning curve on the horizontal. Petroleum Frontiers 6 (4), 52 pp.

Comprehensive review of the geology, stratigraphy, tectonics, reservoir engineering, drilling practices and horizontal drilling activity of the Bakken formation, Williston basin.

653. Montgomery, S. L. (1990). Techniques of horizontal drilling: Concepts and consequences. Petroleum Frontiers 7 (1), 1-48.

Review of the applications and methods of horizontal drilling including: completions, cementing, logging, and drilling fluids.

654. Montgomery, Scott (1990). Horizontal Drilling: A Survey of Domestic Case Histories. Petroleum Frontiers 7 (2), 1-46. Published by Petroleum Information Corporation.

Review of horizontal drilling in: Prudhoe Bay field; the Michigan basin; Elk Hills field; and U.S. Department of Energy tight gas formations projects. Projects include high-angle drilling, inverted wells (Prudhoe Bay), and tight gas formations (Mesaverde group).

655. Montgomery, Scott (1990). Horizontal drilling in the Austin Chalk. Part 1: Geology, drilling history and field rules. Petroleum Frontiers 7 (3), 1-44.

A review of the geology, history and recent drilling activity in the Austin Chalk trend of Texas. Field rules for horizontal drilling are also discussed.

656. Moore, H. W. (1982). An overview of the Norman Wells Expansion Project, Norman Wells, Northwest Territories. Paper presented at the Engineering Institute of Canada, 8th Annual Western Region Conference, November 17-18, 1982.

Not examined.

657. Moore, Steven D. (1983). Esso expands Norman Wells production. Petroleum Engineer International 55 (1), 26, 28.

Esso Resources Canada Ltd. plans to add 190 wells to the current 50 wells producing from Norman Wells field on the McKenzie River in Canada's Northwest Territories. The expansion will boost production from 3,000 BOPD to a planned 28,300 BOPD. Part of the field expansion is a waterflood which is expected to increase recovery to 42% of oil in place. Norman Wells field was discovered in 1918.

658. Moore, Steven D. (1986). MWD tools improve drilling performance. Petroleum Engineer International 58 (2), 49, 51-52.

MWD is most commonly used to monitor directional drilling parameters, such as drift angle, azimuth, tool-face heading, and gravitational and magnetic field strengths. Some MWD tools are available with gamma-ray and resistivity tools, which allow real-time evaluation of lithology. Editor's Note: MWD is used on most horizontal wells for the directional drilling parameters.

659. Moore, Steven D. (1986). The hows and whys of downhole drilling motors. Petroleum Engineer International 58 (8), 38, 42.

Downhole motors are used in a wide variety of drilling situations, are cost effective, may increase penetration rates, reduce casing wear, and help control deviation. Downhole motors consist of a bypass valve, motor unit, and bearing assembly. They are either positive displacement or turbine motors. Both types of motor are powered by the flow of mud through the motor. A new development is the steerable motor which can be used in conjunction with MWD to drill directional holes without frequent trips to change bent subs.

660. Moore, Steven D. (1987). High-angle drilling comes of age. Petroleum Engineer International 59 (2), 18-20, 22. Abstract in The Engineering Index Annual (1987), 86, Part IV, 6499.

High-angle drilling (40 to 80 degrees) can increase the area accessible, from a single location, thirty-fold. Modern drilling technology makes high-angle drilling a routine matter; horizontal drilling is becoming common. New technology, such as MWD, downhole motors, and steerable directional tools, combined with better drilling fluids, allow wells to reach over 15,000 feet, horizontally, from the surface location. Slant-hole rigs allow horizontal holes to be drilled in very shallow reservoirs (260-660 feet below ground level in the Athabasca Oil Sands). Deep horizontal wells have been drilled and completed in France (Lacq field), Italy (Rospo Mare field), and the United States (Empire-Abo Unit). Horizontal wells have been drilled in Prudhoe Bay field, Alaska, to improve ultimate recovery of oil.

661. Moore, Steven D. (1987). Horizontal drilling a success at Norman Wells. Petroleum Engineer International 59 (9), 19-20, 22. Abstract in The Engineering Index Annual (1987), 86, Part IV, 6499.

Norman Wells field was discovered in 1918 and is still undergoing development. Esso Resources Canada Ltd. is now using horizontal wells to reach the 60% of reserves under the 3 mile wide Mackenzie River. Esso classifies a horizontal well as one with a hole angle of 86 degrees or greater. Four horizontal wells have been drilled along with several high-angle wells (less than 86 degrees). The longest horizontal well to date has a 4,013 ft long horizontal section. Engineering details are given in the article.

662. Moore, Steve (1988). Horizontal wells prove their worth. Petroleum Engineer International 60 (2), 18-19.

An interview with Jacques Bosio, deputy research director of the Elf Aquitaine Group. Horizontal wells are advantageous in vertically fractured reservoirs, in thin pays, and in increasing ultimate recovery. Horizontal wells are also good in cases where water coning or invasion is a possibility. The idea behind horizontal drilling is to drill where the oil is instead of letting the oil flow through the formation to a vertical well bore. Elf has drilled horizontal wells in Lacq field and Rospo Mare field.

663. Moore, Steven D. (1988). Alberta pushes ahead with underground drilling. Petroleum Engineer International 60 (2), 28-30.

Steamflood and horizontal drilling from tunnels were tested in the McMurray formation heavy-oil sands in Alberta at the Underground Test Facility (UTF) of the Alberta Oil Sands Technology and Research Authority (AOSTRA) in Alberta, Canada. The facility is composed of 3-meter diameter shafts with tunnels 14 meters below the reservoir. Horizontal boreholes were drilled up and into the reservoir and used to inject steam into the oil sands. The mobilized oil is gravity-drained into a manifold and pumped to the surface.

664. Moore, Steven D. (1989). Meridian Oil finds success with horizontal wells. Petroleum Engineer International 61 (11), 17-20, 22. Abstract in The Engineering Index Monthly (June, 1990), 28 (6), 671.

Meridian Oil is a world leader in horizontal drilling with more than 43 wells drilled, so far, in North Dakota, Montana, South Dakota, Alabama, Oklahoma, New Mexico, and Texas. Meridian has had 100% success mechanically in drilling its wells. Meridian's management practices, allowing drilling and completion decisions to be made quickly, frequently at well site, have kept well costs down. Meridian has drilled and completed 20 Bakken shale wells in the Williston basin, with an average horizontal displacement of 2,200 ft, with well bore angle of 86 degrees or greater in the pay zone. The first 20 Bakken wells have average reserves of 237,000 BOE in place, and have an average production of 239 BOE/day. Meridian's after-tax rate of return on the Bakken wells is 36.6%. Meridian has the leases to drill more than 15,000 Bakken wells, and plans to apply its technology to other formations, as well.

665. Moore, Steven D. (1990). Oryx develops horizontal play. Petroleum Engineer International 62 (4), 16-22. Abstract in The Engineering Index Monthly (August, 1990), 28 (8), 599.

Oryx Energy drilled its first horizontal well in the fractured Austin Chalk reservoir in Pearsall field, South Texas, in 1985. The well was designed to drill 240 feet horizontally into the vertical fractures of the reservoir so as to intersect more fractures than would a vertical well. The well was a modest success at 107 BOPD. Oryx has perfected its drilling and completion techniques, and now routinely completes Austin Chalk wells that test over 1,000 BOPD. Now, instead of 240 feet of horizontal displacement, Oryx drills out to over 4,000 feet. Drilling methods include using steerable downhole motors and MWD, while drilling under balanced to let the well flow during drilling. Oryx plans at least 85 Austin Chalk wells.

666. Moore, Steven D. (1990). Horizontal drilling activity booms. Petroleum Engineer International 62 (11), 15-16.

In 1990, 650 horizontal wells are expected to be completed in the Austin Chalk and 130 horizontal wells will be completed in the Bakken shale of the Williston basin. Approximately 70 horizontal wells will be drilled in other plays in the United States. Tight supplies of equipment, rigs, and personnel will limit the number of horizontal wells drilled. Approximately 200 horizontal wells will be drilled by international operators, with Canada having almost half of the wells. The record for total measured depth (15,357 ft) and depth (14,614 ft) for a horizontal well is held by a Smako Ltd. well in the Huxford field, Alabama, producing from the Smackover formation.

667. Moore, Steve (1991). Horizontal well technology moves forward. Petroleum Engineer International 63 (1), 19-21.

Interview with Ted Stagg, a drilling engineer with BP Exploration in Alaska.

668. Moore, Steven D. (1991). The coiled tubing boom. Petroleum Engineer International 63 (4), 16-18, 20.

Coiled-tubing services are increasing in demand at a very high rate. Coiled-tubing is the optimum way to log, complete, and workover horizontal and deviated wells. Coiled-tubing is a high-strength, ductile steel tube usually 1-inch or 1.25-inches in outside diameter. The coiled-tubing is transported on reels and fed into a well through an injector into the tubing string. Coiled-tubing may be run into a hole at 60 to 100 ft/minute. Some units can feed in tubing at 240 ft/minute. Maximum pumping rate for coiled-tubing is 1 or 2 barrels/minute. Some coiled- tubing operations are: paraffin clean out, acid or cement squeezes, low-pressure kills, sand washing, perforating underbalanced, drill-stem testing/mud displacement, fishing, lateral drilling, setting and retrieving bridge plugs, logging, and perforating.

669. Moore, W. D. (September 15, 1980). ARCO drills horizontal drainhole for better reservoir replacement. Oil & Gas Journal 78 (37), 139-142, 147-148.

Horizontal drainholes have been drilled in the Empire Abo Unit of New Mexico to alleviate a gas coning problem. The drainholes were drilled 106 ft horizontally into the reservoir (Abo Reef). The well was turned from vertical to horizontal in 29 ft of drilling. Details of the equipment and drilling methods are given in the text. Drainholes were used as early as 1882 at Santa Barbara, California, where the drainholes were mined rather than drilled. In the 1930's, the Zublin Co. provided tools for drilling horizontal wells. High-angle multiple-drainholes were drilled in the Los Angeles basin in the 1950's. This project was killed by legal problems before the technical and economic feasibility could be determined. Sinclair, in the 1960's, tried drainhole drilling, but the project was not economically feasible. The current project was drilled with wiggly drill collars provided by consultant Don R. Holbert.

670. Morgan, Craig D. and Thomas C. Chidsey (1991). Horizontal drilling potential of the Cane Creek shale, Paradox formation, Utah. A.A.P.G. Bulletin 75 (6), 1133-1134.

Abstract. The Cane creek shale member of the Paradox formation is a naturally fractured, organic-rich, marine shale. The Cane Creek shale is similar to the Bakken formation. Structural trends in Utah define potential fractured reservoirs in the Cane Creek.

671. Morgridge, Dean L. and William B. Smith, Jr. (1972). Geology and discovery of Prudhoe Bay field, Eastern Arctic Slope, Alaska. in King, Robert E. (editor), *Stratigraphic Oil and Gas Fields -Classification, Exploration Methods, and Case Histories.* A. A. P. G. Memoir 16, pp. 489-501. A.A.P.G., Tulsa.

Prudhoe Bay field, Alaska, is the largest field in North America, with an area of 125,000 acres. The field is a stratigraphic trap which produces oil and gas from the Sadlerochit, Sag River, Kuparuk, Shublik, and Lisburne formations. This paper discusses the field history and geology.

672. Morita, N., S. P. Singh, H. S. Chen, and D. L. Whitfill (1990). Three-dimensional well model pre-processors for reservoir simulation with horizontal and curved inclined wells. Proceedings 1990 SPE Annual Technical Conference & Exhibition, Reservoir Engineering volume, September 23-26, 1990, New Orleans, pp. 49-64. Society of Petroleum Engineers, Richardson, TX. SPE paper 20718.

Peaceman's well model produces significant errors when applied to certain cases for horizontal wells. Proper local mesh refinements can minimize the errors of Peacemen's model. Three 3-Dimensional well model processors, given in this paper, give accurate well connection factors and can be applied to any inclined, curved, horizontal, and partially penetrating vertical well.

673. Moritis, Guntis (February 27, 1989). Worldwide horizontal drilling surges. Oil and Gas Journal 87 (9), 53-55, 58-59, 62-63.

Review of the horizontal drilling projects of 1988 and projects planned for 1989.

674. Moritis, Guntis (March 13, 1989). More information from horizontal well survey. Oil & Gas Journal 87 (11), 63-65.

The article is a list of horizontal well projects. Projects and operators are indexed.

675. Moritis, Guntis (1990). Horizontal drilling scores more successes. Oil and Gas Journal 88 (9), 53.

Oil & Gas Journal survey of horizontal drilling projects; 130 wells listed. The list excludes wells with less than 1,000 feet of horizontal displacement.

676. Moritis, Guntis (March 11, 1991). Horizontal drilling technology keeps advancing. Oil & Gas Journal 89 (10), 49-52, 54.

Review of the horizontal drilling activity of the previous year. Horizontal records for the year include: Elf Aquitaine's C-2 offshore Netherlands Block L-7 well which was the deepest TVD horizontal well, at 13,322 ft; longest horizontal displacement was Meridian's Pearsall field well No. 1-337H with a horizontal lateral of 5,208 ft; maximum angle reached in 1990 was 121 degrees in the Amoco Canada Sturgeon Lake South No. 2. A table of 1991 horizontal drilling plans is included.

677. Moseley, L. M. (1986). Discussion of Measurement while drilling - a new tool. Journal of Petroleum Technology 38 (5), 590.

Critical discussion of article by John E. Fontenot "Measurement while drilling - A new tool," Journal of Petroleum Technology 36 (1), 128-130.

678. Moslow, Thomas F. and Roderick W. Tillman (1989). Characterization, distribution of Frontier formation reservoir facies of Wyoming fields. Oil and Gas Journal 87 (22), 95-102, 104.

The Cretaceous Frontier formation, Green River basin, Wyoming, is predominantly a deltaic sandstone. Gas production in the Moxa and Whiskey Buttes fields is mainly from distributary channel sands. These reservoirs are heterogeneous with permeability barriers in both vertical and horizontal directions. Production from Moxa field is from sandstones with "normal" porosities. Additional reserves are expected from tight sands after the "normal" reservoirs are depleted.

679. Mroz, Thomas H. and William A. Schuller (1990). Results of horizontal well site evaluation in fractured Devonian shale reservoirs. A.A.P.G. Bulletin 74 (8), 1308.

Abstract. U.S. Department of Energy Morgantown Energy Technology Center has conducted studies of four horizontal wells drilled in fractured Devonian shales. The studies have resulted in three-dimensional geologic models of the fractured reservoirs.

680. Mueller, F. S. and R. E. Grayson (June/July, 1990). Worksheet helps control slanted hole kicks. Drilling Contractor, pp. 68-69.

Not examined.

681. Mueller, M. D., J. M. Quintana and M. J. Bunyak (1990). Extended reach drilling from Platform Irene. Proceedings of the 22nd Annual Offshore Technology Conference, Houston, v. 1, pp. 217-226. OTC paper 6224. Also in SPE Drilling Engineering 6 (2), 138-144.

Union Oil Co. of California has drilled two wells extending over two mile from Platform Irene in the Point Pedernales field, offshore California. The average angles from the wells are 70.9 degrees and 68.2 degrees. The term "negative width well" is introduced to indicate a compressive negative force used to push the drill string down the hole to overcome the forces of drag and gravity which resist downward movement of the pipe in a high-angle hole. Details of drilling the two wells are given. Some of the conclusions reached in this project are: (1) extensive drag analysis must be made in high-angle wells; (2) liners can be rotated into wellbore in which they can not slide; and (3) shallow reservoirs offshore in state waters can be drilled from shore using extended-reach technology.

682. Mukherjee, Hemanta and Michael J. Economides (1989). A parametric comparison of horizontal and vertical well performance. Proceedings 1989 SPE Annual Technical Conference & Exhibition, Reservoir Engineering volume, October 2-5, 1988, Houston, pp. 411-419. Also in SPE Formation Evaluation 6 (2), 209-216. Society of Petroleum Engineers, Richardson, TX. SPE paper 18303.

This paper compares the performance of a fully completed horizontal well with that of a hydraulically fractured well, and the performance of a hydraulically fractured horizontal well with that of a hydraulically fractured vertical well. Horizontal wells can be viable alternatives to vertical wells in some cases, including cases where Bhd (B = vertical anisotropy; h = reservoir height). Horizontal wells are useful in naturally fractured reservoirs. Horizontal wells may not outperform vertical wells where vertical permeability is greater than, or equal, to 1.5.

683. Muller-Link, D. and A. Steenbock (1987). Alternative drain hole drilling system - Theoretical background and practical experience. Paper presented at the 1987 SPE Offshore Europe Conference, September 8-11, 1987, Aberdeen. Society of Petroleum Engineers, Richardson, TX. SPE paper 16564.

Not examined.

684. Mungan, N. (1986). Horizontal wells for enhanced oil recovery. Paper presented at the 20th Anniversary of Petroleum Recovery Institute, July, 1986.

Not examined.

685. Murphy, Patrick J. (1990). Performance of horizontal wells in the Helder field. Journal of Petroleum Technology 42 (6), 792-800. SPE paper 18340 presented at the 1988 SPE European

Petroleum Conference, London, October 16-19. Abstract in The Engineering Index Monthly (February, 1990), 28 (2), 628.

Helder field, on the Dutch continental shelf, was "redeveloped" using eight horizontal wells sidetracked from existing wells. Advantages of horizontal wells include: lowering water cut; increasing recovery of oil; and lower pressure drawdown. Operating costs were lowered and oil production was increased.

686. Mutalik, P. M., S. P. Godbole, and S. D. Joshi (1988). Effect of drainage area shapes on the productivity of horizontal wells. Paper presented at the 1988 SPE Annual Technical Conference & Exhibition, October 2-5, 1988, Houston. Society of Petroleum Engineers, Richardson, TX. SPE paper 18301. Not printed in the Proceedings volume.

A method is presented of calculating the pressure responses and productivity of centered and off-centered horizontal wells.

687. Nation, Larry (1990). Oil on tap in room in Illinois field: Downhole operations - literally. A.A.P.G. Explorer 11 (11), 16-17.

Gravity draining through horizontal pipes is the production method in an oil mine in Siggins field, Illinois, operated by Three Star Drilling and Producing Corp. Thirty-three lateral drainholes were drilled into the walls of a 24-ft diameter room 425 ft below the surface of the field. The drainholes are from 300 to 2,000 ft long, have 8 ft of casing, and are openhole completions. The drainholes have a cumulative penetration of the reservoir of 34,780 ft. A similar operation in Kaycee, Wyoming, operated by the Kaycee Oil Co., produces 100-250 barrels of oil per day from 70,000 ft of drainhole.

688. Nazzal, Greg (October 8, 1990). Horizontal wells-2. Planning matches drilling equipment to objectives. Oil & Gas Journal 88 (41), 110, 112,114-118.

Drilling successful horizontal wells requires careful pre-well planning, properly trained wellsite personnel, contingency plans for correcting problems encountered in drilling, and meaningful post-drilling appraisal. Pre-drilling planning should include design of the drillstring, analysis of expected torque and drag, formation evaluation, geology of the target formation, and reservoir management. One of the common problems encountered is missing the target zone. Suggestions are given in the article for planning and appraisal of horizontal wells. Editor's Note: Many of the published accounts of horizontal drilling emphasize the reduction in costs with experience. Proper planning and post-drilling appraisal should significantly lower costs by identifying problems and presenting workable solutions.

689. Newman, Harry E., II (1981). Greater Green River basin stratigraphy as relates to natural gas potential. Proceedings of the 1981 SPE/DOE Symposium on Low-Permeability Gas Reserves, May 27-29, 1981, Denver, pp. 177-192. Society of Petroleum engineers, Richardson, TX. SPE/DOE paper 9845.

The Green River basin is now the site of horizontal drilling for Niobrara formation and Mesaverde formation targets. Horizontal drilling has been restricted to the Sand Wash sub-basin, so far, but other formations will probably be horizontal drilling targets in the future. This article is a good review of the stratigraphy of the basin with emphasis on tight gas reservoirs.

690. Newman, Kenneth R., Mark Corrigan, and John B. Cheatham (1989). Safely exceed the 'critical buckling load' in highly deviated holes. Paper presented at the 1989 SPE European Offshore Conference: SPE Offshore Europe 89, September 5-8, 1989, Aberdeen. Society of Petroleum Engineers, Richardson, TX. SPE paper 19229.

The critical buckling-load of coiled-tubing may be exceeded in highly-deviated holes in some circumstances. Coiled-tubing is sometimes used for conveying logging and completion tools in horizontal and high-angle wells. Several case histories are presented.

691. Newton, Graham (1991). World Horizons: Global trends in horizontal drilling. Popular Horizontal 1 (2), 4-5.

General article on horizontal drilling.

692. Nice, Sid B. and W. H. Fertl (1990). New logging, completion techniques boost horizontal well productivity. Petroleum Engineer International 62 (11), 20, 22-24, 26-29.

Discussion of pipe-conveyed and coiled-tubing-conveyed logging and completion system used in horizontal, extended-reach, and high-angle wells.

693. Nice, Sid B. (1991). Leading edge logging: Well logging and completion technology for horizontal wellbores. Popular Horizontal 1 (1), 4-13. Reprinted from Transactions G.C.A.G.S.

Discussion of pipe-conveyed and coiled-tubing-conveyed logging systems for use in horizontal wells. Examples are given from the Austin Chalk, Texas, and Bakken formation, North Dakota.

694. Nice, Sid N. and W. H. Fertl (1991). Logging, completing extended-reach and horizontal wells. World Oil 212 (3), 49- 50, 52-53, 55-56.

Pipe-conveyed and coiled-tubing-conveyed logging and completion tools are available for use in horizontal wellbores. These systems are discussed and examples are given for the Appalachian basin, North Sea, Gulf of Mexico, and onshore Texas.

695. Nicholson, John Taft (February 4, 1991). Algorithm facilitates preplanning well bore trajectories. Oil & Gas Journal 89 (5), 45-49.

An algorithm is discussed that facilitates planning of wellbore trajectories in horizontal, extended-reach, relief, and non-colinear multi-target wells. Examples of each case are given.

696. Nini, C. J. and G. W. Owen (1983). Successful high-angle gravel packing techniques. in Preprints, 1983 SPE Annual Technical Conference & Exhibition, October 5-8, 1983, San Francisco. Society of Petroleum Engineers, Richardson, TX. SPE paper 12105.

Chevron successfully tested gravel packing procedures, in high-angle wells (70-80 degrees), in Garden Banks block 236, Gulf of Mexico. The criteria established for successfully packing 100% of the annular volume include: (1) use clean completion fluids; (2) gravel and screen design based on absolute stoppage and high permeability; (3) all perforations placed in the lower (bottom) portion of the hole using the maximum economic perforation density; (4) use medium viscosity carrier fluid and sand concentration of 4 lbs/gal; (5) tailpipe O.D. to liner I.D. ratio should be a minimum of 0.6; (6) no acid wash should be used; sand volume emplaced should be at least 100% of annular fill up; (7) and design criteria of gravel and screen should be verified on site.

697. Noblett, Bruce R. and Walter H. Fertl (1987). Circumferential acoustic log finds vertical fractures. World Oil 205 (5), 37-38.

Circumferential acoustic logging can locate and determine the orientation of fractures in a borehole. The tool also has excellent thin-bed resolution. Examples are given from Williston basin wells.

698. Noblett, Bruce R. and L. Pat Howard (1991). Using the circumferential borehole imaging log to identify and orient vertical fractures in the Austin Chalk. in Chuber, Stewart (editor), Abstracts and Short Papers of the South Texas Geological Society Austin Chalk Exploration Symposium, February 24-26, 1991, San Antonio, pp. 105-109. South Texas Geological Society, San Antonio.

The Circumferential Borehole Imaging Log is an acoustic tool that provides an "image" of the borehole, including natural fractures. The orientation of natural fractures can be determined from the log. Details of the log and its interpretation are given in the paper with examples from the Austin Chalk in Pearsall field and Gonzales County, Texas.

699. Nobs, Louis and Ronald Ingersoll (1991). Downhole PDM motors: New designs improve horizontal drilling efficiency. Popular Horizontal 1 (2), 8, 10.

General article about positive displacement motors (downhole motors).

700. Nolen-Hoeksema, Richard C. (1989). Fracture development and mechanical stratigraphy of Austin Chalk, Texas: Discussion. A.A.P.G. Bulletin 73 (6), 792-793. See Corbett, K., M. Friedman, and J. Spang (1987). Fracture development and mechanical stratigraphy of Austin Chalk, Texas. A.A.P.G. Bulletin 71 (1), 17-28.

Discussion of the article by Corbett et al. (1987). A different approach to linear regression analysis, than that of Corbett et al., is that confining pressure is a better predictor of rock strength than strain at failure.

701. Norel, G. and C. Dubois (1988). Test bench checks cement in horizontal holes. Petroleum Engineer International 60 (11), 54-59.

Elf Aquitaine and Institut Francais du Petrole (IFP) conducted extensive tests on cementing casing in horizontal wells. The tests showed that the main problems were centralization of the casing, cement slurry composition, and rate of cement and mud displacement. An ultrasonic tool (ARTEP) was developed as a cement-bond logging tool.

702. Norris, S. O. and L. D. Piper (1990). Modeling fluid flow around horizontal wellbores. Proceedings 1990 SPE Annual Technical Conference & Exhibition, Reservoir Engineering volume, September 23-26, 1990, New Orleans, pp. 65-75. Society of Petroleum Engineers, SPE paper 20719.

Horizontal well flow was simulated using a hybrid grid model incorporating radial gridblocks with spherical grid blocks to model fluid flow around the end of the well.

703. Norris, Stephen O., James L. Hunt, and Mohamed Y. Soliman (1991). Predicting horizontal well performance: A review of current technology. Pacific Oil World 83 (3), 42. Abstract of a paper presented at the 61st Western Regional Meeting of the SPE, March 20-22, 1991, Long Beach, California.

Abstract. No data given in the abstract.

704. Northrop, D. A., A. R. Sattler, and J. K. Westhusing (1983). Multi-Well Experiment: A field laboratory for tight gas sands. Paper presented at the SPE/DOE Symposium on Low Permeability Reservoirs, March 14-16, 1983, Denver. Society of Petroleum Engineers, Richardson, TX. SPE/DOE paper 11646.

The U.S. Department of Energy established the Multi-Well Experiment Site (MWX), Colorado, to study recovery methods from tight gas sands in the western United States. The site has closely spaced wells that provide good geological and engineering information about the reservoirs. The reservoirs under study are lenticular sands. In-situ stress field has been measured and hydraulic fracturing experiments planned. Editor's Note: Horizontal and slanthole wells have been drilled in the MWX.

705. Nydegger, Gary L. (1991). Overview of the Rocky Mountain fractured Niobrara play. in Chuber, Stewart (editor), Abstracts and Short Papers of the South Texas Geological Society Austin Chalk Exploration Symposium, February 24-26, 1991, San Antonio, pp. 111-118. South Texas Geological Society, San Antonio.

The Niobrara formation in the Western Interior Cretaceous Seaway was penecontemporaneous with the Austin Chalk of Texas. Both formations produce oil and gas from naturally fractured reservoirs. The Niobrara has four facies, of which three are productive from fractured reservoirs. This paper describes the Niobrara facies and their relationship to oil production.

706. Ocean Industry (March, 1986). Survey gyro tested in Grand Canyon well. Ocean Industry 21 (3), 49.

An extended-reach borehole was drilled in the Grand Canyon National Park for the emplacement of a water pipe to supply potable water to the Indian Garden campground in the Grand Canyon. Eastman Whipstock did the directional surveys of the hole with their Seeker-1 gyro system. The tools use a rate-integrating, or north-seeking gyroscope for a survey reference. Both of the Seeker-1 systems used gave bottom hole locations within 7 ft of the location, as measured by land survey instruments (the hole exited the side of the canyon above the campground).

707. Ocean Industry (1988). Operators active in several new Australian fields. Ocean Industry 23 (1), 19-20.

Brief mention of horizontal drilling in North Herald and South Pepper fields in the Bass Strait, Australia.

708. Ocean Industry (1988). Record horizontal reach well slated for UK North Sea. Ocean Industry 23 (4), 18.

Conoco proposes to drill an extended-reach well with up to 32,000 ft of horizontal displacement at a TVD of 12,000 ft and an angle of inclination of 72 to 75 degrees.

709. Ocean Industry (1990). Using MWD in short-radius horizontal wells. Ocean Industry 25 (8), 36-37.

Halliburton Geodata and Preussag Erdoil und Erdgas have formed a partnership to provide short-radius MWD drilling for horizontal boreholes. The equipment used includes a flexible MWD BHA combined with Preussag's Varidril articulated downhole motor. The system has been used in the Vega field, Italy, and the Reitbrook field, Germany.

710. Odeh, Aziz S. and D. K. Babu (1990). Transient flow behavior of horizontal wells: Pressure drawdown and buildup analysis. SPE Formation Evaluation 5 (1), 7-15. SPE paper 18802. Also in Proceedings 1989 SPE California Regional Meeting, April 5-7, Bakersfield, pp. 567-580.

Four flow periods may occur during transient behavior of a horizontal well: Early radial, early linear, late pseudo-radial, and late linear. Equations for buildup and drawdown are given along with illustrative examples.

711. Offshore (1981). Pipeline enhances Norman Wells potential. Offshore 41 (14), 124, 126-128.

A pipeline has been approved to connect Norman Wells field, Northwest Territories, Canada, to a pipeline at Zama, Alberta. The pipeline will be a 12 inch, 560 mile long underground pipeline. With the new pipeline in place, new development drilling will raise Norman Wells production to 25,000 BOPD, or 7% of Canada's imports. Details of the project are given in the article. Editor's Note: The project includes drilling 160 development wells, many of which will be extended-reach or horizontal wells.

712. Offshore (1988). Southeast Asia Report: Arco drills high angle wells on Bima field. Offshore 48 (2), 27-28.

Atlantic Richfield Indonesia, Inc. has drilled sixteen high-angle and horizontal wells in the Bima field Batu Raja reservoir. Horizontal sections of wells have reached 2,175 ft into the reservoir, with 19 ft of TVD variation. Cost of the horizontal wells is 12% higher than for vertical wells.

713. Offshore (1988). Amoco completes record North Sea directional well. Offshore 48 (11), 30, 32.

Amoco Norway Oil Co. has drilled a world record extended-reach well in Valhal field, offshore Norway. The well has a horizontal displacement of 13,954 ft at a TVD of 6,952 ft with a maximum angle of 76.26 degrees of inclination. The well was drilled with the steerable motor system, Eastman Christensen's Navigation Drilling System. A 9,660 ft section of the hole was drilled in a single run.

714. Offshore (1989). Six-mile well deviation has far-reaching economic implications. Offshore 49 (1), 46-47, 49.

Conoco has developed plans for drilling wells with a horizontal displacement of six miles at a TVD of 12,000 ft with a maximum angle of 75 degrees. These wells will allow an offshore structure to reach 6-10 times more area. The original project, called Extended Reach Drilling Program, failed because of lack of industry support.

715. Offshore (1989). Independent brings profitability to Philippines field. Offshore 49 (2), 45-47.

Alcorn International Inc., a Houston-based company, was created to pursue petroleum opportunities in southeastern Asia. Alcorn acquired the properties of Marathon and Occidental in several Philippines fields. Alcorn drilled development wells in Galoc, Nido, and Matinloc fields. A horizontal well is planned in Galoc field in an unconsolidated turbidite sand. Additional horizontal wells will depend on the results of the first horizontal well.

716. Offshore (1989). Multiple directional wells from single borehole developed. Offshore 49 (7), 29-30.

Horizontal Production Systems (HPS) has developed the technology to drill multiple offset radial drainholes. The system was used by Gardes Directional drilling to drill an "ultra-high level" well with a inclination angle of 150.5 degrees (60.5 degrees above horizontal). The drilling and angle-build ended when the top of the reservoir was reached. The system is used to drill multiple lateral boreholes from a single horizontal borehole. Each lateral is drilled upward to the top of the reservoir and is drilled to the left and right of the horizontal borehole (see diagrams in the article). The system increases the borehole exposure to the reservoir many times that achieved in a single horizontal wellbore. The system was used in a Horwell Energy well in the Annona Chalk of Louisiana.

717. Offshore (1989). Statoil using new technology to drill satellite fields. Offshore 49 (9), 29-30.

Statoil is using horizontal and extended-reach drilling technology to develop Gullflaks South and Statfjord fields in the North Sea. Using the same technology Texaco has drilled an extended-reach well in Main Pass, offshore Louisiana, Shell drilled at Cormorant field, Esso in the Bass Strait, Australia, and Conoco in Hutton field, North Sea.

718. Offshore (1989). Horizontal systems growing more specialized. Offshore 49 (10), 36-39, 42-43.

This article has descriptions of five horizontal drilling systems: Baker Hughes Drilling System, Becfield Horizontal Drilling Services, Bentomatic, Drilex Systems Inc., and Eastman Christensen.

719. Offshore (1989). New tools, techniques reduce high-angle drilling costs. Offshore 49 (11), 47-49, 52-54, 56.

This article gives descriptions and case histories for six horizontal drilling systems: ENSCO Technology Co., Gardes Directional Drilling, Petrolphysics Inc., Smith International, Sperry-Sun Drilling Services, and Andergauge Ltd.

720. Offshore Engineer (July, 1988). No shortage of lateral thinking at Rospo Mare. Offshore Engineer, p. 26.

Elf Italiana plans to convert the deviated well in Rospo Mare field, Italy, to a horizontal well. Three additional horizontal drainholes will be drilled in the field. The one vertical well will be re-entered and three short-radius lateral drainholes will be drilled in three different directions.

721. Offshore Engineer (September, 1988). Pioneering Cyrus well shows productivity promise. Offshore Engineer, p. 11.

The world's first horizontal well drilled from a semi- submersible has been drilled in Cyrus field by BP. The well has a 565-meter horizontal section.

722. Offshore Engineer (June, 1989). Majors trumpet horizontal feats. Offshore Engineer, p. 13.

Amoco China has drilled a well with a 697-meter horizontal section in Liuhua field, South China Sea, People's Republic of China. The well was completed with an un-cemented 7-inch perforated liner. Shell Expro has drilled a well with a horizontal displacement of 4,760- meters in Cormorant field, U.K., North Sea.

723. Offshore Engineer (July, 1990). BP pilots its way ahead to put Cyrus on the level. Offshore Engineer, pp. 30, 32.

BP has drilled two horizontal wells in Cyrus field, U.K., North Sea. Cyrus field, with only a 12-ft oil column, would not be economic without using horizontal wells.

724. Oil & Gas Journal (August 31, 1981). First sandstone reservoir tapped off Philippines. Oil & Gas Journal 79 (35), 40.

Cities Service - 1 Galoc is the first successful well in a sandstone offshore Palawan, Philippines. The well flowed at the rate of 1,850 BOPD of 35 gravity oil through a 3/4-inch choke from 7,338-7,285 feet.

725. Oil & Gas Journal (February 1, 1982). Norman Wells project to include island drilling. Oil & Gas Journal 80 (5), 74.

Esso plans up to 190 additional wells in the Norman Wells field, Northwest Territories, Canada. These wells will be drilled from both natural and artificial islands in the Mackenzie River. Editor's Note: Some of the proposed wells are horizontal and extended-reach wells.

726. Oil & Gas Journal (March 14, 1983). Horizontal drilling taps coal seam gas. Oil & Gas Journal 81 (11), 35.

Consolidation Coal and Conoco have developed a mobile horizontal drilling system that can be used to produce natural gas from coal seams. A combination of rotary drilling and a downhole motor allows holes of 3-4 inches to be drilled more than 2,000 feet into coal seams.

727. Oil & Gas Journal (March 28, 1983). Apache expands Williston holdings, drilling. Oil & Gas Journal 81 (13), 52.

Apache Corp. plans to increase its activity in the Williston basin where it holds extensive acreage. The target formations are the Bakken shale, Red River formation, and Duperow formation.

728. Oil & Gas Journal (August 1, 1983). Elf tries horizontal drilling to tap heavy oil. Oil & Gas Journal 81 (31), 48.

Elf Aquitaine drilled a horizontal well to improve production of heavy oil from Rospo Mare field in the Italian Adriatic Sea. The well produced 3,140 BOPD of water free oil in the first 10 days of production. A nearby vertical well produced 691 BOPD with 2% water cut. The reservoir is a karst pay zone with 1% porosity.

729. Oil & Gas Journal (August 29, 1983). Light oil mining project scheduled in North Louisiana. Oil & Gas Journal 81 (35), 37.

An oil mining, gravity drainage project is scheduled by Petroleum Mining Corp. for the Caddo-Pine Island field, Louisiana, Annona Chalk reservoir. The project will use a combination of mining and horizontal drainholes to produce oil from the naturally fractured Annona Chalk.

730. Oil & Gas Journal (December 26, 1983). Elf to attempt deep horizontal well. Oil & Gas Journal 81 (52), 48-49.

Elf Aquitaine plans a 14,760 ft well into a reservoir under the City of Pau in southwest France. The drilling target will be 1.24 miles horizontally from the surface location. The well is Elf's fifth horizontal well with the others being in Lacq field and Castera Lou field, France, and Rospo Mare field, Italy. Elf believes that one horizontal well may take the place of ten vertical wells in certain situations.

731. Oil & Gas Journal (December 26, 1983). Second Galoc well tests oil off Philippines. Oil & Gas Journal 81 (52), 51.

A second well has been completed on the Galoc structure off Palawan, Philippines. No test results were announced. Editor's Note: Horizontal drilling is now being done in this field.

732. Oil & Gas Journal (October 22, 1984). Exploration pace quickens in Paris basin of France. Oil & Gas Journal 82 (43), 66.

North Sea Sun Oil Co. has obtained a license on 128,000 acres in the Paris basin and has spudded its first exploration well in the basin. Jean C. Ferrand and Joseph L. Hargrave, two individuals from the United States, have obtained a permit on 500,000 acres. Exxon has permits for another large tract.

733. Oil & Gas Journal (December 31, 1984). Soviets to expand heavy oil mining. Oil & Gas Journal 82 (52), 52.

The U.S.S.R. is planning an expansion of the oil mining operation at Yarega. Editor's Note: Soviet agencies have entered into a joint venture with Anglo-Suisse to drill horizontal wells in Yarega field.

734. Oil & Gas Journal (June 17, 1985). Elf group plans development of heavy oil field in Adriatic. Oil & Gas Journal 83 (24), 72.

Elf Italiana SpA group plans a horizontal drilling program as part of the development of the Rospo Mare heavy oil field, offshore Italy, in the Adriatic Sea. A pilot project proved that horizontal wells in the naturally fractured reservoir will avoid water breakthrough. Optimum production rate is expected to be 2,835 BOPD for 2 years.

735. Oil & Gas Journal (November 11, 1985). Devonian shale due horizontal hole, multifracs. Oil & Gas Journal 83 (45), 42.

U.S. DOE let a $1.2 million contract to BDM Corp. for a project to boost natural gas recovery from Devonian shales by using horizontal drilling and multiple hydraulic fracture treatments. The project involves drilling a 4,000 ft TVD well with the deepest portion deviated 60-90 degrees from vertical in a naturally fractured reservoir in the Appalachian basin. The well is part of the Eastern Gas Shales Project of the DOE.

736. Oil & Gas Journal (December 2, 1985). Chinese, U.S. crews drill deep slant hole. Oil & Gas Journal 83 (48), 51.

China's deepest directional well was drilled in Shengli field by Chinese and U.S. crews. The Gubei GB 1-1 well was drilled to 16,306 feet measured depth with maximum drift angle of 40.5 degrees.

737. Oil & Gas Journal (January 13, 1986). Paris basin boosts French oil production. Oil & Gas Journal 84 (2), 28.

In 1985, production in the Paris basin rose 77% from 1984 levels to 22,602 BOPD. Editor's Note: Extended- reach wells are being drilled in the Paris basin to reach reservoirs under villages, and to minimize the amount of land required for surface locations.

738. Oil & Gas Journal (February 17, 1986). Prudhoe Bay horizontal well yields hefty flow. Oil & Gas Journal 84 (6), 42-43.

The first horizontal well drilled at Prudhoe Bay, JX-2, by Sohio has produced for sixty days at a rate of 12,400 BOPD. Conventional vertical wells in the same field typically produce 3,000 BOPD. The JX-2 well was completed with slotted-liner in the 1,615 ft horizontal section of the well in the Permian Sadlerochit reservoir. The oil column is 220 ft thick.

739. Oil & Gas Journal (April 21, 1986). Horizontal well set in Appalachian area. Oil & Gas Journal 84 (16), 35.

BDM International Inc. won a $1.2 million contract to evaluate a horizontal well drilled in the Devonian shale in the Appalachian basin. The well will be drilled horizontally for 2,000 feet in the shale before being evaluated by geophysical logging and borehole television.

740. Oil & Gas Journal (January 19, 1987), New tool measures fracture direction, distance. Oil & Gas Journal 85 (3), 22-23.

Sandia National Laboratories has developed a directional fracture mapping wireline tool. The tool has detected simulated fractures 30 ft from a test hole in a granite quarry near Albuquerque. Sandia believes the tool will detect fractures up to 90 ft from a borehole in dry, resistive rock. The tool uses radar pulses in the 30-300 Mhz VHF band. A prototype tool has been built by Southwest Research Institute, San Antonio.

741. Oil & Gas Journal (February 2, 1987). Paris basin paces French exploratory/development work. Oil & Gas Journal 85 (5), 23.

Paris basin production now exceeds that of the Aquitaine basin. In 1986, 79% of the exploration wells drilled in France were in the Paris basin. Editor's Note: The Paris basin is now the site of several extended-reach drilling projects. See index.

742. Oil & Gas Journal (March 9, 1987). Horizontal holes may unlock Devonian shale. Oil & Gas Journal 85 (10), 17-18. Abstract in The Engineering Index Annual (1987), 86, Part IV, 6499.

A horizontal well drilled in the Devonian shale in Wayne County, West Virginia, is expected to produce seven times the gas produced from conventional vertical wells. Productivity is expected to increase further after multiple hydraulic fracturing is completed. The horizontal wells cost about four times as much as a vertical well. The well was drilled by BDM under contract by the U.S. Department of Energy (DOE). The well was drilled vertically to 2,024 ft, then kicked-off, and angle was built to 75-90 degrees. The horizontal segment is 2,000 ft long.

743. Oil & Gas Journal (May 18, 1987). Elf completes Adriatic horizontal drilling. Oil & Gas Journal 85 (20), 24.

Elf Italiana SpA has completed six horizontal wells on Platform B in Rospo Mare field, offshore Italy. The six wells were drilled in the time scheduled for four wells. The platform will reach a peak of 18,900 BOPD in the next few months of 1987. The heavy-oil is produced by long-stroke pumps. Elf has 61.717 % of the project; Agip SpA and Sarom have the remaining interest.

744. Oil & Gas Journal (June 15, 1987). Soviets launch horizontal drilling in western Siberia. Oil & Gas Journal 85 (24), 26- 27. Abstract in The Engineering Index Annual (1987), 86, Part IV, 6499.

The U.S.S.R. plans a series of horizontal wells in Salym field to improve horizontal drilling technology as well as to improve production from the field. Salym field produces oil from the Bazhenov formation, a tight shale. Previous stimulation methods include nuclear explosion, hydraulic fracturing, and acidizing; none of which had an effect on production. Waterflooding has also been ineffective. The first horizontal well in the field intersected the proposed reservoir but no production was established. The horizontal well was sidetracked at 9,022 ft from a vertical well. The inclination reached 88 degrees. The horizontal portion of the well was drilled with an electric drilling assembly.

745. Oil & Gas Journal)March 7, 1988). Meridian sets records for horizontal drilling. Oil & Gas Journal 86 (10), 24.

Meridian Oil Co. has set half a dozen records in drilling horizontal wells in Elkhorn Ranch field, North Dakota. The wells were drilled into the Bakken shale, a naturally fractured reservoir. The records include: deepest kickoff point; first horizontal well without intermediate casing; longest casing string in a horizontal well; first 8.5 inch wellbore in a horizontal well; thinnest formation penetrated by a horizontal wellbore; longest PDC bit run in the Williston basin; and first frozen horizontal core of the Bakken.

746. Oil & Gas Journal (June 13, 1988). Horizontal oil wells on production off Australia. Oil & Gas Journal 86 (24), 27.

The first horizontal wells completed in Australia are in the marginal North Herald and South Pepper fields in the Indian Ocean. Horizontal wells were used to reduce gas and water coning and to increase oil production rates. Of the two horizontal wells at South Pepper, one has gas coning even with very low drawdown. Details of the three North Herald wells: 4,625 ft MD, 3,916 TVD, 1,322 ft of horizontal offset at 89 degrees. It is believed that South Pepper wells are the first in which 9 5/8-inch casing was set at 90 degrees.

747. Oil & Gas Journal (September 5, 1988). BP completes, tests first horizontal well from semi off U.K. Oil & Gas Journal 86 (36), 28.

BP Petroleum Development Co. completed the first North Sea horizontal well to be drilled from a semisubmersible rig, in Cyrus field. The well tested 6,000 BOPD through a 1-inch choke from the Paleocene Andrew formation at 8,265 ft TVD. The horizontal section was parallel to, and 52.5 ft above the oil/water contact at 8,390 feet TVDss.

748. Oil & Gas Journal (September 26, 1988). Canada horizontal/steam project 'encouraging'. Oil & Gas Journal 86 (39), 32.

Tangleflags pilot steamflood project was producing 140 BOPD of 13-14 gravity oil from a horizontal well. The well has a 1,300-ft long horizontal section in the reservoir. The well was completed through an uncemented 7-inch slotted casing in the unconsolidated Lloydminster sand.

749. Oil & Gas Journal (October 10, 1988). Horizontal drilling program moves ahead. Oil & Gas Journal 86 (41), 24.

Meridian Oil Inc. set horizontal drilling records in the Williston basin Bakken shale play.

750. Oil & Gas Journal (January 30, 1989). Elf Aquitaine starts oil production in Greater Paris area. Oil & Gas Journal 87 (5), 35.

Elf Aquitaine has started production from a well less than nine miles from the southern city limits of Paris. The well, Ivry 101D, produces 189 BOPD from the Upper Triassic Keuper sandstone. Editor's Note: Horizontal drilling is to be used in the Paris basin to limit drillsite area and to reach under inhabited and agricultural areas.

751. Oil & Gas Journal (March 6, 1989). Devonian slant well, air MWD projects funded. Oil & Gas Journal 87 (10), 25.

U.S. Department of Energy will subsidize an experimental slant-hole to be drilled in the Devonian shale in Tariff field, West Virginia. The well will be drilled by Sterling Drilling & Production Co. with McCormick Resources, Inc., and Gas Research Institute as partners. The well is planned to be drilled vertically to 1,960 ft, then angle built to 74 degrees, and the hole drilled through the Devonian shale at that angle. The combination of a slant-hole and multiple stimulations should increase the area of exposure of the reservoir and thereby increase productivity. DOE also let contracts to Geoscience Electronics to improve the reliability of MWD sensors used in air-, air mist-, or air foam-drilling. The improved tool is expected to operate to 10,000 ft in large diameter holes at temperatures of up to 135 degrees C.

752. Oil & Gas Journal (April 17, 1989). Horizontal well tested off Philippines. Oil & Gas Journal 87 (16), 19.

Alcorn International Inc. drilled and tested a horizontal well in Galoc field, offshore Philippines. The well has a measured depth of 9,336 ft of which 2,061 ft are horizontal. The well tested 3,226 BOPD after hydraulic fracturing and acid treatment.

753. Oil & Gas Journal (June 12, 1989). Horizontal well to tap thin Troll oil pay zone. Oil & Gas Journal 87 (24), 26, 27.

A horizontal well is planed for Troll field, Norwegian North Sea. The well will test a thin oil zone. The well will be drilled from the Deep Sea Bergen semisubmersible rig in 1,098 ft of water. Norsk Hydro will be the operator. Troll field is the North Sea's largest gas field.

754. Oil & Gas Journal (June 12, 1989). Chalk zones target of horizontal wells. Oil & Gas Journal 87 (24), 57.

Ultrashort-radius lateral drilling is being done in the Annona Chalk of the Caddo-Pine Island field, Caddo Parish, Louisiana, by Flextech Inc. of Shreveport. The drilling system used is one developed by Petrolphysics Inc, and uses water jets to turn the wellbore 90 degrees within a 1-ft radius. The plan is to drill four 150-200-ft long radial-drainholes from each vertical well. A similar project, operated by JWR Explorations Inc, calls for drilling drainholes in the Saratoga Chalk of the Zwolle field, Sabine Parish, Louisiana.

755. Oil & Gas Journal (July 31, 1989). Horizontal wells help Oryx reduce finding cost. Oil & Gas Journal 87 (31), 105.

In the first half of 1989, Oryx Energy Co. has reduced its finding, development, and acquisition costs by more than $2/bbl, and has replaced more than 90% of its production. Horizontal drilling has played a large, but unspecified, role in the cost reductions and reserve replacement. In addition to the horizontal drilling in South Texas Austin Chalk fields, Bakken shale of the Williston basin, Ellenburger limestone of New Mexico, and Niobrara of the Denver basin, Oryx has conventional discoveries in the Gulf of Mexico. Oryx plans to try horizontal drilling in other plays, such as the Niagaran reefs in the Michigan basin and the Monterey shale in California.

756. Oil & Gas Journal (August 14, 1989). Field rule changes sought for horizontal wells. Oil & Gas Journal 87 (33), 55.

Oryx Energy Co. has requested field rule changes to apply to horizontal wells in Pearsall field of South Texas. The requested changes would increase the size of the proration unit and the allowable for horizontal wells. Meridian Oil Inc. asked for temporary special field rules to apply to its 2 Keystone Cattle Co. well in Flying W Wolfcamp field, Winkler County, Texas. Pinnacle Royalty & Operating Co., with more than 15,000 acres in the Pearsall area, has completed its 1 Hatch horizontal well, which tested 1,000 BOPD and 1 MMCFGPD. Oneok Exploration Co. participated in a Lee County, Texas, horizontal well which tested 224 BOPD and 110 MCFGPD.

757. Oil & Gas Journal (September 11, 1989). 85 horizontal wells due in Austin Chalk if regulations allow. Oil & Gas Journal 87 (37), 23.

Oryx Energy plans 85 additional horizontal wells in the Austin Chalk of South Texas, if regulatory approval is received. Oryx's drilling program will be in Atascosa, Dimmit, Frio, LaSalle, and Zavala counties. Test results on seven wells with lateral extensions of 140-2,650 ft are: initial flow rates of 80 to 1,607 BOPD and 62 to 946 MCFGPD. Oryx owns more than 100,000 net acres in the Austin Chalk trend.

758. Oil & Gas Journal (1989). Horizontal drilling grows in Williston. Oil & Gas Journal 87 (45), 22, 24-25.

Twenty-five oil wells, with no dry holes, have been completed in the Devonian Bakken shale of the Williston basin. The dominant operators in the Bakken play are Merdian Oil Inc., and a joint venture of BWAB Inc. and Canadian Hunter Exploration Ltd. Well costs for the horizontal wells are about double that of a vertical well, with a two to five fold increase in reserves. Meridian estimates that a profit can be made at $10/barrel of oil. Spacing rules in North Dakota usually allow up to four horizontal wells per section. Most wells are being drilled on 320-acre units.

759. Oil & Gas Journal (November 13, 1989). Texas horizontal well program to start. Oil & Gas Journal 87 (46), 26.

Oryx Energy will drill 85 horizontal Austin Chalk wells in Pearsall field and vicinity, by the end of 1990. The wells will qualify for a maximum allowable of 1,312 BOPD on 320-acre, or larger proration units. Seven wells have been completed with test results in excess of the allowable.

760. Oil & Gas Journal (December 11, 1989). Extended reach well sets record in gulf. Oil & Gas Journal 87 (50), 32.

Mobil has drilled a well with a measured depth of 19,705 ft and horizontal displacement of 12,115 ft in Green Canyon Block 988 in the Gulf of Mexico. The well flowed 3,221 BOPD and 3.8 MMCFGPD through a 20/64-inch choke.

761. Oil & Gas Journal (December 25, 1989). Horizontal well slated in Colorado tight sands. Oil & Gas Journal 87 (52), 26.

U.S. DOE plans a $4.3 million slant/horizontal well in the Mesaverde tight gas sands of Colorado. The well will be a technical and economic evaluation of directional drilling to stimulate tight gas reservoirs.

762. Oil & Gas Journal (1990). Oryx's Hauptfuhrer: Big increase due to U.S. horizontal drilling. Oil & Gas Journal 88 (3), 28.

This article is an account of an interview with Robert Hauptfurer, Chairman of Oryx Energy Co. Oryx plans to drill 105 horizontal wells in 1990. Oryx's first 12 Austin Chalk horizontal wells all have sustained production of 1,000+ BOPD. Oryx Energy chairman Robert Hauptfuhrer predicts that horizontal wells will account for 30% to 50% of the total U.S. wells drilled within ten years.

763. Oil & Gas Journal (1990). BP finishes another N. Sea horizontal well. Oil & Gas Journal 88 (5), 32.

BP Exploration successfully drilled a second horizontal well in Cyrus field. No gauge was announced for the well. The previous horizontal well at Cyrus flowed about 6,000 BOPD through a 1-inch choke. The horizontal section of the hole is 825 meters long; the total measured depth of the hole is 3,800 meters below the rotary table.

764. Oil & Gas Journal (August 6, 1990). Horizontal well venture expanded in U.S.S.R. Oil & Gas Journal 88 (), 32.

Anglo-Suisse Inc. (ASI) and Varyeganneftegaz (VOGA) of the Soviet Ministry of Oil & Gas have committed to study the expansion of their horizontal drilling agreement to other Siberian fields. The new agreement may include Roslavl field where as many as 50 horizontal wells may be drilled. The earlier agreement was to drill as many as 400 wells in West Varyegan and Tagrin fields.

765. Oil & Gas Journal (December 10, 1990). Austin Chalk horizontal drilling pace continues to sizzle. Oil & Gas Journal 88 (5), 32.

This article reviews the Austin Chalk horizontal drilling activity in Texas. Most of the activity reported is in the Pearsall area (Pearsall and Thirteen fields) and Giddings field, with some activity in Gonzales County and wildcats in South Texas.

766. Oil & Gas Journal (May 7, 1990). More horizontal drilling projects show up. Oil & Gas Journal 88 (19), 38-40.

Union Pacific Resources Co. plans 50 horizontal wells in Giddings field, Texas, six or seven wells in the Niobrara formation of Wyoming, and will participate in five horizontal wells in the Bakken shale in North Dakota and Montana. Belden & Blake drilled a high-angle well under a waste dump in Ohio. Merrion Oil & Gas is using horizontal drilling in a watered out San Juan basin field. The U.S Department of Energy is drilling deep slant/horizontal wells in the Piceance basin and in the Bartlesville sands in Oklahoma. In Argentina horizontal drilling has been used to produce from the low-permeability tuffaceous sands of Piedras Coloradas field.

767. Oil & Gas Journal (1990). Horizontal chalk well blowout killed. Oil & Gas Journal 88 (21), 22-23.

The GLG Energy Inc.-#1 H.J. Shorp horizontal well exploded and burned on May 8, 1990. The well will be replaced with another well in La Salle County, Texas. The #1 Shorp was a horizontal well that was drilled 1,027 feet into the Austin Chalk when the blowout occurred. The Exploration Co. announced a horizontal well in Chittim field, Maverick County, Texas, to produce gas condensate from the Pearsall formation. This well, the 1-112 La Paloma, will be one of the first commercial horizontal gas wells in the United States. Smacko Ltd. will redrill a vertical well as a horizontal well in Huxford field, Alabama, which produces from the Jurassic Smackover formation. This well will extend 800 feet horizontally from the vertical well bore.

768. Oil & Gas Journal (1990). Meridian logs hefty Chalk test. Oil & Gas Journal 88 (24), 17.

Meridian Oil Inc. successfully completed a 12,254 BOPD well in Gonzales County, Texas. The 1H E.E. Smith well was completed in the Austin Chalk.

769. Oil & Gas Journal (1990). Oryx to step up horizontal drilling campaign. Oil & Gas Journal 88 (26), 22-23.

Oryx Energy Co., in a farmout agreement with Snyder Oil Corp., will earn a 50% interest in 38,500 net acres in Maverick and Dimmitt counties, Texas, as part of the Austin Chalk horizontal drilling play in South Texas. Oryx recently had two successful Austin Chalk wells in Dimmitt County, Texas, near Big Wells. The Oryx - 1 C. Jasik flowed 1,084 BOPD and 643 MCFGPD through a 34/64- inch choke with 340 psi tubing pressure. The second discovery is the Oryx - 1 D. Cross et al., which flowed 728 BOPD and 359 MCFGPD through a 28/64-inch choke, with flowing tubing pressure of 220 psi. Winn Exploration Co. completed their #25 Glasscock in the Austin Chalk in Zavala County. The 25 Glasscock tested 19,560 BOPD and 11.7 MMCFGPD. Serfco, of Pampa, Texas, announced it will drill its #14S well in the Texas Panhandle to test the Permian Brown dolomite.

770. Oil & Gas Journal (July 9, 1990). Horizontal well boosts Crawford field oil flow in British North Sea. Oil & Gas Journal 88 (28), 32.

Hamilton Bros. Oil & Gas Ltd drilled a horizontal well in Crawford field, one of the smallest in the U.K. North Sea. The well, 8,450 ft. TVD and 11,200 ft. measured depth, flowed 6,900 BOPD through a 72/64-inch choke. The well has 2,000 feet of horizontal hole in a Triassic reservoir.

771. Oil & Gas Journal (July 16, 1990). Horizontal wells boost south Texas oil flow. Oil & Gas Journal 88 (29), 26.

Horizontal drilling in the South Texas Austin Chalk trend has boosted production 100 fold in Pearsall field. Recoveries from horizontal wells are significantly higher than vertical wells. Proven reserves in the horizontal wells is 450,000 barrels of oil as opposed to 125,000 barrels of oil actually recovered from a nine year old vertical well.

772. Oil & Gas Journal (July 30, 1990). Austin Chalk boosts oil reserves for Oryx. Oil & Gas Journal 88 (31), 39-40.

Oryx Energy Co. has added 42 million barrels of oil equivalent to its reserves from the Austin Chalk horizontal drilling program in South Texas. Oryx has 7 rigs operating in the Austin Chalk play.

773. Oil & Gas Journal (August 6, 1990). Horizontal well venture expanded in U.S.S.R. Oil & Gas Journal 88 (32), 32.

Anglo-Suisse Inc. (ASI) of Houston, and Varyeganneftegaz [Varyegan Oil & Gas Association] (VOGA) of the U.S.S.R. Ministry of Oil and Gas, have added Roslval field to West Varyegan field and Tagrin field in their White Night horizontal well joint venture. As many as 450 horizontal wells could be drilled in the three fields.

774. Oil & Gas Journal (August 13, 1990). Austin Chalk play keeps setting lively pace. Oil & Gas Journal 88 (33), 32.

Clayton Williams, Jr. Inc. has completed seventeen horizontal wells in Pearsall field of South Texas and four horizontal wells in Giddings field. The company plans 107 more horizontal wells through

1991. Union Pacific Resources Co. completed a horizontal well in Giddings field. The well tested 3.8 MMCFGPD through a 26/64-inch choke at 1,410 psi flowing pressure. Credo Petroleum Corp. has acquired 6,000 acres in Atascosa County, Texas.

775. Oil & Gas Journal (October 22, (1990). Pearsall oil well completed with dual drainholes. Oil & Gas Journal 88 (43), 37.

Pearsall Partners has drilled an Austin Chalk horizontal well with two drainholes. The well, 1 McDermand, had an IP of 2,291 BOPD and 1.65 MMCFGPD. The combined length of the two drainholes is 5,700 ft.

776. Oil & Gas Journal (December 17, 1990). BP overcomes problems in high angle well. Oil & Gas Journal 88 (51), 30.

BP's seven well high-angle drilling program in the U.K. North Sea Miller field is planned to start production in early 1992. Platform installation will be in the summer of 1991. The wells, drilled from a semisubmersible, have experienced unprecedented torque levels. The rig has a top drive torque capacity of 38,000 ft-lbs. Drilling has experienced torque of 30,000-32,000 ft-lbs in the high-angle wells (56 degrees). The torque problems have been overcome; future wells will be drilled with a top drive capacity of 44,000 ft-lbs.

777. Oil & Gas Journal (March 4, 1991). Horizontal drilling technology advances. Oil & Gas Journal 89 (9), 63.

Review of current horizontal drilling activity. Projects mentioned include: Austin Chalk - Peach Creek field, Texas (UPRC); Jurassic in Texas and Arkansas (Great Southwest Energy Co. and American Exploration Co.); Oklahoma (Union Oil Co. of California, Conoco Inc., ESCO Exploration Inc., Jolen Operating Co., Meridian Oil Co., and Bogert Oil Co.); Niobrara (Gerrity Oil & Gas Corp., UPRC, Snyder Oil Corp., Meridian Oil Co.); Bakken (Columbia Gas Development Corp., Slawson Exploration Inc., Conoco Inc., Meridian Oil Co.). The UPRC well in Peach Creek field, Gonzales County, Texas, is believed to be a record length medium-radius horizontal wellbore (horizontal displacement is 6,142 ft.).

778. Oil & Gas Journal (April 1, 1991). Austin Chalk maintains brisk drilling pace. Oil & Gas Journal 89 (13), 35-36.

This article reviews horizontal drilling activity and drilling plans in the Texas Austin Chalk play.

779. Oil & Gas Journal (April 1, 1991). Horizontal drilling spreads out in several states. Oil & Gas Journal 89 (13), 34.

This article reviews horizontal drilling projects in the United States. Notable projects include: Midway field (Smackover formation), Arkansas; Vidor Ames field (Yegua formation), Texas; East Fort Trinidad field (Buda formation), Texas; Soso field (Washita-Fredricksburg), Mississippi; Selma Chalk, Mississippi; and Ojo Encino field, New Mexico. Also mentioned are Bakken formation, North Dakota, Bartlesville sand, Red Fork formation, and Misener formation, Oklahoma.

780. Oil & Gas Journal (April 15, 1991). Statoil sets another mark for extended reach. Oil & Gas Journal 89 (15), 31.

Statoil has drilled a well with a horizontal displacement of 19,905 ft, a measured depth of 23,786 ft, and a TVD of 8,850 ft. The well is in Statfjord field, Norwegian North Sea. Later this year a horizontal water injection well will be drilled in Statfjord field, with a planned horizontal reach of 18,040 ft.

781. Oil & Gas Journal (April 15, 1991). Long production test due off Philippines. Oil & Gas Journal 89 (15), 96.

Alcorn (Production) Philippines Inc. plans a long production test of its 1A West Linapacan well, offshore Palawan, Philippines, later this year. The well flowed 6,708 BOPD last year.

782. Oil & Gas Journal (April 22, 1991). Horizontal drilling yields more good wells. 89 (16), 36-37.

This article reviews recent horizontal drilling successes, including the Sooner Trend in Oklahoma, Paradox formation in Utah, and Austin Chalk in Giddings and First Shot fields, Texas.

783. Oil & Gas Journal (May 20, 1991). U.S. horizontal drilling action still spreading. Oil & Gas Journal 89 (20), 23-25.

This article reports on recent horizontal drilling projects in the United States. Projects discussed include: Soso field (Washita-Fredricksburg), Mississippi; Knox dolomite wells in Kentucky; Austin Chalk wells in Giddings field and in East Texas; Pendleton-Many field (Saratoga Chalk), Louisiana; Midway field (Smackover formation), Arkansas; Bartlesville sand and Sooner Trend in Oklahoma; wells in the Niobrara, Mancos shale, Bakken formation, Amsden formation, and Stevens sand (Miocene).

784. Oil & Gas Journal (May 20, 1991). Unocal redevelops Helder field with horizontal wells. Oil & Gas Journal 89 (20), 24.

Unocal Netherlands BV has increased production from Helder field, Netherlands North Sea, by 160% using five horizontal wells and drilling lateral drainholes from nine existing vertical wells.

785. Oil & Gas Journal (May 20, 1991). Horizontal well will be employed in hydraulic fracturing research. Oil & Gas Journal 89 (20), 45-48.

The Gas Research Institute plans to drill a horizontal well at its Hydraulic Fracture Test Site in the Fort Worth basin. The well will be used to test hydraulic fracturing techniques and effectiveness of hydraulic fracturing on the productivity of tight gas sands. The target formation is the Davis sand. Details of the project are given in the article.

786. Oil & Gas Journal (June 3, 1991). Union Pacific expanding Austin Chalk role. Oil & Gas Journal 89 (22), 54, 56.

Union Pacific Resources Co. has purchased Santa Fe Minerals Holdings Inc. leases in Giddings field, Texas. The acreage includes Austin Chalk, Buda formation, and Georgetown formation production.

787. Oil & Gas Journal (June 10, 1991). Horizontal drilling boosts Devonian gas flow. Oil & Gas Journal 89 (23), 16-17.

Columbia Natural Resources Inc. and Columbia Gas System Service Corp. have completed a horizontal well in the Devonian Huron shale in Big Sandy field, Kentucky. The well produces 3.1 MMCFGPD from six zones in the Lower Huron. The well has a horizontal displacement of 2,872 ft, measured depth of 6,263 ft, and TVD of 3.754 ft. The U.S. Department of Energy subsidized the well as part of a three well Devonian shale project. The well cost $1.4 million. Half of the well cost was attributed to research and learning-curve costs.

788. Oil & Gas Journal (June 10, 1991). Horizontal wells may tap North Sea field. Oil & Gas Journal 89 (23), 27.

Horizontal drilling is being used to develop Eocene discoveries in the U.K. North Sea. BP Exploration plans a horizontal well in Forth field. The well will have a 984 ft horizontal section.

789. The Oilman (April, 1988). Horizontal endeavors. The Oilman, p. 18.

Brief discussion of the horizontal drilling done by Maersk in Dan field, Danish North Sea.

790. Oilweek (February 20, 1984). Special application program for deviated wells. Oilweek 35 (3), 16.

Virtual Computing Services Ltd., Calgary, Canada, has compiled a large horizontal well data base, DISDA. This data base will eventually include all Canadian wells having at least one directional survey and any that were reported deviated.

791. Oilweek (1984). Esso drills World's longest horizontal well. Oilweek 35 (44), 27-28.

Esso Canada drilled and completed the Cold Lake horizontal well pilot No. 2 to a measured depth of 1,773 meters with 1,016 meters of horizontal displacement at a TVD of 475 meters. The well was completed in the Clearwater formation oil sands. Steamflood will be used to recover the heavy-oil.

792. Oil Week (March, 1986). Texaco mothballs oil sands pilot. Oilweek, .

Not Examined.

793. Okrajni, Slavomir S. and J. J. Azar (1986). The effects of mud rheology on annular hole cleaning in directional wells. SPE Drilling Engineering 1 (4), 297-308. SPE paper 14178 presented at the 1985 SPE Annual Technical Conference and Exhibition, Las Vegas, September 22-25.

The University of Tulsa has conducted extensive experiments on the effects of mud rheology on hole cleaning (cuttings transport), the results of which are reported in this paper. The experimenters used water and bentonite/polymer mud in holes with inclinations of 0 degrees to 90 degrees. Three regions of hole inclination, that affect cuttings transport, have been identified : 0-45 degrees, 45-55 degrees, and 55-90 degrees. In low-angle holes (0-45 degrees), laminar flow dominates cuttings transport. In Medium-angle wells, laminar flow and turbulent flow have generally similar effects. In high-angle holes, turbulent flow effects dominate. In laminar flow, higher mud yield values and yield point/plastic-viscosity ratio provide better cuttings transport. In turbulent flow, cuttings transport was not affected by mud rheological properties.

794. Oliver, S. J. P., V. A. Kuuskraa, C. F. Brandenburg, J-M. P. Coates, and J. R. Kelafant (1989). The Antrim shale of the Michigan basin: An active Devonian shale gas play. Proceedings 1989 SPE Gas Technology Symposium, June 6-7, 1989, Dallas, pp. 487-494. Society of Petroleum Engineers, Richardson, TX. SPE paper 19104.

The Devonian Antrim shale of the Michigan basin is estimated to contain up to 76 TCF gas-in-place. The Antrim shale gas production rates are relatively low, in spite of increased drilling and massive stimulation of wells. This paper addresses the possibility that Antrim gas production rates can be increased by better stimulation practices, proper well spacing, and better interval selection. The geology of the Antrim is also outlined.

795. Ong, Tee S. and Roger M. Butler (1989). Wellbore flow resistance in steam-assisted gravity drainage. Journal of Canadian Petroleum Technology 28 (2), 38. CIM paper 89-40-59. Paper presented at the 40th Annual Technical Meeting of The Petroleum Society of CIM.

Abstract. Scaling of the pressure gradient and radius of horizontal wells in physical models are not proportional to the geometric ratio of the laboratory models. A computer program has been developed to account for the discrepancies in modeling.

796. Overbey, W. K., Jr., R. S. Carden, and J. B. Williams (1987). Computer applications in the planning and drilling of a 2,000-ft horizontal well in Wayne County, West Virginia. Presented at the Petroleum Industry Applications of Microcomputers Conference, June 23-26, 1987, Del Lago, Lake Conroe, Montgomery County, Texas. Society of Petroleum Engineers, Richardson, TX. SPE paper 16501.

A personal computer, using commercial software, was used to plan, design, and simulate the drilling of a 2,000-ft horizontal well in West Virginia. The computer was also used to monitor the drilling and to track well costs. The software packages used were OPEN ACCESS and WELLPLAN. The drillstring analysis software was credited with saving $250,000. Details of the system used are given in the article.

797. Overbey, W. K., Jr., L. E. Yost, and A. B. Yost, II (1988). Analysis of natural fractures observed by borehole video camera in a horizontal well. Proceedings Gas Technology Symposium: Gas Supply - A Critical Balance of Politics, the Market Place, and Technology, June 13-15, 1988, Dallas, pp. 461-468. Society of Petroleum Engineers, Richardson, TX. SPE paper 17760. Abstract in The Engineering Monthly Index (July, 1989), 27 (7), 549.

The horizontal section of an air-drilled well in West Virginia was successfully examined using a video camera. The 2,217 ft long horizontal wellbore had more than 200 features interpreted to be natural fractures. Measurement of the orientation of these features showed good correlation with surface features, seismic trends, and fractures in an oriented core taken from the well. The well was drilled in the Appalachian basin Devonian shale in Wayne County, West Virginia, by BDM Corporation under contract to the U.S. Department of Energy.

798. Overby, W. K., Jr., A. B. Yost, A. B., II, and D. A. Wilkins (1988). Inducing multiple hydraulic fractures from a horizontal wellbore. Proceedings of the 1988 SPE Annual Technical Conference and Exhibition, Production volume, October 2-5, 1988, Houston, pp. 673-687. Society of Petroleum Engineers, Richardson, TX. SPE paper 18249. This paper was also presented at the SPE Eastern Regional Meeting, Charleston, West Virginia, November 1-4, and is printed in the Proceedings SPE Joint Rocky Mountain Regional/Low Permeability Reservoir Symposium & Exhibition, Denver, March 6-8, 1989, pp. 27-41. Abstract in The Engineering Monthly Index (November, 1989), 27 (11), 723 and (January, 1990), 28 (1), 614.

Multiple hydraulic fracturing was done to a horizontal test well (RET #1) in Wayne County, West Virginia, as part of a Department of Energy project. The test well was drilled horizontally 2,000 ft into a naturally fractured reservoir. The horizontal section of the well was hydraulically fractured in eleven zones which had six to 72 natural fractures. Using radioactive tracers, flow paths were traced. The tracers indicated that the natural fractures were selected at injection rates between 5 and 10 barrels-per-minute, and induced fractures selected at injection rates greater than 25 barrels-per-minute. Production improvement in the hydraulically fractured zones ranged from 4.1 to 25 times that of the naturally fractured zones. Details of the test are reported in the article.

799. Overhuls, Bob (1991). Overview: An overview of horizontal (lateral) drilling and development. Popular Horizontal 1 (1), 26, 28-29.

Brief review of the history of horizontal drilling.

800. Overhuls, Bob (1991). Ever present dangers. Popular Horizontal 1 (1), 38.

This article is a brief discussion of safety in drilling horizontal wells.

801. Ozkan, E. (1988). Performance of Horizontal Wells. Ph. D. Dissertation, The University of Tulsa, Tulsa, Oklahoma.

Not examined.

802. Ozkan, E. and R. Raghavan (1988). Performance of horizontal wells subject to bottomwater drive. Proceedings 1988 Eastern Regional Conference, November 1-4, 1988, Charleston, West Virginia, pp. 235-255. Society of Petroleum Engineers, Richardson, TX. SPE paper 18545. Abstract in The Engineering Index Monthly (February, 1990), 28 (2), 628. Also in Proceedings SPE Symposium on Energy, Finance, and Taxation Policies, Washington, D.C., September 19-20, 1988.

Analytical solutions are given for the pressure distribution created by horizontal wells in bottomwater reservoirs. When the drainage area of a horizontal well is equal to a critical value, increases in well spacing have no impact on productivity of the wells. Water-free production of horizontal wells increases as the well spacing increases up the critical value mentioned previously.

803. Ozkan, Erdal and Rajagopal Raghavan (1990). Supplement to SPE 16378, horizontal well pressure analysis. SPE manuscript. SPE paper 20271. Abstract in The Engineering Index Monthly (June, 1990), 28 (6), 725.

This paper presents a derivation of the long-time approximation and pseudoskin factor for horizontal well pressure analysis.

804. Ozkan, Erdal, Rajagopal Raghavan and Sadanand Joshi (1989). Horizontal-well pressure analysis. SPE Formation Evaluation 4 (4), 567-575. SPE paper 16378 presented at the 1987 SPE California Regional Meeting, April 8-10, Ventura. Abstract in The Engineering Index Monthly (March, 1990), 28 (3), 670.

Production performance over the past few years confirms the advantages of horizontal wells and drainholes. Drainholes are short horizontal boreholes drilled from existing vertical wells and horizontal wells are new wells drilled usually with horizontal displacements of over 1,000 feet. Horizontal-well pressure responses are a function of a dimensionless well length (discussed in the paper) and the dimensionless well radius. For a single horizontal well, the infinite-conductivity idealization is the only viable boundary condition. For two horizontal boreholes drilled in diametrically different directions from a single vertical well, either the infinite-conductivity or the uniform-flux idealization is appropriate. The pressure response of the horizontal well and pseudoskin factors are independent of the well location in the vertical plane of the reservoir. If dimensionless length is less than 25, the pressure response will be essentially identical to that of fully penetrating vertical fractures; if equal to or less than 50, the pressure response is indistinguishable from that of a vertically fractured well.

805. Paces, R. S. and Liu DeFu (1990). South China Sea extended well testing program: Implementation and results. Proceedings of the 22nd Offshore Technology Conference, Houston, v. 1, pp. 9-17. OTC paper 6200.

Extensive testing was conducted on a vertical well, a horizontal well, and a slanted well in the Liuha field, People's Republic of China. Test results indicated that horizontal wells would most efficiently recover the 16.4 degree API gravity oil. Details of the project are given in the article.

806. Pacific Oil World (1990). K & M Enterprises: A key player for horizontal wells. Pacific Oil World 82 (7), 29.

K & M Enterprises, a California-based horizontal drilling service company, has assisted Unocal in drilling a record-setting offshore well near Point Pedernales, California. They have also assisted in drilling horizontal wells in Midway-Sunset field and South Belridge field.

807. Pacific Oil World (1990). New Solum horizontal simulator reveals clues to production gains. Pacific Oil World 82 (7), 25.

Solum Oil Tool Corp. has developed a horizontal drilling simulator which creates above ground the conditions found in the subsurface when drilling a horizontal well. A description of the simulator is given.

808. Pacific Oil World (1990). Horizontal wells at Elk Hills. Pacific Oil World 82 (12), 17.

The first two horizontal wells in Elk Hills field, California, are producing 1,000 BOPD each. Additional horizontal wells are planned for the field. Some details of the project are given.

809. Pain, D. D., E. T. Johnson, R. H. Byers, and E. R. Vader (1985). Drilling with PDC bits on rotary vs. downhole motor: Southeast Central Texas Austin Chalk fields. Paper presented at the 60th Annual SPE Technical Conference and Exhibition, September 22-25, 1985, Las Vegas. Society of Petroleum Engineers, Richardson, TX. SPE paper 14384.

Drilling with PDC bits is done at higher rotating speeds than other bit types, and consequently has higher bit torque and drillpipe wear. In spite of these problems, PDC bits are cost effective. One solution to the torque and wear problems is to use a downhole motor. PDC bits have proven to be cost effective in the Austin Chalk trend of Texas. Use of downhole motors with PDC bits in the Austin Chalk reduces drillstring wear and allows faster drilling than other combinations.

810. Paptzacos, P., S. A. Gustafson, and S. M. Sljaeveland (1988). Critical time for cone breakthrough in horizontal wells. Seminar on Recovery from Thin Oil Zones, April 21-22, 1988, Stavanger. Norwegian Petroleum Directorate.

Not examined.

811. Papatzacos, P., U. Rogaland, T. R. Herring, R. Martinson, and S. M. Skjaeveland (1989). Cone breakthrough time for horizontal wells. Proceedings 1989 SPE Annual Technical Conference & Exhibition, Reservoir Engineering volume, March 8-11, 1989, San Antonio, pp. 535-550. Society of Petroleum Engineers, Richardson, TX. SPE paper 19822.

Horizontal wells can improve oil recovery from a an oil zone underlying a gas cap, overlying an aquifer, or sandwiched between an aquifer and a gas cap. An analytical solution is given in this article for predicting the time to breakthrough versus rate for horizontal wells in the above situations. Examples are given for Troll and Helder fields.

812. Papatzacos, P.(1989). Gas Coning by a horizontal well. Rogaland Research Institute Report No. K-63/89, a project sponsored by Amoco, Fina, Norsk Hydro, Phillips, and Statoil. Well-Test Analysis.

Not examined.

813. Papatzacos, P. (1989). Gas coning by a horizontal well as a moving boundary problem. Proceedings European Conference on Mathematics of Oil Recovery, July 25-27, 1989, Robinson College, Cambridge University. Oxford University Press, Oxford.

Not examined.

814. Parcevaux, Philippe (October 19, 1987). Guides emerge for cementing horizontal strings. Oil & Gas Journal 85 (42), 25- 37, 40-41.

The most important aspect of cementing horizontal wells is mud displacement, which is connected to mud properties and proper centralization of casing. The best cement slurry will have a density as close to the mud as possible, low yield stress, no free water development, and good compressive strength development.

815. Parker, Cecil D., Dick Weber, David Garza, and Susan Swaner (1982). Austin Chalk stimulations require thorough planning. Petroleum Engineer International 54 (10), 58-60, 64, 66, 68, 72, 76.

Austin Chalk stimulations in Gidding field of Texas depend on several factors, such as: well position, fracture height, pump rates, type of fracture fluids, proppant concentrations, and leak-off controls. The most critical factor affecting the productivity of an Austin Chalk well is the position of the wellbore with respect to faults and fractures in the Austin Chalk. Wells drilled directly into the fractured zones of the chalk are usually the best producers. If a well is not in a good fractured zone, hydraulic fracturing may help make a good well. Hydraulic fracturing methods used in Austin Chalk wells include gelled water fracs with sand, and conventional cross-linked polymer fracs. Fracture heights are usually assumed to be 10-15% higher than the perforated interval. The fracturing is usually designed to be limited to the lower "high resistivity" zone of the chalk, normally 200 to 300 feet thick. Hydraulic fracturing is usually designed to penetrate 750 to 1,000 feet from the wellbore so that as many natural fractures as possible can be intersected. When sand is used as a proppant, concentrations of 1.0 to 2.0 lb/square foot of propped-fracture area are used.

816. Parker, Marcia A. (June 17, 1985). Oil production building in Paris basin. Oil & Gas Journal 83 (24), 80-82.

Oil production from the Paris basin has increased to about 30% of the total production in France. Current targets, if met, will have Paris basin production equal that from the Aquitaine basin by the end of 1986. Active exploration and development drilling are being conducted by Esso, Elf Aquitaine, Sun, Total-Triton, BP, Windsor Gas, Bula Resources, and others.

817. Parra, J. and H. Villasmil (October, 1984). Perforacion Direccional en la Costa Bolivar [Directional drilling on the Bolivar coast {Venezuela}]. Colaper Congreso, October, 1984, Venezuela. In Spanish.

Not examined.

818. Parsons, R. S. and R. W. Fincher (1986). Short-radius lateral drilling: A completion alternative. Proceedings 1986 SPE Eastern Regional Meeting, November 12-14, 1986, Columbus, Ohio, pp. 239-250. Society of Petroleum Engineers, Richardson, TX. SPE paper 15943.

Short-radius lateral drilling is the drilling of a horizontal borehole from a vertical borehole with a curve build rate of 1.5 to 3 degrees per foot drilled. Short-radius horizontal drilling is usually used to drill relatively short drainholes (single or multiple) into reservoirs to improve productivity of the well. This paper describes the short-radius system developed by Eastman Christensen with an example of its use in a Niagaran reef in Michigan.

819. Parsons, Randy (1987). Horizontal drilling in oil and gas applications. in Trevail, R. A. (chairperson), Ontario Petroleum Institute 26th Annual Conference, October 25-27, 1987, Toronto.

Not examined.

820. Partain, Bruce (1989). Meridian heads sideways into Permian. Southwest Oil World 37 (3), 8, 11.

Meridian Oil Inc. and other operators are using horizontal drilling in the Permian-Midland basin of Texas. The targets are the Fusselman, Spraberry, and Wolfcamp formations. Meridian has completed horizontal wells in Acetate field (Fusselman) and Flying W field (Wolfcamp). Oryx Energy is drilling a horizontal well in Sugg field, Texas. Brief discussion of the projects are in the article.

821. Peaceman, Donald W. (1990). Discussion of productivity of a horizontal well. SPE Reservoir Engineering 5 (2), 252-253.

Critical discussion of Babu and Odeh (1989), Productivity of a horizontal well, SPE Reservoir Engineering 4 (4), 417-421.

822. Peaceman, Donald W. (1991). Further discussion of productivity of a horizontal well. SPE Reservoir Engineering 6 (1), 149-150. SPE paper 21611.

Critical discussion of Babu and Odeh (1989), SPE Reservoir Engineering 4 (4), 417-426.

823. PEI Staff Report (1988). Getting to the bottom with slant- hole logging tools. Petroleum Engineer International 60 (2), 32-34.

Getting logging tools to the bottom, in boreholes inclined more than 50 degrees, by gravity, is nearly impossible. Better techniques include drillpipe-conveyed tools and pumping tools through drill pipe. In the late 1970's, the side-entry sub made possible better methods of using drill pipe-conveyed logging tool.

824. Pelensky, P. J. (1984). Project development innovations of slant hole drilling and computerized production transmission. Paper presented at the 35th Annual Technical Meeting of The Petroleum Society of CIM: Recovery and Beyond, June 10-13, 1984, Calgary. CIM, Montreal. CIM paper 84-35-10. Abstract in Meeting Report: The Journal of Canadian Petroleum Technology (1984), 23 (2), 38.

Abstract. A brief description is given of a gas drilling project in Saskatchewan.

825. Pendleton, Larry E.and A. Behmooz Ramesh (1985). An innovative method of drilling horizontal boreholes. Paper presented at the Heavy Oil Sands Technical Symposium, February 20, 1985, University of Calgary, Calgary, Canada.

Not examined.

826. Pendleton, L. E. and A. B. Ramesh (May 27, 1985). Bechtel develops innovative method for horizontal drilling. Oil & Gas Journal 83 (21), 95-99.

Bechtel Investments has developed an ultra-short-radius horizontal drilling method for drilling lateral drainholes from a vertical wellbore. The system has been used to drill 350 horizontal drainholes for both experimental and production purposes. A description of the system is given in the article.

827. Perrine, R. L. (1955). Well productivity increase from drain holes as measured by model studies. Journal of Petroleum Technology 7 (2), 30-34. Also in Transactions, A.I.M.E. (1955) 204, 30-34.

An electrolytic tank model was used to obtain data on the increase in productivity expected from using drainholes instead of conventional vertical wells. Horizontal drainholes were found to increase productivity in proportion to their length, but not their number. The model results had a maximum of 2.5 times more productivity from a horizontal drainhole, in an undamaged reservoir, than a vertical well did. Theoretically, ten-fold increases are possible with drainholes. Drainholes must extend an appreciable distance beyond the damaged zone to be effective. For drainholes extending 40-60 ft into the reservoir, the arrangement of the drainholes is not important. With drainholes shorter than 40 ft, the drainholes are more effective if arranged in pairs 180 degrees apart. Details of the model and analysis of the data are given in the article.

828. Perry, C. J. (1986). Directional drilling with PDC bits in the Gulf of Thailand. Presented at the 61st Annual SPE Technical Conference & Exhibition, October 5-8, 1986, New Orleans. Society of Petroleum Engineers, Richardson, TX. SPE paper 15616.

Unocal Thailand Inc. has used PDC bits in directional drilling in the Gulf of Thailand. These bits show individual directional characteristics. Unocal's analysis of the bit characteristics led them to the conclusion that knowledge of these characteristics must be used during drilling to optimize the directional drilling and drilling operations. Details of these studies are given in the article.

829. Perry, John H., Leonard J. Prosser, and Joseph Cervik (1980). Methane drainage from the Mary Lee coalbed, Alabama, using horizontal drilling techniques. Proceedings 1980 SPE/DOE Symposium on Unconventional Gas Recovery, May 18-21, 1980, Pittsburgh, pp. 335-339. Society of Petroleum Engineers, Richardson, TX. SPE/DOE paper 8967.

Coalbed methane may be drained from coalbeds by drilling horizontal wells from underground locations. Drainage of coalbed methane is dependent on the amount of methane present, in-situ pressure of the coalbed, and permeability of the coalbed. Horizontal drilling to extract methane from the Mary Lee coalbed, in Alabama, was successful with 40 MMCF of methane recovered in the first year of operation.

830. Peterson, Susan K. (1990). An Investigation of Factors Influencing the Productivity of A Horizontal Borehole. Ph. D. dissertation, Texas A & M University, College Station, Texas.

Not examined.

831. Peterson, S. K. and S. A. Holditch (1990). Effects of skin damage upon the productivity of a well containing a horizontal borehole. Proceedings Ninth SPE Symposium on Formation Damage Control, February 22-23, 1990, Lafayette, Louisiana, pp. 143-154. Society of Petroleum Engineers, Richardson, TX. SPE paper 19415. Abstract in The Engineering Index Monthly (August, 1990), 29 98), 575.

Horizontal wells may be used to improve production in some low-permeability reservoirs when long hydraulic fractures can not be created safely. Skin damage should be avoided in horizontal wellbores, if possible. If severe skin damage occurs, acidizing or hydraulic fracturing may be used to stimulate the well. Details are given of a study of the effects of skin damage on horizontal wellbores.

832. Petit, H. J. M., G. Renard, and E. Valentin (1989). Technical and economic evaluation of steam injection with horizontal wells for two typical heavy-oil reservoirs. Proceedings 1989 SPE Annual Technical Conference & Exhibition, Reservoir Engineering volume, March 8-11, 1989, San Antonio, pp. 619-629. Society of Petroleum Engineers, Richardson, TX. SPE paper 19828. Abstract in The Engineering Index Monthly (January, 1990), 28 (1), 663.

Horizontal wells may be an economic alternative for developing thin, viscous-oil reservoirs using steam stimulation/drive EOR methods. Simulations of horizontal and vertical wells using two models (a Venezuelan reservoir and a Canadian reservoir) were conducted to determine the optimum number of steam cycles, use of both injector and producer wells for steamflood, spatial distribution of wells, performance of steamflood depending on reservoir character, and conversion of steamflood to water flood. Details of the simulation, economic analysis, and conversion to waterflood are also addressed. The simulation indicated that steamflood, using horizontal wells, may be economically advantageous by lowering the cost of the produced oil and increasing the productivity significantly.

833. Petrie Parkman & Co. (July 11, 1989). Field Trip Notes: Horizontal Drilling Seminar. Petrie Parkman & Co., Denver. FTN 02.

Not examined.

834. Petrie Parkman & Co. (August 31, 1989). E & P Notes: Horizontal Drilling Revitalizes the Austin Chalk. Petrie Parkman & Co., Denver. EPN 01.

Not examined.

835. Petrie Parkman & Co. (September 18, 1989). E & P Perspectives: Horizontal Drilling Unlocks the Bakken Shale. Petrie Parkman & Co., Denver. EPP 01.

Not examined.

836. Petroleum Engineer International (1984). Multi-drain drilling ups production. Petroleum Engineer International 56 (15), 34. Abstract in The Engineering Index Annual (1985), 84, Part IV, 6203.

Elf Aquitaine has developed a method of drilling multiple drainholes from a single vertical wellbore. The method was used to obtain a five-fold increase in production from a well in the Eschau field, Alsace, France. The system allows for the drilling of three drainholes; is applicable in low-permeability reservoirs, low-pressure reservoirs, and with reservoirs containing viscous crude.

837. Petroleum Engineer International (1987). PEI initiates horizontal drilling awards. Petroleum Engineer International 59 (6), 55.

Petroleum Engineer International announces awards to horizontal drillers. Anyone who participates in drilling, engineering, or completing a horizontal well is eligible for an award. Esso Resources Canada Ltd. holds the record for the longest horizontal borehole section, with a well in the Norman Wells field of Canada that has a 4,013-ft long horizontal section.

838. Petroleum Engineer International (1988). Horizontal drilling records recognized. Petroleum Engineer International 60 (2), 35.

PEI Horizontal Drilling Awards to: Peter Edlund, ARCO and Dwight Goolsbay, Gearhart Industries for the ARCO W.M. Schrock 26 No. 8 well with horizontal displacement of 1,821 ft; an Oil and Natural Gas Commission of India well in Bombay High field, with 1,568 ft of horizontal displacement; Lacq 91 well with 1,540 ft of horizontal displacement; and Meridian Oil Inc. with a North Dakota well with 2,603 ft of horizontal displacement.

839. Petroleum Engineer International (1988). PEI issues more horizontal drilling awards. Petroleum Engineer International 60 (10), 31-32.

Petroleum Engineer International horizontal drilling awards went to Gulf Canada and BP for horizontal wells in Pelican Lake pilot project and Cyrus field, respectively.

840. Petroleum Engineer International (1989). Horizontal drilling stays hot. Petroleum Engineer International 61 (4), 24, 26. Abstract in The Engineering Monthly Index (August, 1989), 27 (8), 757.

Petroleum Engineer International magazine "Horizontal Drilling Awards" were presented to the DOE/BDM/Eneger Corp Recovery Efficiency Test No. 1 well, Wayne County, West Virginia, which was the first horizontal well to have a borehole video camera survey run and the first to have multiple hydraulic fractures induced.

841. Petroleum Engineer International (1989). Manufacturers offer many choices in MWD systems. Petroleum Engineer International 61 (5), 29-30, 32.

This article is a list of companies offering MWD services and gives the parameters of each system.

842. Petroleum Engineer International (1989). PEI announces more horizontal drilling awards. Petroleum Engineer International 61 (11), 40.

Petroleum Engineer International horizontal drilling award went to BHP Petroleum, Meridian Oil, and Anders Energy Corp. for horizontal wells in Texas and North Dakota.

843. Petroleum Engineer International (1990). Many MWD choices available. Petroleum Engineer International 62 (5), 37-38, 40.

This article has a list of MWD specifications and manufacturers.

844. Petroleum Engineer International (1990). Horizontal drillers earn PEI awards. Petroleum Engineer International 62 (6), 32.

Maxus Energy Co. drilled its first horizontal well in Billings County, North Dakota. The Maxus State 14-32H well measured depth was 12,540 ft, TVD was 10,364 ft, with a horizontal section 1,900 ft long.

845. Petroleum Engineer International (1990). PEI honors horizontal drillers. Petroleum Engineer International 62 (12), 34.

Horizontal Drilling Awards went to Saskatchewan Oil & Gas Corp. (2 long-radius wells at Plover Lake), Atwood Oceancis Australia Pty., and Western Mining Corp. Ltd. (1 well at South Pepper field, Australia), and GLG Energy for several wells.

846. Petroleum Engineer International (1991). Horizontal drillers receive PEI awards. Petroleum Engineer International 63 (2), 33.

PEI horizontal drilling awards were given to Citation Oil & Gas Corp. and B.P. Petroleum Development Ltd.

847. Petroleum Gazette (1990). Exploration increasingly depends on high-tech methods to find oil. Petroleum Gazette 26 (4), 7-12.

This is a brief discussion of "high-tech" exploration methods, including horizontal drilling methods.

848. Petzet, G. Alan (July 18, 1983). Research effort aims to trim drilling costs, boost recovery. Oil & Gas Journal 81 (29), 41-44, 46.

Horizontal drainhole drilling is expected to be a cost effective way to increase production. Arco is using horizontal drainholes in the Empire Abo Unit, New Mexico. Drainholes in the Empire Abo Unit are expected to recover 550,000 barrels of oil, or twice the recovery of the same number of vertical wells. Texas Eastern, through a subsidiary, has a tool that aids in drilling horizontal holes with conventional rigs. Horizontal drainholes have been drilled in Grassy Trail Creek field, Utah, and have been commercial where vertical wells were not. Texas Eastern has also drilled horizontal wells in the Eastern Devonian shales, Miocene Monterey shale in Cat Canyon field, California, and Austin Chalk in Texas.

849. Petzet, G. Alan (February 17, 1986). Prudhoe Bay horizontal well yields hefty flow. Oil & Gas Journal 84 (7), 42-43.

Sohio's first Prudhoe Bay horizontal well, JX-2, has sustained production in its first 60 days of 12,400 BOPD of oil, as compared to 3,000 BOPD for a conventional vertical well. Horizontal wells are believed to drain twice the area as does a vertical well. The JX-2 produces through a slotted-liner spanning a horizontal distance of 1,615 feet in the Permian Sadlerochit formation. Drilling history and engineering data are given in the article.

850. Petzet, G. Alan (June 1, 1987). Canadians combine steam drive, horizontal well. Oil & Gas Journal 85 (22), 22.

Sceptre Resources Ltd. and Murphy Oil Ltd. plan to try a combination of steamflood, gravity drainage, and horizontal wells to improve production in Tangleflags North heavy-oil field, Saskatchewan, Canada. The project will recover 13-14 gravity oil from the Lloydminster sand. The project may recover 40-50% of the oil-in-place. The project is patterned after Esso Canada's Cold Lake field and Norman Wells field projects.

851. Petzet, G. Alan (April 11, 1988). Operators make wider use of horizontal drilling technology. Oil & Gas Journal 86 (15), 15-17.

An estimated 110 to 150 horizontal wells and drainholes have been drilled worldwide. Drilling and completion costs have dropped from 5-10 times that of a vertical well, to as low as 2 times the cost of a vertical well. Horizontal wells have been drilled in Prudhoe Bay field, Empire Abo Unit, Grassy Trails Creek field, Spraberry Trend, and other U.S. fields and plays. Still unknown is whether horizontal wells increase ultimate recovery as well as production rate. Wells are now being drilled to test the effectiveness of horizontal drilling in several plays such as the Michigan's Antrim Shale and Silurian reefs and the Mesaverde group in Colorado.

852. Petzet, G. Alan (May 14, 1990). Buda strikes may boost South Texas action. Oil & Gas Journal 88 (20), 25.

Two Frio County, Texas, wells have established production from the Buda formation, an Upper Cretaceous limestone. The wells are in the Pruitt (Navarro, Buda) field, 15 miles east of Pearsall, Texas. The Fossil Bay Petroleum Inc-#1 Tschirhart well was drilled vertically and started to flow 70 ft into the Buda at 6,577 ft. The Fossil Bay Petroleum Inc.-#1 Williams well was drilled as a horizontal well in the same fault block as the #1 Tschirhart. The #1 Tschirhart tested 842 BOPD and 360 MCFGPD and produced 5,317 barrels of oil during the first 7 days; The #1 Williams tested 36 BOPD. The Buda is 11- ft thick, in this area.

853. Petzet, G. Alan (October 22, 1990). U.S. operators expand horizontal programs. Oil & Gas Journal 88 (43), 36-37.

Additional horizontal drilling projects are underway in Texas, Louisiana, Mississippi, and Alabama. Horizontal wells are now being drilled (or planned) for the Anacacho formation, Olmos formation, Wilcox (Eocene) formation, Smackover formation, and Saratoga Chalk. Many of these projects are in older fields such as Pendleton-Many field and Olla field of Louisiana. Horizontal wells have been drilled in Panhandle, Huxford, Ina-Taylor, Clear Springs, Fort Jessup, Fairfield, Clear Springs, Center, and Grosebeck Creek fields.

854. Petzet, G. Alan (October 29, 1990). More horizontal drilling under way in U.S. Oil & Gas Journal 88 (44), 24-27.

List of new horizontal drilling projects by many companies in New Mexico, North Dakota, Oklahoma, Wyoming, Colorado, Montana, and Michigan.

855. Petzet, G. Alan (November 26, 1990). Horizontal drilling still fanning out in U.S. Oil & Gas Journal 88 (48), 33-34.

This is a review of horizontal drilling projects including: Austin Chalk, Texas - Giddings field (Exxon; Santa Fe Energy Partners LP/Enserch Exploration Partners Ltd), Karnes county (Leede Oil & Gas); Bakken formation, Williston basin - Murphy Creek field (Edisto Resources Corp.); Niobrara formation - Silo field, Wyoming (Union Pacific Resources Co.). Other plays and companies mentioned are: Viola formation, Oklahoma (Esco Exploration Inc.), Saratoga Chalk, Texas (Maxwell Oil & Gas Co.), Northeast Lyle field, Oklahoma (Ceja Corp.), Devil's Basin field, Montana (Golden Globe Engineering Co.).

856. Petzet, G. Alan (December 10, 1990). U.S. horizontal drilling continues to spread. Oil & Gas Journal 88 (50), 31- 32.

This is a discussion of current horizontal drilling projects. Projects and operators are indexed.

857. Pickett, George W. (1967). Techniques and deflection tools in high-angle drilling: Past, present, and future. Journal of Petroleum Technology 19, 469-476.

This article gives a discussion of high-angle, directional drilling techniques.

858. Pidcock, Gary (1988). Gulf Canada improves drilling with MWD techniques. Petroleum Engineer International 60 (9), 16, 19- 22, 24.

Gulf Canada Resources Ltd had developed two MWD interpretation techniques: mechanical efficiency log (MEL) and sticking pipe indicator (SPIN). MEL monitors the condition of the bit; SPIN monitors the frictional losses in the drill string. Both methods run during drilling.

859. Pitts, John Paul (August, 1980). ARCO's Lea County drainhole proves a success. Drill Bit 29 (8), 24-35. Arco Oil & Gas successfully drilled a horizontal drainhole in the Empire Abo Unit, New Mexico. The drainhole was drilled to reduce the high gas/oil ratio caused by gas coning in the reservoir. The technique is believed by Arco to be suitable for use in the Spraberry Trend, Texas, and fractured gas reservoirs in Oklahoma. Some details of the project are given in the article.

860. Plahn, S. V. (1986). Predicting horizontal well performance in solution gas drive reservoir. M.S. Thesis. Texas A & M University, College Station, TX.

Not examined.

861. Plahn, S. V., R. A. Startzman, and R. A. Wattenbarger (1987). A method for predicting horizontal well performance in solution-gas-drive reservoirs. Proceedings 1987 SPE Production Operations Symposium, March 8-10, 1987, Oklahoma City, pp. 157-168. Society of Petroleum Engineers, Richardson, TX. SPE paper 16201.

A type-curve method is presented for predicting the performance of horizontal wells. The method may be used in isotropic, homogeneous, solution-gas drive reservoirs, for wells producing at their maximum rate.

862. Pocovi, Antonio, Hector Sosa, and Miguel Angel Casares (1990). Horizontal wells improve short-term reserves. World Oil 210 (3), 51-61. Abstract in The Engineering Index Monthly (June, 1990), 28 (6), 671.

Yacimientos Petroliferos Fiscales (YPF) has drilled four horizontal wells in two fields of the Neuquen basin, Argentina. The wells were redrilled from vertical wells in Chihuido de la Sierra Negra field and Loma de la Lata field. The project was successful and will probably result in the use of horizontal wells to develop other fields in Argentina. Details of the project, the technology used, and project economics are given in the article.

863. Pocovi, A. S., L. Gustavino, A. Pozzo, and J. A. Musmarra (1991). Comparing cost and performance of horizontal wells. World Oil 212 (3), 39-40, 42-44.

Yacimientos Petroliferos Fiscales (YPF), the national oil company of Argentina, has drilled four horizontal wells in the Nequen basin. Eleven other horizontal wells have been drilled by private companies. Two of the wells were drilled as horizontal wells, while the other 13 wells were horizontal re-completions of existing vertical wells. The horizontal wells were completed in the Cretaceous Avile sandstone, Jurassic Sierra Blanca sandstone, and Cretaceous Quintuco formation. The wells were logged, flushed, and cased. The casing was not cemented; 3.5 inch screen pipes were used in the horizontal sections. The horizontal wells proved to be less expensive than the vertical wells from which they were drilled, with costs running from 25% to 84% of the cost of the original well. Production from the horizontal wells was four times that of the vertical wells. Details of the wells and economics are given in the article.

864. Poon, D. (1990). Decline curves for predicting production performance from horizontal wells. Preprints of CIM/SPE International Technical Meeting, June 10-13, 1990, Calgary, Canada, Paper 41. Petroleum Society of CIM, Montreal. CIM/SPE paper 90-41. Abstract in The Journal of Canadian Petroleum Technology (1990) 29 (2), 33. Paper also printed in Journal of Canadian Petroleum Technology (1991), 30 (1), 77-81.

Production decline-curves for horizontal wells were developed using Green's and source functions. Production performance of a horizontal well over a very long time period approaches two times that of a vertical well. Decline-curves can be used to determine the effective length of a horizontal well, to determine the production forecast, and predict the ultimate recovery.

865. Pope, C. D. and P. J. Handren (1990). Completion techniques for horizontal wells in the Pearsall Austin Chalk. Proceedings 1990 SPE Annual Technical Conference & Exhibition, Production Operations and Engineering volume, September 23-26, 1990, New Orleans, pp. 657-664. Society of Petroleum Engineers, Richardson, TX. SPE paper 20682.

Oryx Energy Co. uses three completion techniques for Austin Chalk horizontal wells in the Pearsall field, Texas: open hole completions; liner completions; and cased-hole completions. Oryx's guidelines for using these techniques are: (1) open hole completions should be used unless the hole is washed out, has severe dogleg problems, depleted fractures exist, or the Anacacho shale is exposed in the borehole; (2) slotted liner should be used if there are washouts or doglegs, and there are no depleted fractures; (3) if there are depleted intervals near the vertical wellbore, partially cemented slotted liner should be used; (4) ECP's and port collars should be used if there are multiple depleted zones present through non-continuous intervals, if water-bearing fractures need to be isolated, or if site selective stimulation or production is desired.

866. Pope, Scott C. (1991). Horizontal potential of the Austin Chalk in East Texas. in Chuber, Stewart (editor), Abstracts and Short Papers of the South Texas Austin Chalk Exploration Symposium, February 24-26, 1991, San Antonio. South Texas Geological Society, San Antonio.

Abstract. Comparison is made of the Austin Chalk in East Texas with the Austin Chalk in Pearsall field. East Texas is defined as east of the Giddings field, and includes Milam County. Milano field, Cherokee field, Brookeland/Scrappin' Valley field, and Hilltop Resort fields are also briefly discussed.

867. Popular Horizontal (1991). News Horizons. Popular Horizontal 1 (1), 30-33.

This article consists of news items, including: a review of horizontal drilling status; horizontal drilling in Wilson County, Texas; and leasing in Lee County, Texas. Of special note is a horizontal well completed in the Georgetown formation, Brazos County, Texas, by Union Pacific Resources Company. The Elaine Hester No. 1 well flowed 374 BOPD and 258 MCFGPD in a 24 hour test.

868. Popular Horizontal; (1991). News Horizons. Popular Horizontal 1 (2), 40-44.

This article consists of brief news items concerning horizontal drilling and related subjects. The items are: Logging horizontal wells; Eastman Christensen establishes a division to provide horizontal drilling services to the environmental remediation industry; Union Pacific Resource Co. completes a record length Austin Chalk horizontal well in Gonzales County, Texas (horizontal displacement 6,142 ft).

869. Powell, John W. (1988). Thixotropic biopolymers: Evaluation of their low shear rate viscosity and its effect on radial slip of cuttings in high angle wells. M-1 Drilling Fluids, In-house publication.

Not examined.

870. Power, Michael M., Roger Chapman, Robert O'Neal (1990). Horizontal well sets depth record: Horizontal drilling below 14, 600 ft. Petroleum Engineer International 62 (11), 36, 38.

Smako Ltd. has drilled a horizontal well in Huxford field, Alabama, to a depth of 14,462 ft (TVD). The well was drilled as a recompletion of a watered-out vertical well. The horizontal section was drilled through a window cut in the casing of the vertical well at 14,075 ft (TVD). The well was completed in the Jurassic Smackover formation. Details of drilling the well are given in the article.

871. Power, Michael M., Roger Chapman, John O. Hughes, and Richard W. Parliman (1990). Horizontal well sets depth record: Completing deep horizontal well. Petroleum Engineer International 62 (11), 37, 40-41.

Smacko Ltd. recompleted a watered-out vertical well in Huxford field by drilling a horizontal well from the old vertical wellbore at 14,075 ft. (TVD). The well was drilled to a measured depth of 15,357 ft. The well was completed with a slotted liner in the horizontal section.

872. Prestwich, S. M. and L. G. Miller (1978). Drain-hole drilling can increase productivity. World Oil 193 (12), 69-70.

Drainhole drilling was used at a geothermal project in the Raft River valley, Idaho, to enhance production of steam for the power plant. The drainholes were planned to intersect major fracture systems in the reservoir rock. The drainholes increased productivity by about 500%.

873. Prevedel, B. (1985). Berkhoepen 2001; Horizontalbohrorojekt. [Berkhoepen 2001: Horizontal drilling project]. Erdoel Erdgas 101 (12), 377-383. In German with English abstract.

Abstract. The Berkhoepen 2001 horizontal well was drilled by Preussag AG as an experimental project.

874. Prevedel, B. (1985). How one operator drilled horizontally through a salt dome. World Oil 201 (7), 69, 73, 76, 80. Abstract in The Engineering Index Annual (1986), 85, Part IV, 6527.

Berkhoepen 2001 is a horizontal drilling research project in Hanover, Germany. The Berkhoepen T1 well was a reentry in a vertical well, and was designed to test horizontal drilling methods. The Berkhoepen 2001 well was drilled in the Oedesse salt dome to a measured depth of 8,184 ft. The well penetrated 2,000 ft into the salt; 700 ft of that was at 90 degrees from vertical. Steerable Navi-Drill downhole motors were used with MWD for the directional part of the hole. A salt-saturated, water-based mud was used throughout, with mud weights in the range 10.4 ppg to 12.7 ppg because of the squeezing salt problem. Bits used were 8.5 inch roller-bearing bits and PDC bits. The article has a detailed description of drilling operations.

875. Prevedel, B. (1986). Berkhoepen 2001 -Horizontalpohrorojekt. [Berkhoepen 2001 - Horizontal drilling project]. Forschung Planung und Betrieb 7, 47-53. In German.

Indexed but not annotated.

876. Prevedel, B. (1987). Preussag Gamma-Steering-Tool Ein neuer Weg zur Optimierung von Richtbohroperationen. Erdol Erdgas Kohle 103 (2), 67-77. In German with an English abstract.

Abstract. Preussag AG has developed an MWD gamma-ray and steering tool called Preussag Gamma Steering Tool, which has been used to drill horizontal wells in Georgsdorf field, Germany.

877. Prevedel, B. (1987). New techniques in horizontal and drainhole drilling optimization: Part 2 -Georgsdorf 503, horizontal drilling project. SPE unsolicited manuscript. SPE paper 17080. Abstract in The Engineering Index Annual (1988), Part IV, 6006.

Two horizontal drainholes were planned for Georgsdorf field, Germany. The two wells were to be parallel and separated laterally by 250 meters. Both wells were to have a horizontal section 650 meters long. One well was to be a production well; the other an injection well. The target formation was the Bentheimer sandstone, a poorly consolidated reservoir that was expected to be underpressured. Drilling problems in the first well resulted in a sidetrack of the entire kickoff section, stuck BHA, etc. The horizontal section was terminated short of the target length. The well was completed with slotted liner. Due to formation damage and the lack of a water injection well, the horizontal well did not produce significant amounts of oil. Details of drilling and completion of the well are given in the paper.

878. Prevedel, B. (1987). New techniques in horizontal and drainhole drilling optimization: Lehrte 41 lateral drilling program. Proceedings 1987 5th Middle East Oil Show, March 7-10, 1987, Manama, Bahrain, pp. 75-84. Society of Petroleum Engineers, Richardson, TX. SPE paper 15694.

Preussag Ag Oil & Gas conducted horizontal drilling experiments in Lehrte field to improve oil recovery from depleted sands. Preussag AG developed a system called the PREUSSAG "Alternative Drilling System" which uses an "articulated drillstring" (ADS). The reservoir is in the Cornbrash formation which dips 30 degrees. The project was designed to recover the attic oil in the updip part of the pinching-out reservoir sand. The first effort missed the reservoir, but the second effort successfully intersected the reservoir sands. After the workover the production increased from 2 cubic meters per day of fluid with 97% water cut to 2 cubic meters per day of pure oil. Details of the project are given in the article.

879. Price, Leigh and Julie LeFever (1991). Horizontal drilling in the Bakken formation - The hunt for an elephant that never left the source system. A.A.P.G. Bulletin 75 (6), 1137.

Abstract. Oil from the Bakken formation makes a minor contribution to other reservoirs in the Williston basin. Oil generated in the Bakken is still in place. The Bakken may have generated as much as 150 billion barrels of oil.

880. Pugh, G. E. (1982). Drilling of three horizontal wells pattern, Fort McMurry, Alberta. Paper presented at the 1982 Annual Technical Meeting of the Petroleum Society of CIM, June 6-9, 1982, Calgary. CIM, Montreal. Paper 82-83-68. Abstract published in The Journal of Canadian Petroleum Technology (1982) 21 (2), 43.

Texaco Canada Resources Ltd. drilled three horizontal wells in the Fort McMurray Pilot Plant, Alberta, Canada. The wells were drilled with horizontal sections of 350 meters, at depths of 125 meters from the surface. The wells were drilled from the same pad, with the non-horizontal sections drilled at 45 degrees inclination. The wells were drilled to test the applicability of horizontal wells for bitumen recovery from the Athabasca Tar Sands.

881. Quang, D., S. M. Farouq Ali, and A. E. George (1990). Scaling criteria and model experiments for horizontal wells. Paper presented at the International Technical Meeting, The Petroleum Society of CIM and the Society of Petroleum Engineers, June 10-13, 1990, Calgary. The Petroleum society of CIM, Montreal. CIM/SPE paper 90-128. Abstract in The Journal of Canadian Petroleum Technology (1990) 29 (2), 50.

Abstract. Equations have been developed to model three- phase flow in the vicinity of a horizontal well. Isothermal models and steam models were used in these experiments.

882. Rahman, S. S. (1989). Drilling horizontal well while maintaining a constant borehole pressure. Proceedings 22nd Drilling Symposium, Twelfth Annual Energy-Sources Technology Conference and Exhibition, January 22-25, 1989, Houston, pp. 71-81. ASME, New York.

Not examined.

883. Rankin, M. D., T. J. Frisenhahn, and W. R. Price (1989). Lightened fluid hydraulics and inclined boreholes. Proceedings SPE/IADC 1989 Drilling Conference, February 28-March 3, 1989, New Orleans, pp. 439-450. Society of Petroleum Engineers, Richardson, TX. SPE paper 18670. Abstract in The Engineering Monthly Index (November, 1989), 27 (11), 760.

In the case of under-pressured formations or severe formation damage, compressible (lightened) fluids may be superior to conventional drilling fluids. In under-pressured horizontal wellbores, stable foams have better hole cleaning characteristics than aerated fluids, mists, or dry gases. Details of experimental results of the use of lightened fluids are given in this article.

884. Ranney, L. (1939). The first horizontal well. Petroleum Engineer X (9), 25-30.

The first horizontal well drilled in the United States was drilled in Morgan County, Ohio. The well was drilled into the outcrop of the First Cow Run sand in Havener Run to avoid having to drill a vertical shaft. The reservoir was a low-permeability gas sand. The purpose of the well was to expose a longer length of borehole to the tight reservoir. Ranney proposed drilling up to sixteen horizontal holes radiating out from a single shaft into the shallow reservoir. The alternate horizontal holes were to be 'shot' with TNT and a noncompressible gelatin to enlarge the drainhole and fracture the formation.

885. Ranney, L. (January 20, 1941). Drilling wells horizontally. The Oil Weekly 100 (7), 12-14.

The first horizontal well was drilled in the First Cow Run sandstone in Ohio. The well was drilled horizontally from a vertical shaft; 7,500 ft of horizontal drainhole was been drilled. The reservoir rock had 15% oil saturation. The method has also been used to recover coalbed methane and to produce water. Details of the drilling methods are given in the article.

886. Ranney, L. (April, 1949). Horizontal drilling increases yield of Australia oil field. Oil & Gas Journal 47 (49), 48, 50.

In 1949, gravity drainage was used to produce heavy-oil from Australia's only oil field. The heavy-oil was drained into a shaft through horizontal drainholes. The drainholes were drilled with a 1.5-inch diamond drill into a glauconitic sandstone at a depth of 1,200 ft below the surface. The reservoir is underlain by an aquifer. The production rate in 1941 was 25 to 30 BOPD of 17 gravity oil. A brief description of the project is given.

887. Rao, M. Vikram and John E. Fontenot (January 25, 1988). MWD poised for future - 1. Many factors determine need for real-time or recorded data. Oil & Gas Journal 86 (4), 65-69.

A review of the status of MWD.

888. Rao, M. Vikram and John E. Fontenot (February 8, 1988). MWD poised for future - 2. MWD gains as formation-evaluation tool. Oil & Gas Journal 86 (5), 44-48.

MWD formation evaluation tools now available are: natural gamma-ray, resistivity, neutron porosity, and formation density logs. MWD formation evaluation tools are especially useful in high-angle wells that require special techniques to log with wireline logging tools.

889. Redford, D. A. (1985). In-situ recovery from the Athabasca oil sands - past experience and future potential. Journal of Canadian Petroleum Technology 24 (3), 52-62.

This article reviews in-situ recovery pilot projects in the Canadian Athabasca oil sands and the Conoco Streeter Ranch project, Texas. The methods reviewed include horizontal drilling, steamflood, mining, electrical heating, and hydraulic fracturing to create horizontal fractures. Horizontal wells, as shallow as 85 meters, were successfully drilled at Texaco's Fort McMurray pilot. Horizontal wells were also used as steam injection wells in the "hot finger" concept. Editor's Note: This article is good review of the methods used to produce the heavy- oil in the Athabasca oil sands and similar reservoirs.

890. Rehm, W. A. (1987). Horizontal drilling as applied in slim holes. Petroleum Engineer International 69 (2), 24-26, 28. Abstract in The Engineering Index Annual (1987), 86, Part IV, 6499.

Slim-hole drilling is less expensive than conventional drilling because of the smaller size of tools, rigs, casing, and mud systems. Medium-radius slim-hole drilling is feasible in horizontal wells, if all drilling is done with a downhole motor, and if there is little or no rotation of the drill string, if diamond bits are used, and if a BHA close to "slick" design is used. Rotation of a slim hole drill string can cause buckling.

891. Rehm, W. A. and A. Garcia (1989). Horizontal drilling in mature oil fields. Proceedings 1989 SPE/IADC Drilling Conference, February 28-March 3, 1989, New Orleans, pp. 755-764. Society of Petroleum Engineers, Richardson, TX. SPE/IADC paper 18709. Abstract in The Engineering Monthly Index (November, 1989), 27 (11), 762.

Using horizontal drilling to recomplete wells in mature fields appears to be economically feasible based on the results of a twenty-five well project. The twenty-five wells used in this report were drilled in Texas (Austin Chalk) and Argentina (sandstone reservoir near Mendoza). The recompletions in Argentina showed an increase in production of 1.68 to 4.69 times that of the original vertical wells. The reservoirs were in thin sands. No comparative data are given for the Texas examples. The Texas wells produced oil from previously undrilled fault blocks. Care must be used to avoid skin and/or reservoir damage when recompleteing with horizontal drilling. Engineering details of the recompletions are given in the article. Editor's Note: The Austin Chalk example is evidently from Giddings field. The recompletions were actually drilled into fractured chalk zones that were not drilled previously.

892. Reid, Stephen A., George S. McJannet and Orville D. Hart (1990). Application of horizontal drilling to the development of a complex turbidite sandstone reservoir, Elk Hills field, Kern county, California. A.A.P.G. Bulletin 74 (5), 747.

Abstract. Horizontal drilling has been used to develop the turbidite reservoirs in Elk Hills field, California. The turbidites are in a deep-sea fan in the Miocene Monterey formation. Horizontal wells were used because the oil-water contact was level, and believed to be stable; water production is low; horizontal wells would produce longer than a vertical well; and several sandstones could be produced from one well. Some details of the project, including unexpected problems, are described in this abstract.

893. Reijers, T. J. A. (1985). Porosity characteristics and evolution in fractured Cretaceous carbonate reservoirs, La Paz field area, Maracaibo basin, Venezuela. in Bartok, P. (editor), Carbonate Petroleum Reservoirs, pp. 407-423. Springer-Verlag, New York.

Not examined.

894. Reiley, R. H., J. W. Black, T. O. Stagg, D. W. Walters, and G. R. Atol (1987). Cementing of liners in horizontal and high-angle wells at Prudhoe Bay, Alaska. Proceedings 1987 SPE Annual Technical Conference & Exhibition, Drilling volume, September 27-30, 1987, Dallas, pp. 303-310. Society of Petroleum Engineers, Richardson, TX. SPE paper 16682.

Cementation of liners in horizontal and high-angle wells, in Prudhoe Bay field, has been accomplished by simultaneous rotation and reciprocation of liners while cementing. This method was used on liners as long as 2,587 ft (789 m) in 90 degree wellbores. Off-bottom cementing of liners was done using a cross-linked polymer plug. Cement evaluation logs can be drillpipe or tubing-conveyed in high-angle wells.

895. Reilley, R. H., J. W. Black, T. O. Stagg, D. A. Walters, and G. R. Atol (1988). Improving liner cementing in high- angle/horizontal wells. World Oil 207 (1), 60-80. Abstract in The Engineering Index Annual (1988), 87, Part III, 5995.

Eight unconventional wells (high-angle and horizontal) were completed using cemented liner for zone isolation. The hole inclination of these wells ranged from 79 degrees to 89 degrees. Liner sizes used were 5.5-inches (3 wells) and 7-inches (4 wells). The cementing operations were successful. Cement evaluation logs were run, either with coiled-tubing or were pipe-conveyed; both methods were successful. Details of the operations are given in the paper.

896. Reiss, Herbert and Francois Giger (December, 1982). Le forage horizontal: premiers realizations en Europe. Petrole et Techniques No. 294, 33-36. In French.

Article indexed but not annotated.

897. Reiss, L. H., A. P. L. Jourdan, F. Giger, and P. Armessen (1984). Offshore and onshore European horizontal wells. Offshore Technology Conference, May 7-9, 1984, Houston, Texas, v. 3, pp. 137-143. OTC paper 4791.

Four horizontal wells have been drilled in Europe: one well in Rospo Mare field, offshore Italy; two wells in Lacq field, France; and one well in Castera Lou field, France. The deepest well (TVD) is 2,800 m (8,400 ft) at Castera Lou; the longest horizontal displacement is 1,800 m (5,400 ft) at Rospo Mare of which 600 m (2,000 ft) is in the reservoir. The reservoirs are all fractured limestones or dolomites. Rospo Mare is a heavy-oil field.

898. Reiss, L. H. (1987). Production from horizontal wells after 5 years. Journal of Petroleum Technology 39 (11), 1411-1416. SPE paper 14338 presented at the 1985 SPE Annual Technical Conference and Exhibition, Las Vegas, September 22-25. Abstract in The Engineering Index Annual (1988), 87, Part IV, 6015.

Elf Aquitaine drilled four horizontal wells (2 in Lacq field, 1 in Castera Lou field and 1 in Rospo Mare field) and has produced them for 2 to 5 years. The wells were designed to test drilling, coring, logging, and completion methods in horizontal wells. In Lacq field, one of the horizontal wells has three times the daily production rates of nearby vertical wells. The other Lacq field well is in a more geologically complex situation, with both fractured carbonate and dolomite reservoirs. This second well has significantly greater production, but is limited by the nature of the reservoir. The well in Castera Lou field has five times the daily flow of the three vertical wells in the field. The Rospo Mare field horizontal well is 20 times more productive with 4.6 times the cumulative production of the vertical wells in this heavy-oil field. Critical rate value is 7.4 times that of the vertical wells.

899. Renard, G. and J. G. Dupuy (1990). Influence of formation damage on the flow efficiency of horizontal wells. Proceedings Ninth SPE Symposium on Formation Damage Control, February

22-23, 1990, Lafayette, Louisiana, pp. 133-142. Society of Petroleum Engineers, Richardson, TX. SPE paper 19414. Abstract in The Engineering Index Monthly (August, 1990), 28 (8), 597.

Skin damage in horizontal wells is less detrimental to productivity than it is in vertical wells. Skin damage will, however, damage productivity to some degree. Methods are discussed in this article for the determination of the effect of skin damage on productivity.

900. Restine, J. L., W. G. Graves and R. Elias, Jr. (1987). Infill drilling in a steamflood operation: Kern River field. SPE Reservoir Engineering 2 (2), 243-248. SPE paper 14337 presented at the 1987 Annual SPE Technical Conference and Exhibition, September 22-25, 1986, Las Vegas.

Kern River field, California, is a stratigraphic trap containing heavy-oil. An infill drilling program has been started in an area containing 3,640 producing wells and 1,875 injection wells. Well spacing was reduced for the 574 infill wells. The infill wells in one area of the field (Canfield area) increased ultimate recovery by 5-8% of pre-displacement oil-in-place. Details of the project are given in the paper. Editor's Note: A horizontal well pilot project has been carried out in Kern River field (see Dietrich, 1988).

901. Rial, R. M. (1984). 3D simulation using a horizontal wellbore for steamflooding. Paper presented at the 1984 SPE Annual Technical Conference & Exhibition, September 16-19, 1984, Houston. Society of Petroleum Engineers, Richardson, TX. SPE paper 13076.

Simulations of horizontal wells used for steam injection in Kern River field, California, show that horizontal wellbores have more consistent aerial sweep than vertical wells; increased recovery of oil-in-place can be expected when horizontal wells are used; and horizontal wells have improved temperature and oil phase saturation. Details of the model are given in the paper.

902. Richardson, A. S. (1985). Slant drilling of Valhalla waterflood project. Journal of Canadian Petroleum Technology 24 (2), 24. CIM paper 85-36-60. Paper presented at the Joint Annual Meeting of The Petroleum Society of CIM and the Canadian Society of Petroleum Geologists, June 2-5, 1985, Edmonton.

Abstract. Slant drilling of twenty-three wells has been done at the Valhalla Doe Creek project.

903. Richter, T. H. and B. R. Bright (1990). A regulatory approach to horizontal drainholes. Proceedings 1990 Permian Basin Oil & Gas Recovery Conference: Positioning for the 90's: Improving Profitability Through Todays Technology, March 8-9, 1990, Midland, pp. 157-160. Society of Petroleum Engineers, Richardson, TX. SPE paper 20101. Abstract in The Engineering Index Monthly (June, 1990), 28 (6), 674.

The Texas Railroad Commission has developed guidelines for permitting horizontal wells. The guidelines do not address the issues of well spacing, acreage and allocation assignments, or the definition of a horizontal drainhole. The guidelines are given in the paper.

904. Rinaldi, Roger E. (1987). Improved reservoir productivity through use of horizontal drilling. A.A.P.G. Bulletin 71 (8), 996.

Abstract. Horizontal drilling, when properly conducted, can: (1) increase production from thin formations; (2) increase production from low-permeability formations; (3) increase production from fractured formations; (4) allow production of heavy-oil reservoirs by improving efficiency of steamfloods; and (5) increase productivity of low-energy reservoirs.

905. Ringgenberg, Paul and R. L. Hinkie (1989). Right combination of technology saves stimulation time. Drilling 50 (3), 18-23.

A significant saving of time in stimulating an offshore horizontal well was made by improved use of technology. In two weeks, five zones of the well were acid stimulated using tubing-conveyed perforating guns, gelled polymer plugs, and a ratcheting reclosable circulating valve. Prior to this method two zones had been stimulated in two weeks using drillable packers.

906. Rintoul, B. (October, 1982). Hot plate' cooking in California. Drilling Contractor, pp. 46-54.

Not examined.

907. Rintoul, Bill (June, 1985). BecWell offers horizontal drilling system. Pacific Oil World 77 (6), 16-18.

BecWell, a subsidiary of Bechtel Investments Inc., was formed to market a proprietary ultra-short-radius lateral drilling method developed in conjunction with Petrolphysics Ltd of San Francisco. The system has been successfully used in the heavy-oil sands of Kern River field and Midway-Sunset field, California. A brief description is given of the project.

908. Rintoul, Bill (1985). BecWell introduces horizontal well technology. Well Servicing 25 (4), 24-25.

BecWell, a Bechtel group company, has developed a horizontal drilling system that can be used on double-stand workover rigs. The system is for drilling ultra-short-radius wells with a 12-inch turning radius. A brief discussion of the method is given in the article.

909. Rintoul, Bill (1991). Horizontal drilling in California. Pacific Oil World 83 (5), 4-6, 9.

Review of horizontal drilling projects in California. Projects described include: Midway-Sunset field (Unocal, Mobil, Oryx Energy Co., Berry Petroleum Co.); Elk Hills field (Bechtel Petroleum Operations); Yowlumne field (Arco Oil & Gas Co., Texaco); Kern River field (Texaco, Fenix & Scission, Ladd Petroleum, Cornell Heavy Oil Process, Barber Heavy Oil Process Inc.).

910. Roberts, J. D., A. S. Mathieson, J. M. Hampson (1987). Statfjord field. in Geology of Norwegian Oil and Gas Fields, pp. 319-340. Norwegian Petroleum Society, Graham & Trotman, London.

This paper reviews the geology of Statfjord field, North Sea, Norway.

911. Robinson, L. H., Jr., J. M. Speers, L. A. Watkins, A. Barry, and J. F. Miller (1980). New telemetry system employs unique inside wireline. World Oil 190 (5), 57-61.

Telemetry of downhole data, such as downhole pressure, mud resistivity, or temperature, can be achieved using a wireline inside the drillpipe. Exxon Production Research Co. has developed a system that can be rotated in the hole, allowing continuous drilling while recording data.

912. Roemershauser, A. E. and M. F. Hawkins, Jr. (1955). Effect of slant hole drainhole and lateral hole drilling on well productivity. Journal of Petroleum Technology 7 (2), 11-14.

Slantholes, drainholes, and lateral (horizontal) boreholes normally have greater productivity than vertical wells that fully penetrate the reservoir. The increased productivity is due to the increase in cross-sectional area exposed to the reservoir. Model studies conducted for each of the cases are discussed in the article.

913. Roland, M. (December, 1987). Workshop on extended reach and horizontal drainage wells. Drilling and Pumping Journal, pp. 48-56. Schlumberger and Cambridge Research Ltd.

Not examined.

914. Rolf, E. Gerald and G. A. Stone (May, 1990). The Austin Chalk - horizontal drilling. South Texas Geological Society Bulletin XXX (10), 35-38.

The Austin Chalk production in Pearsall field of South Texas is primarily from fractures in clean chalk zones. The natural fractures are attributed to tensional, stress relaxation. The entire area appears to be potentially productive, so seismic data, while helpful, is not required. Part of the field has been partially depleted and, as a result, has multiple drilling and production problems. Drilling is usually done with NaCl brine drilling fluids. Wells that potential less than 100 BOPD are usually acid treated.

915. Rosa, A. J. and R. de S. Corvalho (1986). An infinite conductivity model of a horizontal well producing from an anisotropic reservoir. Bull. Techn. Petrobras 29 (3), 211-220.

Not examined.

916. Rosa, Adalberto J. and Renato de Souza Carvalho (1989). A mathematical model for pressure evaluation in an infinite-conductivity horizontal well. SPE Formation Evaluation 4 (4), 559-566. SPE paper 15967.

A mathematical model is presented for transient pressure behavior in a well with an infinite-conductivity horizontal drainhole in an infinite slab reservoir.

917. Rosenzweig, J. J., D. C. Korpics, and G. E. Crawford (1990). Pressure transient analysis of the JX-2 horizontal well, Prudhoe Bay, Alaska. Proceedings 1990 SPE Annual Technical Conference & Exhibition, Formation Evaluation and Reservoir Geology volume, September 23-26, 1990, New Orleans, pp. 851-864. Society of Petroleum Engineers, Richardson, TX. SPE paper 20610.

Pressure transient tests were carried out on the JX-2 horizontal well, Prudhoe Bay field, to determine skin factor, vertical permeability, and horizontal permeability. The simulations failed to match the data until gas cap data were incorporated. Analytical and parametric simulations showed that the gas cap exerts a significant effect during the pseudo-radial flow period. Multiphase numerical simulations are required for accurate interpretation of pressure data where a gas cap is present.

918. Rountree, R. (1981). Powder River basin: Conoco digs second oil mine in Johnson County. Western Oil Reporter 38 (8), 144-145.

Conoco Inc. has started work on its second "oil mine" in Wyoming. The North Tisdale Gravity Drainage Project consists of a 14-ft diameter, 235-ft deep shaft from which four horizontal drainholes will be drilled from a collection chamber 60 ft long and 18 ft wide. The project will produce from the Lakota formation. The horizontal drainholes will be 3.25 inches in diameter, and up to 2,000 ft long.

919. Rourke, W. E (March 15, 1971). Marathon Oil sets big kick record with Cook Inlet well. Oil & Gas Journal 69 (11), 55- 61.

Marathon drilled a record extended-reach well in McArthur field, Cook Inlet, Alaska. The well had a horizontal displacement of 12,584 ft (17,010 ft MD and 10,517 ft TVD). Details of the project are given in the article.

920. Ruhl, W. (1989). Schwerkraft-Entolung im Erdol-Bergbau Wietze [Gravity drainage in the Wietze oil mine]. Erdol Erdgas Kohle 105 (2), 54-59. German with an English abstract.

Abstract. The Wietze oil mine utilizes gravity drainage to recover heavy-oil from the Wietze Wealden Oilsands. The field was first developed in 1902 and contains 480 vertical wells. The field recovered 12% of the oil-in-place using vertical wells. Gravity drainage may recover an additional 30% of the oil. Thermal methods may boost total recovery to 60% of the original oil-in- place.

921. Sabins, Fred L., R. C. Smith, M. D. Broussard, K. L. Talbot, and S. R. Olaussen (1990). An Investigation of factors contributing to the deposition of cement sheaths in casing under highly deviated well conditions. Proceedings 1990 IADC/SPE Drilling Conference, February 27-March 2, 1990, Houston, Texas, pp. 203-210. SPE paper 19934.

Experiments were conducted to discover a solution to the problem of cement-sheath formation inside casing during cementing operations. The experiments show the critical values of shear rate rheology and yield point of the cement slurry that must be maintained to reduce or eliminate the promotion of a cement sheath. Details are given in the article.

922. Sabins, Fred L. (1990). Problems in cementing horizontal wells. Journal of Petroleum Technology 42 (4), 298-400. SPE paper 20005.

Problems found in cementing casing in horizontal boreholes are: hole cleaning; drilling-fluid displacement; centralization of pipe; cement slurry design; and evaluation with acoustic tools.

923. Sahuquet, B. and B. Corre (1982). Injection de vapeur en milieu carbonate fissure pilote de Lacq Superieur. Proceedings 2ne European Symposium on Enhanced Oil Recovery, November 8-10, 1982, Paris, France, pp. 611-623. Editions Technip, Paris.

This paper discusses the steamflood EOR pilot project at Lacq Superieur field, France, prior to the horizontal drilling conducted in the field. Lacq Superieur field is a fractured carbonate reservoir containing heavy-oil.

924. Sahuquet, Bernard C. and Jerome J. Ferrier (1982). Steam- drive pilot in a fractured carbonated (sic) reservoir: Lacq Superior field. Journal of Petroleum Technology 34 (4), 873- 880. SPE paper 9453. Abstract in The Engineering Index Annual (1982) 81, Part IV, 5262.

In 1977, a pilot steamflood project was initiated in Lacq Superieur field, France. Lacq Superieur field produces from a fractured carbonate reservoir. This paper discusses the laboratory studies and steamflood project initiated in the field before horizontal drilling was attempted. Editor's Note: Lacq Superieur field was the first field in which "modern" horizontal drilling was tried. This paper is helpful in understanding the later horizontal drilling project.

925. Sahuquet, B. (1985). Pilot project for recovery of oil by steam injection (Upper Lacq field). in Debouw, R., E. Millich, J. P. Joulia, and D. Van Asselt (editors), New Technologies for the Exploration and Exploitation of Oil Gas Resources, v. 2, pp. 880-889. Graham & Trotman, London. ISBN 0 86010 676 4. A similar article was published by Sahaquet, titled "Pilot installation for the recovery of heavy crude by steam injection into the Lacq field," in New Technologies for Exploration and Exploitation of Oil and Gas Resources, volume 2, pp. 1111-1122, Graham & Trotman Ltd., London, (1979); ISBN 0 86010 159 2.

The Upper Lacq field steam injection project is the first such project in a fractured carbonate reservoir. Steam injection caused the recovery of an additional 35,600 tons of oil from the reservoir. Upper Lacq field was 25 years old at the time of pilot project initiation. The project showed that if natural fracturing is sufficient, heat transfer from the steam to the reservoir occurs, ensuring success without steam breakthrough. Oil expelled from the matrix can move to the production wells. Thermal expansion is the predominant mechanism in steam-drive in a fractured- medium, and tritium water is the best tracer.

926. Sahuquet, B. C., A. M. Spreux, B. Corre, M. Guittard (1990). Steam injection in a low-permeability reservoir through a horizontal well in Lacq Superior field. Proceedings 1990 SPE Annual Technical Conference & Exhibition, Enhanced Oil Recovery/General Petroleum Engineering volume, September 23- 26, 1990, New Orleans, pp. 559-566. Society of Petroleum Engineers, Richardson, TX. SPE paper 20526.

Lacq Superieur field, France, is a low-permeability, fractured-carbonate reservoir discovered in 1948. The field was produced naturally until 1977, after which, steamflood was applied to the fractured portion of the field. The steamflood was effective in the fractured portion of the field, but not in the low-permeability portion. A horizontal well has been drilled to attempt to improve productivity by injecting steam at a high rate. The injection well has a 350-meter long horizontal section in the reservoir. The new injection well boosted production from five wells to 55 from 9 cubic meters of oil per day. Details of the reservoir model and drilling of the horizontal steam injection well are given in the article.

927. Saleh, S. T. (1990). Review of hole cleaning requirements for directional/horizontal wells. Preprints of CIM/SPE International Technical Meeting, June 10-13, 1990, Calgary, v. II, Paper 11. Petroleum Society of CIM, Montreal. Paper withdrawn. Abstract in The Journal of Canadian Petroleum Technology (1990) 29 (2), 28. CIM paper No. 90-11.

This paper reviews hole cleaning requirements in horizontal and directional wells. The abstract has no data. The paper was withdrawn from presentation and not published in the Preprints.

928. Salamy, S. P., B. S. Barddji, C. O. Okoye, J. D. Mercer, and A. B. Yost, II (1987). Recovery efficiency aspects of horizontal well drilling in Devonian shale. Proceedings SPE/DOE Low-Permeability Reservoir Symposium, May, 1987, Denver, pp. 217-226. Society of Petroleum Engineers, Richardson, TX. SPE/DOE paper 16411.

As part of the Department of Energy Eastern Gas Shales Project, tests were conducted to determine the relative recovery efficiencies of vertical and horizontal wells in Devonian shales. The two test sites chosen are Wayne County, West Virginia, and Meigs County, Ohio. The two sites were chosen on the basis of differences in permeability anisotropy: Wayne County has no determinable anisotropy and Meigs County has a calculated permeability anisotropy of 8:1. Simulations showed that a 2,000-ft horizontal wellbore has a drainage area comparable to that of two vertical wells (72 acres). Production histories showed that the horizontal well was three times more efficient in recovering gas from a fixed volume of rock. The conclusions reached were: (1) horizontal wells are more economical than vertical wells for infill drilling in a field; (2) horizontal wells are more effective in draining a reservoir; (3) horizontal wells need to be drilled orthogonal to the preferred flow direction of the reservoir in order to maximize production.

929. Salazar, J., J. Parra, E. Alturo, and H. Villasmil (February, 1983). Perforacion Direccional de Pozos desde una Localizacion Comun [Drilling directional holes from a common location]. Maraven, S.A., Venezuela. Technical Report IT-9267. In Spanish.

Not examined.

930. Saldivar-Sali, Arthur, Hans G. Oesterle and Dennis N. Brownlee (November 30, 1981). Geology of offshore northwest Palawan, Philippines - 1. Oil & Gas Journal 79 (48), 119- 120, 122, 124-126, 128.

Part 1 of a three part discussion of the geology in the South China Sea near Palawan, Philippines.

931. Saldivar-Sali, Arthur, Hans G. Oesterle and Dennis N. Brownlee (December 7, 1981). Geology of offshore northwest Palawan, Philippines - 2. Oil & Gas Journal 79 (49), 234, 239-240, 242.

Part 2 of a three part discussion of the geology of the South China Sea near Palawan, Philippines. Horizontal wells are now being drilled in this area.

932. Saldivar-Sali, Arthur, Hans G. Oeserle and Dennis N. Brownlee (December 14, 1981). The geology of offshore northwest Palawan, Philippines-3. Oil & Gas Journal 79 (50), 196-197, 199.

Part 3 of a discussion of the geology of the south China Sea near Palawan, Philippine. Horizontal wells are now being drilled in this area.

933. Santley, David J. and William E. Ardrey (1987). Improve MWD data interpretation. Drilling 48 (1), 22, 24. MWD was initially used to provide directional drilling information. Recent developments now allow MWD to provide real-time measurement of formation resistivity, natural gamma-ray radioactivity, weight-on-bit, torque, and other drilling parameters.

934. Santos, Otto L., Eric E. Maldla and Julius P. Langllnais (1989). Field application of computer graphics for monitoring borehole trajectories. Journal of Petroleum Technology 41 (12), 1254-1258. SPE paper 20097, "Supplement to SPE 17782, Field Application of Computer Graphics for Monitoring Borehole Trajectories", available from SPE Book Order Department. Originally presented at the 1988 SPE Symposium on Petroleum Industry Application of Microcomputers, June 27-29, 1988, San Jose, California.

Interactive computer graphic programs used with microcomputers enable the user to view the trajectory of a wellbore and its relationships to other nearby wellbores immediately, without time consuming drafting.

935. Schaef, D. G. (1983). Review - Esso Resources Canada Ltd., Norman Wells Expansion Project drilling program. Proceedings IADC/SPE Drilling Conference, February 20-23, 1983, New Orleans, pp. 115-122. Society of Petroleum Engineers, Richardson, TX. IADC/SPE paper 11371.

Esso Resources Canada Ltd. plans to drill 150 development wells in Norman Wells field, Northwest Territories, Canada, over a three year period. Norman Wells field is located partially under the Mackenzie River, and will require extended-reach drilling from onshore drilling locations and from islands (natural and artificial) in the river. The target reservoir is the Devonian Kee Scarp formation reef. Editor's Note: This article predates the decision to use true extended-reach and horizontal drilling methods in the development of Norman Wells field.

936. Schmoker, James W. and Timothy C. Hester (1983). Organic carbon in Bakken formation, United States portion of Williston basin. A.A.P.G. Bulletin 67 (12), 2165-2174.

The upper and lower members of the Bakken formation, in the Williston basin, are organically rich black shales. The total organic carbon in the upper member averages 12.1 % and in the lower member 11.5 % based on 260 analyses of samples from North Dakota and Montana. It is possible that 132 billion barrels of oil has been expelled from the United States portion of the upper and lower members of the Bakken.

937. Schmoker, James W. and Timothy C. Hester (1990). Formation resistivity as an indicator of oil generation -Bakken formation of North Dakota and Woodford shale of Oklahoma. The Log Analyst 31 (1), 1-9. Paper originally presented as Paper H at the 1989 SPWLA Symposium, Denver, Colorado; abstract in The Log Analyst (1989) 30 (2), 100. See also Schmoker and Hester (1990), A.A.P.G. Bulletin 74 (8), 1345, for the abstract of a similar paper presented at the A.A.P.G. Rocky Mountain Section Meeting, Denver, September 16-19, 1990. The abstract of this paper is also reprinted in Hansen, William B. (editor) (1991), 1991 Guidebook to Geology and Horizontal Drilling of the Bakken Formation, p, 115, Montana Geological Society, Billings, Montana.

The Bakken formation, Williston basin, and Woodford shale, Anadarko basin, are organic-rich, low-porosity shales. Core studies show that volatile hydrocarbons increase in both formations when the resistivity exceeds 35 ohm-m. Resistivity maps can be used to locate the potentially productive areas for these formations. Crossplots of the thermal-maturity index and formation resistivity can show changes in kerogen composition. Calibration of mathematical indices of time-temperature exposure can be done for the onset of oil generation by crossplotting against formation resistivity.

938. Schmoker, Jim (1990). Formation resistivity as an indicator of oil generation in the Bakken shale. in Hansen, William B. (editor), Bakken Formation Short Course, p. 23. Montana Geological Society, Billings. Study Notes No. 1. Excerpted from U.S. Geological Survey Open-file Report 83-717.

Abstract. Thermal maturity maps of the Bakken shale can be made by mapping resistivity values from well logs.

939. Schnerk, G. C. and C. N. Madeen (1990). The Austin Chalk: Simulation of horizontal wells in a heterogeneous formation. Proceedings 1990 SPE Annual Technical Conference & Exhibition, Reservoir Engineering volume, September 23-26, 1990, New Orleans, pp. 25-36. Society of Petroleum Engineers, Richardson, TX. SPE paper 20716.

The Austin Chalk is a low-permeability, low-porosity formation that produces only from natural fractures. Oryx Energy Co. began experimenting with drilling short-radius lateral drainholes to recomplete wells in Pearsall field Austin Chalk reservoir in 1985. The drainholes proved to be up to 25 times more productive than vertical wells. Later, two long-radius horizontal wells were drilled that have now produced 291 MBO and are expected to produce another 790 MBO. Cores taken in Pearsall field show that the effective porosity is concentrated in "fracture swarms," trending NE-SW, and composed of vertical fractures. Pressure transient tests show that effective fracture permeability is usually in the range 0.1 to 10 md. Production histories indicate that knowledgeable operators can expect a five-fold increase in production from horizontal wells over that of vertical wells. Horizontal wells penetrate more of the "fracture swarms" than do vertical wells.

940. Schroeter, D. R. and H. W. Chan (1989). Successful application of drilling technology extends directional capability. SPE Drilling Engineering 4 (3), 230-236. SPE paper 17660. Abstract in The Engineering Index Monthly (January, 1990), 28 (1), 662.

High-angle, long-reach drilling is being done in the Malaysian South China Sea, Malay basin, using MWD, PDC bits, top-drive drilling systems, and low-toxicity oil-based drilling mud. The high-angle boreholes are inclined 60-73 degrees, and have horizontal displacements of 2,000 to 3,000 ft. Details of the project are given in the article. Estimated net saving from using the improved technology was: MWD $1.5 million; PDC bits $2.5 million; top-drive $2.3 million; oil-based-mud $1.0 million.

941. Schuh, Frank J. (1989). Horizontal drilling: Where we've been and where we're going. Drilling 50 (3), 10-14.

Schuh reviews the brief history of horizontal drilling and gives "20 Tips for Failure." Schuh's point is that careful planning and a clear purpose are prerequisites for success when drilling horizontal wells instead of vertical wells.

942. Schuh, Frank J. (1991). Horizontal well planning. A.A.P.G. Bulletin 75 (3), 668.

Abstract. A multidisciplinary team approach to planning horizontal wells to avoid the cost of trial and error resolution of technical problems.

943. Seeberger, M. H., R. W. Matlock, P. M. Hanson (1989). Oil muds in large-diameter, highly-deviated wells: Solving the cuttings removal problem. Proceedings 1989 SPE/IADC Drilling Conference, February 28-March 3, 1989, New Orleans, pp. 161-170. Society of Petroleum Engineers, Richardson, TX. SPE/IADC paper 18635.

Conventional oil-based muds are deficient in their ability to transport cuttings, in high-angle wells,because of their rheological profile. Experiments and field tests were used to develop and test a new additive that improved the cuttings transport ability of the oil-based muds. The experiments, field tests and additive are described in this paper.

944. Seheult, M., L. Grebe, J. E. Traweek, and M. Dudley (1990). Biopolymer fluids eliminate horizontal well problems. World Oil 210 (1), 49-53.

Kelco Oil Field Group has developed a high-suspension, natural polymer drilling fluid system that improves cutting transport and hole cleaning. Their fluid and its uses are described in the article.

945. Seines, K., I. Aavatsmark, S. C. Lien, and P. Rushworth (1990). Important reservoir considerations for planning horizontal wells; a field example. Paper presented at the 1990 SPE Latin American Petroleum Engineering Conference, October 14-19, 1990, Rio de Janeiro. Society of Petroleum Engineers, Richardson, TX. SPE paper 21124.

A method is presented for implementing friction loss in horizontal wellbores in a simulation study. The simulation was done to determine the effect, for planning purposes, of horizontal wellbores in Troll field, Norwegian North Sea. Troll field has a thin (22-26 meters) oil zone. Horizontal wells were planned to produce this oil with minimum gas or water coning.

946. Shaef, D. G. (1983). Review - Esso Resources Canada Limited, Norman Wells Expansion Project Drilling Program. Proceedings 1983 IADC/SPE Drilling Conference, February 20-23, 1983, New Orleans, pp. 115-122. Society of Petroleum Engineers, Richardson, TX. IADC/SPE paper 11371.

Esso Resources Canada Ltd. has embarked on a development drilling program in Norman Wells field, Northwest Territories, Canada. Norman Wells, located partially under the Mackenzie River, requires drilling of some wells from natural and artificial islands. High-angle, extended-reach wells (60-73 degrees; 1,100 meters from surface location) are required to reach under the river from the drill sites. Details of the development project and of the extended-reach wells are given in the article.

947. Shah, P. C., D. K. Gupta, Lehmbar Singh,B. G. Deruyek (1990). A field application of the methodology for interpretation of horizontal well transient tests. Proceedings 1990 SPE Annual Technical Conference & Exhibition, Formation Evaluation and Reservoir Geology volume, September 23-26, 1990, New Orleans, pp. 865-876. Society of Petroleum Engineers, Richardson, TX. SPE paper 20311.

Simultaneous measurement of downhole pressure and flow rate of wells in Bombay High field, India, enabled identification of the early- and late-time pseudo-radial flow regimes and, to a lesser extent, of the intermediate linear rate. Analysis of these data yielded estimates of the anisotropic formation permeabilities and the mechanical skin factor. The second step of the analysis is to compare the measured pressure or flow rate response with that predicted by a model of the horizontal well. An algorithm is presented for history-matching of the measured downhole flow rate using wellbore pressure as an input. The methodology presented here provides estimates of the horizontal and vertical permeabilities and the true skin factor of the formation.

948. Shah, S. N. and D. L. Lord (1989). Hydraulic fracturing slurry transport in horizontal pipes. Proceedings 1989 SPE Eastern Regional Conference, October 24-27, Morgantown, West Virginia, pp. 7-20. Society of Petroleum Engineers, Richardson, TX. SPE paper 18994. Also published in SPE Drilling Engineering (1990) 5 (3), 225-232 and Proceedings 1989 SPE Joint Rocky Mountain Regional Low-Permeability Reservoirs Symposium & Exhibition, Denver, March 6-8, pp. 615-628. Abstract in The Engineering Index Monthly (January, 1990), 28 (1), 684.

Three transparent horizontal pipes were used to study the transport of various hydraulic fracture slurries. Results of the experiments indicate that critical deposition and resuspension velocities were lower for non-Newtonian fracturing gels than for water; critical deposition and resuspension velocities are dependent on pipe size; if proppant density is greater than sand, higher velocities are required; and, for less viscous fluids, the velocities increase slightly with increased proppant concentration.

949. Shaw, Michael (1987). Redrilling high angle gas wells in the Bass Strait: Case study: Snapper A-14B. Drilling 48 (6), 12, 14-15.

Esso/BHP joint venture re-drilled watered out oil and gas wells in Bass Strait, Australia, to reach new geological targets. The redrilling was done with high-angle wells (62-72 degrees). The redrilled wells were turned 45 degrees to the left from their original azimuth. Drilling fluid was flocculated gel/seawater/native solids mud to 500 ft above the reservoir, where mud weight was raised to 10.8 lb/gal. In the pay zone, trip frequency was every four hours of bottom hole time. Caving was troublesome in the pay zone, which was partially thick coal beds, because the borehole was parallel to bedding planes. The redrilling took 27 days to drill 9,750 ft in the Snapper A-14B well. Logging was done with drillpipe-conveyed tools. The well was completed with perforations through cemented liner. Perforating was done overbalanced because the perforating gun had to be pumped down, due to the high inclination of the borehole.

950. Shaw, M. N. and W. G. Sutherland (1988). Innovative approaches to drilling high-angle wells in Bass Strait. Paper presented at the IADC/SPE 1988 Drilling Conference, February 28-March 2, 1988, Dallas, Society of Petroleum Engineers, Richardson, TX. IADC/SPE paper 17233.

High-angle (72 degrees) wells were drilled from the Snapper platform in Bass Strait, offshore Australia, by Esso/BHP. The wells were drilled to reach new targets after the original reservoirs watered out. The high-angle drilling proved to be 7% less expensive than the original directional drilling, even though the wells had longer reach and higher angles. Details of the drilling and completion of the redevelopment project are given in the paper.

951. Sheikholeslami, B. A., B. W. Schlottman, F. A. Seidel, and D. M. Button (1989). Drilling and production aspects of horizontal wells in the Austin Chalk. Proceedings 1989 SPE Annual Technical Conference & Exhibition, Reservoir Engineering volume, March 8-11, 1989, San Antonio, pp. 575-590. Also in Journal of Petroleum Technology (1991) 43 (7), 773-779. Society of Petroleum Engineers, Richardson, TX. SPE paper 19825. Abstract in The Engineering Index Monthly (January, 1990), 28 (1), 663.

Amoco has used two different drilling methods to drill ten horizontal wells in Giddings field Austin Chalk reservoir. The first three wells in the program were short-radius wells, with a total of five horizontal sections; the remaining seven wells were medium-radius wells. The horizontal sections of the wells ranged from 475 ft to 2,200 ft. Productivity was found to increase over that of vertical wells by factors of 2.5 (500 ft horizontal section) to 7 (2,000 ft horizontal section). The ten wells drilled in this program were all successful. One lesson learned in this program was the necessity for using a "specialist team" of engineers and geologists to design and implement the drilling plans.

952. Sheikholeslami, B. A., B. W. Schlottman, F. A. Siedel, and D. M. Button (1990). Drilling and production aspects of horizontal wells in the Austin Chalk. A.A.P.G. Bulletin 74 (5), 762.

Abstract. Amoco, using both medium- and long-radius horizontal wells, has improved production from the Austin Chalk [Giddings field] in Texas by 2.5 to 7 times that of vertical wells.

953. Shepard, Warren (1990). LANDSAT tectonic analysis, some applications for Bakken exploration in the Williston basin. A.A.P.G. Bulletin 74 (8), 1345. Abstract also published in Hansen, William B. editor (1990), Bakken Formation Short Course, Montana Geological Society Study Notes No. 1, p. 23.

Abstract. LANDSAT data analysis can show fracture trends, tectonic relationships, and other anomalies that can help in locating Bakken formation fractured reservoirs. A plate tectonic model is suggested for thermal maturation and fracturing in the Williston basin.

954. Shepard, Warren (1990). Landsat analysis techniques, some applications for Bakken exploration in the Williston basin. in Hansen, William B. (editor), Bakken Formation Short Course, p. 23. Montana Geological Society, Billings. Study Notes No. 1.

Abstract. Landsat data can be used to map fracture zones in the Bakken formation. The Laramide orogeny caused fracturing, diagenesis, and hydrocarbon expulsion in the Bakken formation.

955. Sherrard, Dave W., Bradley W. Brice and David G. MacDonald (1987). Application of horizontal wells at Prudhoe Bay. Journal of Petroleum Technology 39 (11), 1417-1425. Paper SPE 15376 presented at the 1986 SPE Annual Technical Conference and Exhibition, New Orleans, October, 5-8, 1986. Abstract in The Engineering Index Annual (1988), 87, Part IV, 6015.

Three horizontal wells drilled in Prudhoe Bay field have improved production by two to four times that of vertical wells. The wells cost 50% more than a vertical well, and it is hoped that this will be offset by higher ultimate recoveries. Horizontal wells should delay water breakthrough. Only long term observation will show if the experiment was a success, but initial results are encouraging. Details of the drilling and completion method and design of the test are given in the paper.

956. Shirley, Kathy (1990). Going out is going good: New targets emerge for horizontal drilling. A.A.P.G. Explorer 11 (10), 1, 6.

This article discusses horizontal drilling targets and activity in Oklahoma.

957. Shirley, Kathy (1990). Complex targets force geologist to face the fracs: Horizontal drilling sparks interest in fractured reservoirs. A.A.P.G. Explorer 11 (11), 1, 6-7.

This article is a brief discussion of fractured reservoirs and the role of horizontal drilling.

958. Shirley, Kathy (1990). Austin Chalk play still raises eyebrows: Big flows spice 'old' news. A.A.P.G. Explorer 11 (11), 22-23.

The Austin Chalk play in Texas accounts for 10% of the rigs operating in the United States. The play was initiated by Oryx Energy Co. in Pearsall field. The play has expanded to Giddings field and other areas. Another major player in the Austin Chalk, Meridian Oil Co., is concentrating on Gonzales County, Texas. Winn Exploration Co. will not discuss its spectacular wells in the Austin.

959. Shryne, R. and J. E. Smith (1986). New tool joint improves high-angle drilling. Petroleum Engineer International 58 (9), 44-46. Abstract in The Engineering Index Annual (1987), 86, Part IV, 6498.

East Wilmington field in Long Beach, California, harbor has been drilled for twenty years by deviated wells drilled from artificial islands. More than 50% of these wells are at inclinations of greater than 50 degrees. Replacement of tool joints, with a new joint of larger inside diameter and an additional shoulder, has improved performance. The larger inside diameter improved hydraulics and facilitated the passage of wireline tools.

960. Simison, H. (1988). Discussion of stability of highly inclined boreholes. SPE Drilling Engineering 3 (4), 418. SPE paper 18596.

Critical discussion of paper by Aadnoy and Chenevert (December, 1987), "Stability of highly inclined boreholes," SPE Drilling Engineer, 364-374. See also Aadnoy and Chenevert (December, 1988), "Authors' reply to discussion of stability of highly inclined boreholes," SPE Drilling Engineering 3 (4), 418.

961. Simorangkir, P., O. S. Adisapoetra, and J. Handoyo (1986). Rotary drilling of a horizontal well, West Java Sea. Paper presented at the 15th Annual Indonesian Petroleum Association Convention, October 20-22, 1986, Jakarta, Indonesia.

Not examined.

962. Siu, Alan (1990). Numerical modeling of a thermal horizontal well. Preprints of CIM/SPE International Technical Meeting, June 10-13, 1990, Calgary, v. III, Paper 122. The Petroleum Society of CIM, Montreal. CIM/SPE paper no. 90-122.

Numerical modeling of a coupled reservoir and well model in a steam-assisted gravity drainage system for a bitumen/heavy-oil reservoir. The model demonstrates that heat transfer between annulus and tubing affects the shape of the sweep zone; uneven fluid exchange between the horizontal wellbore and the reservoir can be caused by the frictional pressure drop.

963. Skopec, R. A., M. M. Mann, D. Jeffers, and S. P. Grier (1990). Horizontal core acquisition and orientation for formation evaluation. Proceedings 65th Annual SPE Technical Conference and Exhibition, September 23-26, 1990, New Orleans, pp. 153-166. Society of Petroleum Engineers. SPE paper 20418.

Newly developed methods and hardware have made coring of medium-radius horizontal wells reliable. Oryx Energy Company has developed techniques which have resulted in nearly 100% recovery of core in horizontal wells. The method includes the use of a slip scribe and Electronic Multishot Instrument for orientation of the core. Improved efficiency has been achieved by collecting shorter core sections and improving stabilization of inner and outer core barrels. The core assembly is aligned by use of a patented Laser Alignment System. A detailed discussion and description of the methods used by Oryx is given in the paper.

964. Sloan, Jeff M. (1990). Fire still burns in chalk play. Gulf Coast Oil World 8 (5), 18-19.

Horizontal drilling in Pearsall field, Texas, and other horizontal plays is briefly reviewed.

965. Sloan, Jeff M. (1990). Horizontal drilling heats up in Rockies. Western Oil World 47 (8), 10-11.

This article reviews horizontal drilling projects in New Mexico, Colorado, Wyoming, and Montana. Fields, operators, and formations are indexed.

966. Smagla, Tom (March 9, 1981). The Cretaceous Niobrara play. Oil & Gas Journal 79 (10), 204, 209-210, 212, 217-218.

The Cretaceous Niobrara formation produces natural gas from shallow fields in the Denver basin. The Niobrara also has potential in the Kennedy and Salina basins of Nebraska, and in parts of Montana, North Dakota, and South Dakota. The Niobrara is a chalk with porosity that ranges from 30% to 45%. Production is from depths of 1,000 to 3,500 ft. Economic production requires hydraulic fracture stimulation. The article reviews the history and general geology of the play. Editor's Note: The Niobrara is now a target of horizontal drilling.

967. Smith, G. W. and R. M. Butler (1979). Studies on the use of tunnels and horizontal wells for the recovery of heavy crudes. Paper presented at the First International Conference on the Future of Heavy Crude and Tar sands, 1979, Edmonton, Alberta, Canada. UNITAR/UBDP. Report No. 62.

Not examined.

968. Smith, G. W., R. M. Butler, R. F. Meyer, and C. T. Steele (1981). Studies on the use of tunnels and horizontal wells for the recovery of heavy crudes. Proceedings of the Conference on the Future of Heavy Crude and Tar Sands, pp. 455-459.

Not examined.

969. Sobey, I. and A. Gavigert (August, 1978). A model for the transport of cuttings in highly deviated wells. Drilling and Pumping Journal, pp. 42-49.

Not examined.

970. Soeder, D. J. and P. Randolph (1984). Porosity, permeability, and pore structure of the tight Mesa Verde sandstone, Piceance basin, CO. Paper presented at the 59th Annual SPE Technical Conference & Exhibition, September 16-19, 1984, Houston. Society of Petroleum Engineers, Richardson, TX. SPE paper 13134.

Report of the results of special core analysis of cores from the Department of Energy Multiwell Experiment Site (MWX), Rifle, Colorado. The MWX site has been used to experiment with drilling and completion techniques in a low-permeability sandstone reservoir (Mesaverde sandstone). Detail descriptions are given of the Mesaverde reservoir rocks.

971. Soliman, M. Y., J. L. Hunt, and W. El Raaba (1988). On fracturing horizontal wells. Paper presented at the 1988 SPE Regional Meeting, November 1-4, 1988, Charleston, West Virginia. Society of Petroleum Engineers, Richardson, TX. SPE paper 18542.

The productivity of horizontal wells in naturally fractured reservoirs, or in artificially fractured reservoirs, depends on the number of fractures intersected or created in the horizontal wellbores. Multiple hydraulic fractures may be created in horizontal wellbores that are drilled parallel to the minimum horizontal stress. A model is presented to determine the optimum number of fractures in a horizontal wellbore.

972. Soliman, Mohamed, Bob Rose, Wadood El Rabaa, and James L. Hunt (1989). Planning hydraulically fractured horizontal completions. World Oil 209 (3), 54-56, 58. Abstract in The Engineering Index Monthly (June, 1990), 28 (6), 679.

Hydraulic fracturing is needed in horizontal wells if the vertical permeability is low or non-existent; in naturally fractured reservoirs where the fractures are not continuous or uniform; cases where vertical wells needed extensive stimulation; or if the reservoir has a low stress difference between it and surrounding zones. Horizontal wells should be placed to maximize production or to minimize water and/or gas coning. If the well is to be hydraulically fractured, the stress field in the reservoir needs to be considered, because the fractures will propagate perpendicular to the direction of the least principal stress.

973. Soliman, M. Y., James L. Hunt and A. N. Rabaa (1990). Fracturing aspects of horizontal wells. Journal of Petroleum Technology 42 (8), 966-973. SPE paper 18542 presented at the 1988 SPE Eastern Regional Meeting, Charleston, West Virginia, November 1-4. Abstract in The Engineering Index Monthly (February, 19990), 28 (2), 637. Paper also in Proceedings 1989 SPE Joint Rocky Mountain/Low Permeability Reservoir Conference, Denver, pp. 85-98.

Hydraulic fracturing and hole stability of horizontal wells is dependent on the relative magnitude of the three principle stresses, and the orientation of the hole with respect to the minimum horizontal stress. Multiple fractures may be created if the horizontal well is drilled parallel to the minimum horizontal stress.

974. Souza, R. F. (April, 1987). Comportamento de Pressao em Pocos Horizontais Nao Surgentes. M.E. Thesis. U. de Federal de Ouro Preto, Minas Gerais, Brazil. In Portuguese.

Not examined.

975. Southwest Oil World (1988-1989). Regional News: Downhole motor sets single-trip world record. Southwest Oil World 36 (11), 9.

The single-trip world record for a steerable downhole motor was set in a 9,660-ft run by Eastman Christensen Navigation Drilling System on an Amoco Norway Oil Co. well in the North Sea. The well had a maximum hole angle of 76.2 degrees and a horizontal displacement of 13,425 ft.

976. Southwest Oil World (1990). Tips for traipsing horizontal holes. Southwest Oil World 37 (7), 20.

An estimated 10 horizontal wells may be drilled in West Texas this year. The target will be Ellenburger formation fractured reservoirs. Halliburton Services recommends, first, drilling a vertical well to evaluate the reservoir; then following up with appropriate horizontal wells.

977. Southwest Oil World (1990). Grace goes after horizontal. Southwest Oil World 37 (7), 26.

Grace Drilling Co. added six rigs to its Odessa based southwestern division in anticipation of an increased demand for horizontal drilling.

978. Southwest Oil World (1991). Horizontal drilling gets hotter in Texas. Southwest Oil World 37 (12), 18.

In November, 1990, the Texas Railroad Commission received applications for 115 drilling permits for horizontal wells. Most of the permits (58) were for Pearsall (Austin Chalk) field; most of the remaining permits were for West Texas counties.

979. Southwest Oil World (1991). Horizontal well permits increase in West Texas as operators improve knowledge and enthusiasm. Southwest Oil World 38 (1), 19.

West Texas horizontal drilling permits now equal in number those for the Austin Chalk play. Some horizontal drilling is being done in East Texas. List of numbers of permits by county in West Texas and discussion of trends in Texas.

980. Souza, R. F. and A. J. Rosa (1987). Slug test in horizontal wells. SPE paper 17136.

This paper is a report of the results of a physical experiment and mathematical model of slug tests, in horizontal wells, under the conditions of an anisotropic reservoir, with impermeable upper and lower boundaries.

981. Sparlin, Derry D., Jr. and Raymond W. Hagen (1988). Controlling sand in a horizontal completion. World Oil 207 (5), 54-55, 57-58, 60.

Control of sand in horizontal wellbores, in poorly consolidated formations, may best be accomplished by using pre-packed with gravel screens, instead of slotted liners or screens.

982. Spencer, A. M. et al. (1987). Geology of the Norwegian Oil and Gas Field. Graham and Trotman, London. ISBN 0 86010 9089.

This volume has separate articles on the geology and hydrocarbon reservoirs of each of the Norwegian fields. Of special interest are articles 28-30 on Statfjord field and its satellites, and article 34 on Troll field. Horizontal drilling has been done in both fields.

983. Spencer, Charles W. (1987). Hydrocarbon generation as a mechanism for overpressuring in Rocky Mountain region. A.A.P.G. Bulletin 71 (4), 368-388.

Most overpressuring in the Rocky Mountain region is caused by recent oil or gas generation in low-permeability source rocks. Overpressureing occurs in rocks of Devonian to Tertiary age, most commonly in Cretaceous and Tertiary rocks. Overpressuring in rocks older than Cretaceous only occurs in units with very highly organic source beds, such as the Bakken formation. Overpressuring in the Greater Green River basin, Piceance basin, Williston basin, Uinta basin, Big Horn basin, Powder River basin, and Hanna basin is discussed.

984. Spencer, Charles W. (1989). Review of characteristics of low-permeability gas reservoirs in western United States. A.A.P.G. Bulletin 73 (5), 613-629.

Low-permeability gas reservoirs (less than 0.1 md), exclusive of eastern gas shales, contain an estimated 574 TCFG. Most of this gas is in the western United States. Western tight gas sands occur in two types or reservoirs: blanket and lenticular. These reservoirs are generally difficult to analyze and stimulate using standard logging and stimulation practices. The geology and engineering characteristics of these tight gas reservoirs are discussed in detail including the use of high-angle and horizontal drilling to exploit them.

985. Spencer, Charles W., John C. Lorenz, and Charles A. Brown (1991). Application of horizontal drilling to tight gas reservoirs. A.A.P.G. Bulletin 75 (3), 675.

Abstract. Horizontal drilling has not been used for tight gas reservoirs in the United States due to low gas prices. U.S. tight gas reservoirs exist in the Piceance, Uinta, Appalachian, and Denver basins. The U.S. Department of Energy has funded a horizontal drilling test in the tight Mesaverde Group of the Piceance basin, Colorado, and one well in the Devonian shales of the Appalachian basin in West Virginia. These projects have provided information about the openness and abundance of deeply buried fractures in tight reservoirs.

986. Sperr, Jay Thomas (1990). Exploration models for Bakken shale reservoirs, Williston basin, North Dakota. A.A.P.G. Bulletin 74 (8), 1345. Complete paper printed in Hansen, William B. (editor) (1991), 1991 Guidebook to Geology and Horizontal Drilling of the Bakken Formation, pp. 143-149, Montana Geological society, Billings, Montana.

The Bakken formation reservoirs have five main trapping types: (1) traps along the depositional edge; (2) tensional fracturing where the Bakken is draped over basement structures; (3) fracturing along regional lineaments; (4) fracturing caused by overpressuring; and (5) "conventional" sandstone reservoirs in the middle member of the Bakken.

987. The Spraberry Area Engineering Committee (1954). Statistical Report 1947-1954 Spraberry Fields. Spraberry Area Engineering Committee, Midland.

This report is a list of monthly production statistics for leases and wells in the indexed fields.

988. Spreux, Alain, Christian Gerges and Jacques Lessi (June 13, 1988). Horizontal well operations - 6. Most problems in horizontal completions are resolved. Oil & Gas Journal 86 (24), 48-52. Abstract in The Engineering Index Annual (1988), 87, Part III, 5995-5996.

Most horizontal well completion problems have been solved. This paper gives a summary of the status of logging, completion methods, cementing, perforating, stimulation, and artificial lift of horizontal wells.

989. Spreux, A. M., F. Foxonet, and L. Bruckert (1988). Improved oil production by horizontal well in fine unconsolidated sand. Seminar on Recovery from Thin Oil Zones, April 21-22, 1988, Stavanger. Norwegian Petroleum Directorate.

Not examined.

990. Spreux, A. M., A. Louis and M. Rocca (1988). Logging horizontal wells: Field practice for various techniques. Journal of Petroleum Technology 40 (10), 1352-1354. SPE paper 16565 presented at the 1987 SPE Offshore European Conference, Aberdeen, September 8-11.

Five techniques for logging horizontal and highly deviated wells are: wireline method; MWD; Simphor; pumpdown stinger method; and coiled-tubing method. Simphor is a trade name for a system of pushing logging tools downhole using drillpipe. Some other trade names for drillpipe-conveyed logging systems are: "Drillpipe-Conveyed Logging;" "Tool Pusher;" and "Tough Logging Conditions." This method of logging is accomplished by attaching the logging sensors to the end of the drillpipe with a cable connector and side-entry sub, through which the logging cable passes into the drill pipe. The zone of interest can be logged down or up. The pumpdown stinger method conveys the tool sensor and wire to the bottom of the hole by pumping them through the drillpipe. This method can only use small-diameter tools. When the tool is on bottom, it is retrieved by pumping fluid into its locomotive, attached to the upper end of the tool. Coiled-tubing-conveyed logging systems use standard logging tools attached to coiled-tubing to push the tools downhole and retrieve them. MWD is used to log the well during drilling.

991. Springer, S. J., P. Pentell, and S. Asgarpour (1990). Enhanced heavy oil production by horizontal drilling case study. Paper presented at the International Technical Meeting, The Petroleum Society of CIM and the Society of Petroleum Engineers, June 10-13, 1990, Calgary. The Petroleum Society of CIM, Montreal. CIM/SPE paper 90-126. Abstract in The Journal of Canadian Petroleum Technology (1990) 29 (2), 50.

Abstract. A black oil simulator was used to predict performance of a Gulf Canada horizontal well drilled in the Edam area.

992. Stagg, T. O., John W. Powell, R. Lee Dewees, and Mike P. Stephens (1986). Horizontal drilling project uses sized-salt fluid. Petroleum Engineer International 58 (7), 31-36. Abstract in The Engineering Index Annual (1986), 85, Part IV, 6526.

Two extended-reach horizontal wells drilled at Prudhoe Bay field, Alaska, used water-soluble sized-salt technology to minimize torque and drag. The system, "Clean Bridge", a patented blend of polymers and sized NaCl salt, was added to a 10 lb/gal NaCl brine. Diesel can be added, up to 30% by volume, to reduce hydrostatic pressure, adjust rheology,, and improve filtrate control. Optimum hole cleaning requires drillpipe rotation and reciprocation, while circulating bottoms up. At Prudhoe Bay, drilled solids should be limited to 40%. Aeration of the mud should be avoided to minimize problems with the MWD-transmitted signal. Various grades of sized-salt can be used for lost circulation pills. Cement can cause problems with the sized-salt polymer system.

993. Stagg, T. O. and R. H. Relley (1990). Horizontal well completions in Alaska. World Oil 210 (3), 37-44. Abstract in The Engineering Index Monthly (June, 1990), 28 (6), 669.

Since 1984, BP Exploration has drilled 20 horizontal wells in Alaska, including two horizontal gas injection wells. These wells have been completed with both fully cemented and slotted liners. Two of the wells have been drilled inverted, or up into the reservoir from the bottom. This method was used to better control gas coning. These methods and the circumstances of their use are discussed in this article.

994. Standen, Eric (April, 1991). Tips for analyzing fractures on electrical wellbore images. World Oil 212 (4), 99-100, 102, 112, 114, 116, 118.

Discussion of the interpretation of wellbore anomalies on electrical wellbore images (Schlumberger's Formation MicroScanner). These anomalies are classified as: (a) drilling-induced fractures; (b) drilling-enhanced fractures; and (c) natural fractures. Wellbore anomalies can be attributed to one of these causes with careful analysis.

995. Stang, Carl W. (November 6, 1989). Alternative electronic logging technique locates fractures in Austin Chalk horizontal well. Oil & Gas Journal 87 (45), 42-45. Abstract in The Engineering Index Monthly (February, 1990), 28 (2), 635.

Oryx Energy Co. used a combination of spectral gamma- ray log (Schlumberger Natural Gamma-ray Tool) and Formation MicroScanner tool (FMS) to detect fractures in Austin Chalk wells in Pearsall field. These tools were run in tandem on the end of the drill pipe. Logging was done on the trip out of the hole. Problems encountered in logging were: erratic dragging of the drill pipe on the high side of the hole causing "streaking" on the FMS log; and loss of pad contact with the side of the well bore, when the pads were on the high side, caused by gravitational forces.

996. Stanislav, J. F., K. Christensen, and A. Chakma (1990). Two-phase flow in highly deviated wells. Preprints of CIM/SPE International Technical Meeting, June 10-13, 1990, Calgary, Canada, v.II, Paper 70. Petroleum Society of CIM, Montreal. CIM/SPE paper 90-70. Abstract in The Journal of Canadian Petroleum Technology (1990) 29 (2), 40.

Report of experiment of two-phase (gas-liquid) systems in inclined pipes modeling high-angle wellbores.

997. Stark, Philip H. (March, 1991). Horizontal drilling - Overview of geologic aspects and opportunities. Houston Geological Society Bulletin 33 (7), 10-11. Abstract of paper presented at the March, 1991, HGS Luncheon Meeting.

Abstract. Subjects discussed include: horizontal drilling in the Austin Chalk, Bakken formation, Ellenberger formation, Niobrara formation, Spraberry Trend, coalbed methane, and Prudhoe Bay field. The examples are categorized to show the various situations in which horizontal drilling is useful: reduction of coning, fractured reservoirs, stratigraphic traps, and paleokarst.

998. Stark, Philip H. (1991). Integrated database and GIS: Examples from analysis of horizontal drilling opportunities. in Abstracts and Schedule Of the Houston Geotech Technical Conference, February 17-22, 1991, Houston, p. 39. Houston Geotech Association, Houston.

Abstract. Use of Geographical Information Systems (GIS) with integrated databases can enhance selection of horizontal drilling opportunities.

999. Stayton, R. J. and S. R. Peach (1990). Horizontal drilling enhances production of Austin Chalk well. Proceedings 1990 IADC/SPE Drilling Conference, February 27 - March 2, 1990, Houston, pp. 629-636. Society of Petroleum Engineers, Richardson, TX. IADC/SPE paper 19984.

Pinnacle Royalty and Operating Co. drilled a long-radius horizontal well in Pearsall field, Texas. Steerable positive-displacement motors were used in the build and horizontal sections of the hole. Casing was set at the top of the Austin Chalk formation. A 1,253-ft horizontal section was drilled under-balanced to minimize formation damage. Two fractured sections were identified in the horizontal borehole. The first fractured zone was 400 ft into the horizontal section and was detected by calcite crystals in the cuttings. At 7,865 ft MD oil flowed to the surface and surface pressure increased to 675 psi. The well was brought back under control and drilling continued. At 8,670 ft MD another fracture system was encountered and oil flow to the surface occurred. The well TD was 8,707. The horizontal section was drilled with a penetration rate of 12.7 ft/hr, and was completed in 101 hours of drilling time. The well tested 1,600 BOPD with a gas/oil ratio of 380:1. The operator feels that one medium-radius well would reduce costs because of the reduced well profile.

1000. Steenbock, A. (January 25, 1985). Horizontalbohrung gelungen-Preussag exprobte neue Methode in der Berkhopen 2000. Erdoel Informationdienst. In German.

Not examined.

1001. Steenbock, A. (1985). Neue Drainhole-Bohrtechnik [A new drainhole drilling tool]. Erdol Erdgas 101 (3), 80-84. In German with an English abstract.

Abstract. Techniques and tools for drilling horizontal drainholes are described.

1002. Steenbock, A. and D. Muller-Link (1987). Alternative drainhole drilling system - theoretical background and practical experience. Paper presented at the SPE 1987 European Offshore Conference: Offshore Europe 87, September 8-11, 1987, Aberdeen. Society of Petroleum Engineers, Richardson, TX. SPE paper 16564/1

Short-radius, lateral drainhole drilling is one method of increasing the formation flow area. Preussag AG used an Articulated Drill String (ADS) with a mud motor to recover attic oil in stripper wells in a German oil field. The equipment and drilling methods are described in this article.

1003. Sterne, W. P. (August 26, 1946). Lateral drilling of limestone reservoirs. The Oil Weekly 122 (13), 34, 37.

Stanolind Oil & Gas Company drilled multiple horizontal drainholes in Pine Island field, McCamey field, and Slaughter field to improve productivity from the fractured limestone reservoirs. At Pine Island field, Louisiana, 54 drainholes were drilled from six vertical wells in the Annona Chalk reservoir. One drainhole was drilled in the San Andres reservoir in McCamey field, Texas. Eleven drainholes were drilled from two vertical wells in the San Andres reservoir Slaughter field, Texas. From the information given in the article, the experiments do not seem to have been economically successful, although some increase in production was obtained. Some details of these projects are given in the article.

1004. Steven, Robert R. (1984). Troll development tests subsea technology. Offshore 44 (1), 76, 78, 80-83.

This article discusses the development plans for Troll field, North Sea, Norway, before the horizontal and extended-reach drilling was done.

1005. Stewart, D. J., J. Goodacre, and C. Cruickshank (1989). Bass Strait: Production decline mitigation. Proceedings 1989 Asia-Pacific Conference, September 13-15, 1989, Sydney, Australia, pp. 17-26. Society of Petroleum Engineers, Richardson, TX. SPE paper 19467.

Declining production in Gippsland basin, Australia, has been offset by application of conventional techniques and, later, applying low-cost, quick response techniques to well workovers. These wells include several high-angle wells. Procedures used in these workovers are described in the article.

1006. Stewart, J. S. (1948). Norman Wells oil field, Northwest Territories, Canada. in Howell, J. V. (editor), Structure of Typical American Oil Fields. Tulsa, A.A.P.G v. III, pp. 86-109.

Norman Wells field, Northwest Territories, Canada, is a stratigraphic trap in a Devonian reefal limestone. The trap is caused by the pinchout of the reef facies. The field is located on the Mackenzie River. The top of the reservoir is at depths of 1,050 to 1,410 ft. In October, 1944, production was 4,300 BOPD from 60 wells. The history and geology of the field are discussed in this article.

1007. Stewart, C. D. and D. R. Williamson (1988). Horizontal drilling aspects of the Helder field redevelopment. Proceedings of the 20th Annual Offshore Technology Conference, May 2-5, 1988, Houston, v. 3, pp. 415-426. OTC paper 5792. Also published in Proceedings International Meeting on Petroleum Engineering, November 1-4, 1988, Tianjin, China, (SPE paper 17886), pp. 933-944.

The first eight horizontal wells drilled in the North Sea were in Helder field, offshore Netherlands. Helder field produces oil from the Vlieland Oil Sand. The reservoir is in a faulted structure, and has a bottomwater drive. Permeability in the 25-40 meter thick reservoir is 6000 md. Horizontal wells were believed to be the solution to several of the problems that developed in Helder field as reserves declined. These problems were: water cut had increased to an average of 94%, resulting in a high gross fluid production; short-lived electric submersible pumps were used; reservoir pressure decline was too high; and carbon dioxide corrosion and erosion affected the tubing. The horizontal wells were designed to reduce these problems by placing a horizontal wellbore in the upper three meters of the reservoir sand. Drilling and completing the Helder field horizontal wells takes 14 days/well. Two trips were required for each well, one for the build section and one for the horizontal section. Unstabilized steerable drilling motors were most effective in drilling the horizontal sections. Good hole cleaning in the horizontal interval was achieved using turbulent flow when circulating, and using a top drive system for mechanical cleaning. Sand control was achieved in the horizontal holes by using prepack gravel pack screens. Costs were comparable to that of conventional directional wells.

1008. Stewart-Gordon, Tom (April 1, 1991). Horizontal drilling boom may be headed to Mexico. The Oil Daily No. 9716, pp. 1, 8.

Smith International Inc. has been awarded a contract for drilling a horizontal well in the Chicontepec Paleo-Canyon, Mexico, by Petroleos Mexicanos (PEMEX). The Chicontepec Paleo-Channel play consists of sand lenses distributed throughout a 120-ft thick section of channel fill. Vertical wells have low productivity because of the limited reservoir exposure in the wellbore. It is expected that horizontal wells may overcome this problem and be more productive, as well as requiring fewer wells.

1009. Stormont, D. A. (August 17, 1953). Increasing drainage of oil into well by drain-hole drilling. Oil & Gas Journal 52 (15), 105-108, 145.

In the year prior to publication of this article, over 50 multiple drainhole projects were drilled in California, the Mid-Continent, Texas, and Venezuela. The number of drainholes per well varied from two to eight. Some of the drainholes were drilled laterally 100 ft into the reservoir. In the early 1930's, lateral drainholes were drilled in approximately 120 wells in the Mid-Continent of the United States. These projects were largely unsuccessful because the reservoirs were near depletion. The recent drainhole drilling projects have been in the Spraberry Trend of Texas, Wilmington field, California, and La Paz field, Venezuela. Drilling and completion methods are discussed in the article. The two major contractors drilling lateral drainholes are Oilwell Drainhole Drilling Co., Long Beach, California, and Zublin & Co., Los Angeles, California.

1010. Stramp, R. L. (1980). The use of horizontal drilling drainholes in the Empire Abo Unit. Paper presented at the 1980 SPE Annual Technical Conference and Exhibition, September 21-24, Dallas. Society of Petroleum Engineers, Richardson, TX. SPE paper 9221.

Two short-radius horizontal drainholes were drilled in the Empire Abo unit, New Mexico, to evaluate horizontal drilling technology, and to determine the effect of drainholes on the gas coning problem. The Empire Abo unit produces from a Permian dolomitized reef. The productive reef is 12.5 miles long and 1.5 miles wide, and produces from a depth of 6200 ft. Production is by gravity drainage with later gas injection. The field was discovered in 1957; by the early 1970's, gas coning had become a problem in certain wells. Wells with gas coning were shut-in and their allowables transferred to other wells until there were no wells remaining with excess gas-free production. In 1979, horizontal drainholes were drilled to test their efficiency in producing oil with no gas coning. The reduced pressure-gradient achieved by using a 200 ft horizontal drainhole, instead of a conventional vertical completion, it was believed, would reduce gas coning. The first drainhole, K-142, was drilled S20E from the surface location so as to intersect the field's major fracture trend. The build section was drilled with a 28 ft radius of curvature. The first drainhole was drilled with a horizontal section of 106 ft, where it was stopped because of increasing torque due to severe corkscrewing. The well was swabbed for several days and then cleaned out using coiled-tubing nitrogen injection. The flowing pressure was insufficient to produce the well, so a hydraulic jet pump was installed. The production tests were 200 BOPD with a GOR of 670 CFG/barrel of oil, which is similar to that of a conventional completion. The second drainhole, J-213, was drilled with a 127 ft horizontal section. The horizontal section had problems with hole climbing. Editor's Note: This well had not been completed at the time of the publication of the article. The effect of horizontal drainholes on gas coning was not known at the time of publication. The feasibility of short-radius drainhole drilling was demonstrated by this project.

1011. Striegler, J. (May 24, 1982). ARCO finishes fourth horizontal drainhole. Oil & Gas Journal 80 (21), 55-61. Abstract in The Engineering Index Annual (1983) 85 Part IV, 4942.

ARCO Oil & Gas Co. drilled four horizontal drainholes in the Empire Abo Unit, New Mexico, to reduce gas coning. The wells were drilled into the lower part of the oil column. Details of the project are given in the article.

1012. Striegler, J. H. (1982). Paper B. Horizontal drainhole drilling update. Paper presented at the 1982 Annual API Production Department Meeting, April 4-7, 1982, San Antonio.

Not examined.

1013. Stright, Dan (1990). Computer modeling of Bakken fracture systems. in Hansen, William B. (editor), Bakken Formation Short Course, p. 24. Montana Geological Society, Billings. Study Notes No. 1.

Abstract. 3D hydraulic fracture simulation models and three-phase dual porosity reservoir simulation models can be used for predicting the production performance of wells in the Bakken formation.

1014. Strubhar, M. K., J. L. Fitch and E. E. Glenn (1975). Multiple vertical fractures from an inclined wellbore. Journal of Petroleum Technology 27 (5), 641-647. SPE paper 5115 presented at the 49th SPE-AIME Annual Fall Meeting, October 6-9, 1974, Houston.

Mobil Research & Development Corp., in field tests, was able to determine that it is possible to carry out a multiple hydraulic fracturing (Multifrac) of a formation from an inclined borehole. The test was conducted in Caddo-Pine Island field, Natchitoches Parish, Louisiana, in the Annona

Chalk reservoir. The target reservoir, Annona Chalk, is a low-permeability reservoir, 150 ft thick, and is at a depth of 1,400 ft. The fracturing was done from a wellbore inclined 52 degrees from the vertical, in the pay zone. The well was fractured so that four vertical fractures were created which intersected natural fractures in the field. The productivity index of the well was 2-3 times that of a conventional well in the field. Details of the project are given in the paper.

1015. Sugianto, S. and R. M. Butler (1990). The production of conventional heavy oil reservoirs with bottom water using steam-assisted gravity drainage. Journal of Canadian Petroleum Technology 29 (2), 78-86. CIM paper 89-40-33. Paper presented at the 40 the Annual Technical Meeting of The Petroleum Society of CIM; abstract printed in Journal of Canadian Petroleum Technology (1989(, 28 (2), 35.

Producing heavy oil economically from a bottomwater reservoir may be possible using parallel horizontal production wells placed near the bottom of the oil zone, and injecting steam at the top of the reservoir from horizontal injection wells. This paper discusses the results of a simulation of such a system, using a Lloydminster-type of reservoir. The Lloydminster model has oil that is mobile at reservoir conditions, unlike the Cold Lake model which requires heating the oil to achieve mobility. The model indicates that from 48% to 87% of the original oil-in-place may be recovered using this technique, with the thickness of the bottomwater zone and well configuration having a significant effect on recovery.

1016. Sulaiman, S. and T. A. Bretherton (1989). Thin oil development in the Bream field. Proceedings 1989 Asia-Pacific Conference, September 13-15, 1989, Sydney, Australia, pp. 245-254. Society of Petroleum Engineers, Richardson, TX. SPE paper 19493. Abstract in The Engineering Index Monthly (October, 1990), 28 (10), 610.

Bream field, Gippsland basin, Australia, produces from a pancake oil zone 13.5 meters thick, overlain by a thick gas cap, and underlain by a basin-wide active aquifer. Horizontal wells are being drilled to evaluate their effect on gas coning in the reservoir. The field is initially being developed as an oil field with the gas reinjected.

1017. Sumpter, Randy (March 23, 1981). Chalk play expands at fast clip in Texas. Oil & Gas Journal 79 (12), 51-55.

The Austin Chalk play in Giddings field is expanding, and now covers parts of five counties in Texas. The Austin Chalk play extends from the Rio Grande River to Burleson County, a length of 150 miles, and includes the Pearsall field. Predictions are that the Austin could account for up to 6,000 wells with average reserves of 75,000 barrels of oil each. Editor's Note: This article was written before the horizontal drilling "boom" in the Austin Chalk and underestimates the activity in the Austin. The article does give a good review of activity and expectations as of 1981.

1018. Sung, W. and T. Eretkin (1987). Performance comparison of vertical and horizontal hydraulic fractures and horizontal boreholes in low-permeability reservoirs: A numerical study. Proceedings SPE/DOE Low Permeability Reservoirs Symposium, May 18-19, 1987, Denver, pp. 185- 193. Society of Petroleum Engineers, Richardson, TX. SPE paper 16407.

A numerical simulator was used to compare the performance of hydraulic fractures in horizontal boreholes in low-permeability gas reservoirs. The simulation showed that, for equal surface area, the horizontal borehole is more productive than hydraulic fractures, and that reservoir thickness is a critical factor in determining the feasibility of a horizontal borehole. Details of the simulator are given in the article.

1019. Suprunowicz, R. and R. M. Butler (1989). Vertical confined water drive to horizontal well. Part I: Water and oil of equal densities. Paper presented at the 3rd Technical Meeting of the South Saskatchewan Section, the Petroleum Society of CIM, September 25-27, 1989, Regina.

Not examined.

1020. Suprunowicz, Ryszard amd Roger M. Butler (1990). The choice of pattern size and shape for regular arrays of horizontal wells. Preprints of CIM/SPE International Technical Meeting, June 10-13, 1990, Calgary, v. III, Paper 129. Petroleum Society of CIM, Montreal. CIM/SPE paper 90-129. Abstract in The Journal of Canadian Petroleum Technology (1990), 29 (2), 50.

Horizontal wells tend to have oval-shaped drainage patterns, so that suitable well arrays should be placed in rectangles of elongated hexagons. Quantitative methods for determining the optimum array have been developed and are discussed in the article.

1021. Sutcliffe, Brod and David Sim (March 25, 1991). Drilling optimized by monitoring BHA dynamics with MWD. Oil & Gas Journal 89 (12), 41-45.

Drilling dynamics measurements can now be made in realtime using advanced MWD equipment. Teleco Oilfield Services Inc. now offers a package, including drilling dynamics with gamma-ray tool, that measures torque and weight-on-bit (WOB). Description of the tool, its operation, and interpretation are given in this article.

1022. Suzuki, K. and T. Nanba (1990). Horizontal well test analysis system. Proceedings 1990 SPE Annual Technical Conference & Exhibition, September 23-26, 1990, New Orleans, pp. 893-904. Society of Petroleum Engineers, Richardson, TX. SPE paper 20613.

A horizontal well test analysis is proposed that addresses two rarely addressed aspects: (1) variable production rates; and (2) early-time behavior. The analysis uses a "robust nonlinear regression algorithm."

1023. Tannenbaum, Elo, Brod Sutcliffe and Andrew Franks (1988). A composite lithology log while drilling. A.A.P.G. Bulletin 72 (2), 253.

Abstract. MWD drilling mechanical data (ate of penetration, torque), MWD geophysical data (gamma-ray, resistivity, neutron-density), and drill-cuttings in combination can be used to produce a composite lithology log of the borehole.

1024. Tarr, B. A., A. F. Kuckes, and M. Ac (1990). Use of new ranging tool to position a vertical well adjacent to a horizontal well. Proceeding 1990 SPE Annual Technical Conference & Exhibition, Drilling volume, September 23-26, 1990, New Orleans, pp. 421-430. Society of Petroleum Engineers, Richardson, TX. SPE paper 20446.

As part of the pilot project to test horizontal wells in the heavy-oil reservoirs at Cold Lake, Canada, three vertical observations wells were drilled. The locations of the vertical wellbores with respect to the horizontal steam injection well was done using a long range, controlled source, electromagnetic ranging method developed by Vector Magnetics. The new method was able to locate the wellbores within the 5 meters (+/- 3 meters) separation required for the experiment. MWD directional surveys were not as accurate and produced an error greater than 10 meters.

1025. The Technical Review (1987). Pioneering horizontal wells in Prudhoe Bay. The Technical Review 36 (1), 26-31.

Not examined.

1026. Terwilliger, P. L., L. E. Wilsey, H. N. Hall, P. M. Bridges, and R. A. Morse (1951). An experimental and theoretical investigation of gravity drainage performance. Petroleum Transactions of AIME 192, pp. 285-296.

This paper reports on experimental and theoretical studies of the production rate of oil in a gravity drainage system. At low rates of production by gravity drainage (oil drainage by virtue of its own weight), recovery of oil-in-place may be 70%-90%. At high rates of production, in a similar system, recovery by gravity drainage may be negligible. This paper describes the experimental apparatus and procedure, as well as the theoretical predictions, based on measured properties of the fluids and rock, of gravity flow in the experimental system. The experimental results and theoretical predictions were in very close agreement. Editor's Note: Gravity drainage of heavy-oil reservoirs may be accomplished using horizontal boreholes, either drilled from the surface or from underground shafts and tunnels, and with or without heating or injecting water to enhance flow. Several such projects are included in this bibliography.

1027. Teufel, L. W., C. M. Hart, A. R. Sattler, and J. A. Clark (1984). Determination of hydraulic fracture azimuth by geophysical, geological, and oriented-core methods at the Multiwell Experiment Site, Rifle, CO. Paper presented at the 59th Annual SPE Technical Conference & Exhibition, September 16-19, 1984, Houston. Society of Petroleum Engineers, Richardson, TX. SPE paper 13226.

Determination of natural and hydraulically-produced fractures in horizontal drilling and wellbores is critical to obtain optimum production. This paper discusses in detail the methods used to determine fracture orientations at the Department of Energy Multiwell Experiment Site (MWX), Rifle, Colorado. The methods discussed include geophysical methods, borehole impression packers, oriented cores, surface fractures, and other geological methods.

1028. Teufel, Lawrence W. and Helen E. Farrell (1990). Distribution of in-situ stress and natural fractures in the Ekofisk field, North Sea: Implications for reservoir permeability. A.A.P.G. Bulletin 74 (5), 777.

Abstract. The chalk reservoir of Ekofisk field, Norwegian North Sea, produces primarily from natural fracture-porosity. In-situ measurements of stress and fracture orientations in the field were made from core studies, hydraulic fractures, and well-bore breakouts. These measurements show that the azimuth of the maximum horizontal stress is not uniform across the field, and tends to be oriented perpendicular to the structural contours. Conclusions reached are: (1) radial fractures have the highest permeability; (2) reservoir permeability will be anisotropic; (3) maximum flow direction will not be uniform, but will vary radially around the flanks of the structure; and (4) production and reservoir depletion will increase effective stress and may create new fractures aligned with the stress. Editor's Note: This article is included to emphasize the kind of information needed to understand the most common type of reservoir that is the target of horizontal drilling, and the dynamic nature of a producing field.

1029. Thomas, S. (October 19-22, 1982). Pipeline crossing of the River Thames by directionally controlled horizontal drilling technique. Petroleum Review.

Not examined.

1030. Thorogood, J. L. (1982). Discussion of MWD North Sea field use, Aug. 1978-Feb. 1979 (with 1982 update). Journal of Petroleum Technology 35 (5), 905-906.

1031. Thorogood, J. L. (1988). Discussion of 'Evaluation of directional survey errors at Prudhoe Bay.' SPE Drilling Engineering 3, ( ), 334-335. SPE paper 18557.

Critical discussion of paper by Dubrule, O. and P. H. Nelson (1988), SPE Drilling Engineering 3, 335.

1032. Thorogood, J. L. (1988). Instrument performance models and their application to directional surveying operations. Proceedings 1988 SPE Annual Technical Conference & Exhibition, October 2-5, 1988, Houston, Drilling volume, pp. 291-303. Society of Petroleum Engineers, Richardson, TX. SPE paper 18051. Also published in SPE Drilling Engineering (December, 1990), 294-298.

This report has a discussion of instrument performance models for evaluation of directional surveys.

1033. Thorogood, John L. (1989). Directional survey operations management. Journal of Petroleum Technology 41 (12), 1250-1252. SPE paper 19462.

Directional surveys are critical in horizontal and other directional drilling. Definition of the geological target boundaries and the potential survey errors are essential in a directional drilling project.

1034. Tillman, Ned (July 12, 1982). Fractured reservoirs and targets of the Appalachian basin and thrust belt. Oil & Gas Journal 80 (28), 153.

Many of the Devonian shale reservoirs of the Appalachian basin are dependent on natural fractures for their effective porosity. This article has a brief description of naturally-fractured Devonian shale reservoirs, which are now targets of horizontal drilling.

1035. Tillman, J. Edward (February 21, 1983). Exploration for reservoirs with fracture-enhanced permeability. Oil & Gas Journal 81 (8), 165-172.

Natural fractures enhance the productivity of many hydrocarbon reservoirs. The two techniques currently used for exploring for fractured reservoirs are lineament studies and high-resolution seismic surveys. Recently, a combination of geochemical surveys and structural analysis has been developed to locate these reservoirs. Examples of fractured reservoirs are: Austin Chalk in Texas, Spraberry Trend (sandstone) in Texas, basement metamorphics in California, and Devonian shales in the Appalachian basin. Fractures may be caused by folding and bending of competent rocks or intra-formational fractures in flat-lying rocks associated with sand pinchouts or other features. Geochemical and structural analysis can determine which fractures enhance production and which do not. Structural data may be obtained from logging tools, such as the borehole televiewer or impression packer, while fluid inclusion analysis can determine thermal history of the matrix and migration history in the matrix and fractures.

1036. Tiraspolsky, Wladimir (1985). Hydraulic Downhole Drilling Motors. Gulf Publishing Co., Houston. ISBN 0-87201-766-4.

This book comprehensively covers all aspects of downhole drilling motors.

1037. Tolle, Glen and Thomas Dellinger (May 26, 1986). Mobil identifies extended-reach-drilling advantages, possibilities in North Sea. Oil & Gas Journal 84 (21), 78-86. Abstract in The Engineering Index Annual (1986), 85, Part IV, 6522.

Extended-reach wells may be used to develop offshore reservoirs not now considered to be economic, to drill under shipping lanes, under environmentally sensitive areas, to accelerate production, and to reduce the number of platforms needed to develop large reservoirs. Mobil drilled an experimental extended-reach well on the West Ranch, Texas, which proved successful in demonstrating the technology. Mobil will use extended-reach drilling in Beryl field and Statfjord field, North Sea. Details of the West Ranch test well are given in the article.

1038. Toma, P., D. Redford, and D. Livesey (1984). The laboratory simulation of bitumen recovery by steam stimulation of horizontal wells. Paper presented at the 1st WRI-DOE TAR-SAND Symposium, June 14-15, 1984, Vail, Colorado. Western Research Institute and U.S. Department of Energy.

Not examined.

1039. Toma, Peter, Decian Livesey and Theodore Heldrick (1988). New sand-control filter for thermal recovery wells. SPE Production Engineering 3 (2), 249-257. SPE paper 15057 presented at the 1986 SPE California Regional Meeting, Oakland, April 2-4.

Severe sand inflow problems were encountered in horizontal wells drilled in the Athabasca Tar Sands pilot plant, Fort McMurray, Alberta, Canada. The sand problem was caused by three conditions: fine-grained cohesionless sand; cyclic steam injection; and the production of water, steam and high-viscosity bitumen. A new type of filter was developed to cope with the sand problem. The new filter was made of compressed metallic wool and a metallic cloth. The new filter (SWC filter) will withstand high-temperature and high-pH conditions.

1040. Tomren, P. H. (1980). The transport of drilling cuttings in an inclined eccentric annulus. M.S. thesis, University of Tulsa, Tulsa.

Not examined.

1041. Tomren, R. H., A. W. Iyoho and J. J. Azar (1986). Experimental study of cuttings transport in directional wells. SPE Drilling Engineering 1 (1), 43-56. SPE paper 12123 presented at the 1983 SPE Technical Conference and Exhibition, San Francisco, October 5-8. Abstract in The Engineering Index Annual (1986), 85, Part IV, 6522.

Large-scale experimental studies of cuttings transport in high-angle wells have been conducted at the University of Tulsa, the results of which are reported in this paper. These studies used 40-ft pipe sections, annulus inclinations varying from 0 degrees to 90 degrees, pipe rotation and eccentricity, and several types of drilling muds and flow regimes. Cuttings transport is affected primarily by drilling fluid velocity, hole inclination, and fluid rheology properties. Conclusions reached in this study are: (1) the customary theoretical transport ratio (for cuttings) is invalid for inclined holes, because of cuttings segregation (by size) which leads to formation of a cuttings bed; (2) when hole angle exceeds 40 degrees, a cuttings bed is formed, which reduces flow area, even with high fluid flow rates; (3) major factors affecting mud carrying capacity, in directional wells, are fluid velocity, hole inclination, and mud rheology; (4) increased hole inclination reduces drilling fluid carrying capacity, when other parameters are kept constant; (5) within a particular flow regime, higher viscosity fluids give better cuttings transport; (6) pipe rotation had only slight effects on cuttings transport in inclined holes; (7) pipe/hole eccentricity affected cuttings transport adversely in all tests; (8) at hole angles of inclinations between 35 and 55 degrees, cuttings beds form and slide downward against the flow; (9) previous reports of cuttings behavior in vertical wells were confirmed.

1042. Torgerson, Tod A. (1991). Metal seals reduce Williston horizontal problems. Western Oil World 48 (5), 36-37

Substitution of metal face seals for rubber elastomer seals in rolling cone bits has increased bearing life of bits in the Williston basin. Bit performance and cost discussed in the article.

1043. Towson, D. E. and R. F. Meyer (1981). Chapter 36. Current in-situ oil sands pilot projects in Alberta, Canada. UNITAR International Conference on Long-term Energy Resources, pp. 499-503. Pittman, Boston.

Not examined.

1044. Towson, D. E. (1984). The Mine Assisted In Situ Project. Proceedings Exploration for Heavy Crude Oil and Bitumen - AAPG/UNITAR Joint Conference, October 29-November 2, 1984, Santa Maria, California, v. II.

Not examined.

1045. Towson, D. E. (1984). The mine assisted in-situ project. 5th Annual Advances in Petroleum Recovery and Upgrading Technology Conference, June 14-15, 1984, Calgary. Session No. 1, Paper No. 4. Session No. 1, Paper No. 4.

The "Mine Assisted in-situ Project", conducted by a consortium of five companies (Husky Oil Operations Ltd., Esso Resources Canada Ltd., Gulf Canada Resources Inc., Canada-Cities Service Ltd., and Petro-Canada) was designed to test the feasibility of extracting bitumen from the Athabasca Oil Sands. The project involved drilling production wells from underground locations. Steam- injection wells were used in conjunction with production wells drilled upward at angle of 3 degrees. The project successfully demonstrated that it is possible to drill long horizontal wells from mine tunnels and shafts, and that the heavy-oil production was possible using this system.

1046. Tracy, P. B. (1988). Lateral drilling technology tested on UCG project. Proceedings 1988 IADC/SPE Drilling Conference, February 28-March 2, 1988, Dallas, pp. 493- 502. Society of Petroleum Engineers, Richardson, TX. IADC/SPE paper 17237.

Six medium-radius horizontal wells were drilled in the Hanna formation coal beds in the Rocky Mountain No. 1 Underground Coal Gasification Project of the Gas Research Institute and U.S. Department of Energy. Details of the project and the wells are given in the article.

1047. Trichel, D. K. and M. P. Ohanian (1990). Unique articulated downhole motor holds promising future for short radius horizontal drilling. Proceedings 1990 SPE Annual Technical Conference & Exhibition, September 23-26, 1990, New Orleans, pp. 137-152. Society of Petroleum Engineers, Richardson, TX. SPE paper 20417.

An articulated downhole motor has been developed that enables drilling short-radius horizontal holes with longer horizontal sections. The articulated motor makes it possible to steer the bit in a short-radius hole. Details of the system and a general discussion of short radius drilling are given in the article.

1048. Underdown, D. R., K. Das and H. Nguyen (1985) Gravel packing highly deviated wells with a cross-linked polymer system. Journal of Petroleum Technology 37 (13), 2197-2202. SPE paper 12481, presented at the 1984 SPE Formation Damage Control Symposium, Bakersfield, February 13-14. Abstract in The Engineering Index Annual (1986), 85, Part IV, 6522.

In highly deviated wells, conventional gravel-packing fluids allow the gravel to fall to the low side of the tubing, causing dunning and resulting in premature sandouts. These problems can be largely prevented by using a cross-linked polymer system such as hydroxethylcellulose (HEC) cross-linked with a heavy metal ion. HEC cross-linked with ZrOCl2 in an aqueous medium will provide a zero gravel-fall rate, stable up to temperatures of 150 degrees F. This fluid can be made in the field and has adequate fluid loss, in most cases, to ensure good gravel packing. The cross-linked HEC gravel pack system causes minimal formation damage.

1049. Varnardo, S., A. Meidinger and M. Bexon (1985). Horizontal drilling techniques - Considerations in planning a horizontal drilling programme. Proceedings of the Oil & Gas Machinery Exhibition, November, 1985, New Delhi, India.

Not examined.

1050. Van Der Vlis, A. C., H. Duns and R. F. Luque (1979). Increasing well productivity in tight chalk reservoirs. Proceedings of the 10th World Petroleum Congress, Bucharest, September 9-14, 1979, Bucharest, v. 3, pp. 71- 78. Heyden & Son, London. ISBN 0 85501 473 3. Abstract in The Engineering Index Annual (1982) 81, Part IV, 5232.

This article presents the results of tests, conducted by Shell, on low-permeability chalks to improve production. Hydraulic fracturing tests indicated that soft chalks require an acid/propped fracturing treatment, while competent chalks require acid-fracturing using a viscous finger technique. Oil production was found to increase two- to three-fold when multiple drainholes with inclinations of 80-85 degrees were used.

1051. van Kruysdijk, C. P. J. and H. Niko (1988). Alternatives for draining tight naturally fractured gas reservoirs: horizontal hole drilling vs massive hydraulic fracturing. Proceedings of the European Petroleum Conference, October 17-19, 1988, London, pp. 47-57. Great Britain Society of Petroleum Engineers. SPE paper 18339.

This article addresses the problem of using horizontal wells or massive hydraulic fracturing (MHF) to develop naturally fractured tight gas reservoirs. Some of the conclusions are: horizontal wells are real alternatives to MHF because they are less prone to coning and have a greater likelihood of intersecting natural fractures; the influence of soluble porosity is minimal in both cases; Green's function is a powerful tool for analysis of transient pressure response. Other conclusions are made concerning well testing for both the horizontal well and MHF cases.

1052. Vanorsdale, Charles R. (1987). Evaluation of Devonian shale gas reservoirs. SPE Reservoir Engineering 2 (2), 209-216. SPE paper 14446 presented at the 1985 Annual SPE Technical Conference and Exhibition, September 22-25, Las Vegas.

Devonian shales occur in the Appalachian basin, Illinois basin, and Michigan basin. They vary in thickness from a few hundred feet, in western Ohio, to 11,000 feet in eastern Pennsylvania. 90% of the wells drilled in Devonian shales are productive, although most require artificial stimulation. Many of the gas wells produce by sequential release mechanisms, and behave as multiple-porosity systems. Several engineering evaluation methods are addressed in this paper, including wireline formation evaluation, pressure/cumulative production, well test analysis, analogy to offset wells, material balance, rate vs. cumulative-production analysis, reservoir simulation, and decline-curve analysis. Conclusions reached are: (1) reliable evaluation of Devonian shale gas wells is best done when continually updated reservoir descriptions are obtained (production-rate/time decline-curve analysis is the most easily grasped method); (2) typical production-decline curves for these wells exhibit rapid drop in production rate, followed by production stabilization; (3) the decline curves follow a hyperbolic pattern and display b exponent values greater than 1.0, resulting from multi-porosity, multi-layer systems; (4) late-life decline rates are very shallow, with production rate forming an almost horizontal line on the rate/time curves; (5) the best well stimulation method is "explosive treatment", closely followed by hydraulic fracturing (natural completions decline more rapidly after the sixth year; some wells have produced for over 50 years); (6) rate/time curves are essentially the same for all parts of the study area in Ohio, Kentucky, and West Virginia.

1053. van Rensburg, W. C. J. and Krishnan A. Malik (1988). Tracking changes in oil exploration and development technology and their economic implications. Journal of Petroleum Technology 40 (1), 71-74. SPE paper 17021.

Improvements in technology, geological interpretation, and geophysical data acquisition and interpretation have greatly improved oil exploration. One of the improvements mentioned is horizontal drilling.

1054. Van Waart, T. L. and R. Hartley (1985). Research on improved hydrocarbon recovery from chalk deposits. DeBouw, R, E. Millich, J. P. Joulia, and D. Van Asselt (editors), New Technologies for Exploration and Exploitation of Oil and Gas Resources, v. 2, pp. 848-853. Graham & Trotman, London. ISBN 0 86010 676 4.

Chalk reservoirs are sometimes characterized by high-porosity and low-permeability, which may make production of hydrocarbons uncommercial. Acid-fracturing, using a proppant, of such reservoirs can improve productivity, particularly when combined with multiple high-angle drainholes. This method is recommended in the North Sea chalks.

1055. Varnado, S. G., C. F. Huff and P. Yarrington (1980). Studies aim at optimum design for PDC hard-formation bits. World Oil 190 (4), 63-66, 70.

Discussion of the development of improved PDC bits by Sandia National Laboratories and General Electric. Bit design and construction are discussed.

1056. Vega, A., L. Montefusco and N. Perdomo (n.d.). Experiencia en Perforacion de Pozos Inclinados [Experience in drilling inclined holes]. Maraven, S.A., Venezuela. Technical Report FPO-0598. In Spanish.

This report was not examined.

1057. Ventre, J. and A. M. Dormigny (1985). Horizontal drilling in Rospo Mare oil reservoir. in DeBouw, R., E. Millich, J. P. Joulia, and D. Van Asselt (editors), *New Technologies for the Exploration and Exploitation of oil and Gas Resources,* v. 2, pp. 827-832. Graham & Trotman, London. ISBN 0 86010 767 4.

Three test wells were drilled in Rospo Mare field, offshore Adriatic Sea, Italy. One well was vertical, one well was deviated, and one was horizontal. They were drilled to test the productivity of each type in a heavy, viscous-oil reservoir. The horizontal well was drilled in 88 days without any difficulties. Productivity proved higher than anticipated. The reservoir is a karstic limestone reservoir, with an underlying active aquifer. The horizontal well produced 212,000 barrels of oil as opposed to 73,000 barrels from the vertical well in the period covered in this report. Daily flow rates for each was 480 BOPD for the horizontal well and 120 BOPD from the vertical well.

1058. Vicanek, J. (1985). Erdolforderung durch horizontale Bohrungen [Oil production with horizontal wells]. Erdol Erdgas 101 (3), 85-90. In German with an English abstract.

Abstract. Examples are given of the production efficiency of horizontal wells.

1059. Vielvoye, Roger (February 16, 1981). New slant drilling technique slated in Irish Sea. Oil & Gas Journal 74 (7), 74- 75.

Slant drilling will be used to develop Morecambe Bay field, Irish Sea, because of the shallowness of the gas reservoir. The reservoir lies at a depth of 3,000 feet. Six to eight drilling platforms and 120 wells will be required to develop the 12 mile long field.

1060. Vielvoye, Roger (April 22, 1991). Danish oil flow rises. Oil & Gas Journal 89 (16), 44.

Denmark is increasing its oil and gas production in the North Sea fields by use of horizontal drilling. Danish North Sea fields include Dan field, Dagmar field, and Kraka field. Production from these fields is now 135,000 BOPD. The operator of the Danish fields is Dansk Undergrund Consortium, a group composed of A. P. Moller, Royal Dutch/Shell Group, and Texaco Inc.

1061. Vieser, H. (1986). Horizontalbohrungen strecken oelreserven [Horizontal drilling extending oil reserves]. Bild der Wissenschaft 23 (9), 11-13. ISSN 0006-2375. In German.

Not examined.

1062. Villalobos, C. and J. Parra (1981). Perforacion Direccional Controlada desde Localizaciones Multipozos [Drilling directionally controlled multiple wells]. M.S. Thesis, Universidad de Zulia, Venezuela. In Spanish.

Not examined.

1063. Vozniak, John P. (1991). Horizontal well personal computer applications. Schedule and Abstracts Houston Geotech Technical Conference, February 17-22, 1991, Houston, pp. 40. Houston Geotech Association, Houston.

Abstract. Maurer Engineering Inc. has developed 18 manuals and 13 personal computer-based programs for reservoir screening, horizontal well planning, and tubular design.

1064. Vrielink, H. J. and A. Hippman (1988). The optimization of slant well drilling in the Lindbergh field. Proceedings 1988 SPE Rocky Mountain Regional Meeting, May 11-15, 1988, Casper, Wyoming, pp. 435-446. Society of Petroleum Engineers, Richardson, TX. SPE paper 17539. Also published in SPE Drilling Engineering (1989) 4 (4), 307-314.

Lindbergh field, Alberta, Canada, is a heavy oil field (API gravity 12.5), with a Cretaceous Lower Mannville formation reservoir at 1,970 ft (600 m). Dome Petroleum has developed the field with slant-well drilling from four pads of 14 or 15 wells each. The wells were placed on primary production for two years, after which, steam injection was used. Wells require two days to drill. The methods used are economic because of improved drilling times, pad design, and lower rod and tubing wear resulting from the "straight" slantholes as opposed to curved deviated wells. Details are given in the paper.

1065. Wagner, Duanne T. (1990). Controls on fracture distribution in the Giddings Austin Chalk. Kinsland, Gary and Tony Cagle (editors), Transactions of the 40th Annual Meeting, G.C.A.G.S. & GSC-SEPM, October 17-19, 1990, Lafayette, Louisiana, p. 597. Gulf Coast Association of Geological Societies. Abstract also published in A.A.P.G. Bulletin (1990), 74 (9), 1512; and in Chuber, Stewart (editor) (1991), Abstracts and Short Papers of the South Texas Geological Society Austin Chalk Exploration Symposium, February, 1991, San Antonio, Texas.

Abstract. Structure and stratigraphy of the Austin Chalk control fracture distribution in Giddings field, Texas. Fracturing mechanisms include normal faulting, bending over buried structures, gravity creep, differential compaction, and aquathermal pressuring.

1066. Walden, William and C. H. Jacoby (1963). Exploration by horizontal drilling at Avery Island, Louisiana. Berksticker, A. C., K. E. Hoekstra and J. F. Hall (editors). Symposium on Salt, pp. 367-378. Northern Ohio Geological Society, Cleveland.

This article discusses the use of horizontal drilling in salt exploration and mining at Avery Island, Louisiana. Editor's Note: This article is included because it appears in several reference lists and bibliographic data bases (GEOREF). It is not applicable to petroleum exploration.

1067. Walker, Bruce H. (1986). Factors controlling hole angle and direction. Journal of Petroleum Technology 38 (1), 1171-1173. SPE paper 15963.

Borehole angle (vertical deviation) and direction (azimuth of horizontal deviation) of deviated wells are affected by many factors, including formation strength and anisotropy, drill bits, BHA (bottom hole assembly), borehole shape and curvature, and operating parameters. Bit deviation is most affected by alternating formation hardness, which also affects penetration rate. Type of bit is important in directional drilling, as roller bits tend to "walk" to the right while PDC bits and diamond bits do not. BHA components affect the weight and rigidity, which affect overall performance. The curvature of the borehole can cause the BHA to be deflected in a complex shape nearly independent of the BHA components. Operating parameters such as bit weight, torque, and rotary speed also affect borehole trajectory. Modeling of these parameters can allow designs that overcome the adverse affects to produce the desired borehole trajectory.

1068. Warpinski, Norman R., Paul Branagan and Roy Wilmer (1985). In-situ stress measurements at U.S. DOE's Multiwell Experiment Site, Mesaverde group, Rifle, Colorado. Journal of Petroleum Technology 37 (3), 527-536. SPE paper 12142 presented at the 1983 SPE Annual Technical Conference and Exhibition, San Francisco, October 5-8. Abstract in The Engineering Index Annual (1985), 84, Part IV, 6221.

In-situ stress measurements were made of the Mesaverde group sands and shales at the DOE Multiwell Experiment Site in Colorado. The tests were made at depths of 7,300 to 8,100 ft (2,225 to 2,470 meters). The tests were small-volume, hydraulic-fracture stress tests through perforations. Marine shales above and below the sands have a fracture gradient greater than 1.0 psi/ft [23 kPa/m], indicating inelastic behavior (creep and fracture are dominant stress controlling mechanisms). Sandstones and siltstones have fracture gradients of 0.85 to 0.9 psi/ft [19 to 20kPa/m]. Containment of hydraulic fractures would therefore be expected.

1069. Wash, Renee March, 1981). ARCO's fourth successful drainhole drilled in cased hole. Drill Bit 30 (9), 35-36.

Arco's fourth horizontal drainhole in the Empire Abo Unit, New Mexico, was drilled from a vertical cased hole. The well was drilled horizontally 190 ft into the reservoir. The casing was milled out at 6,224 ft before kicking off the horizontal segment. This well demonstrated the feasibility of reentering vertical wells and recompleting horizontally.

1070. Weaver, O. D. and Arthur Van Damme (May 23, 1988). France - 2. Renewed interest seen in Aquitaine basin. Oil & Gas Journal 86 (21), 55-58.

The Aquitaine basin has produced 75% of France's oil and 98% of its gas since 1949. This article reviews the history and geology of the basin. Editor's Note: Elf Aquitaine drilled its first horizontal wells in Lacq field and a horizontal well in Pau field in the Aquitaine basin.

1071. Webster, M. B., G. E. Otott, Jr., and D. L. Rice (1987). Cementing high-angle wells using cement-expanded formation packers and/or casing rotation. Proceedings 1987 SPE/IADC Drilling Conference, March 15-18, 1987, New Orleans, pp. 745-754. Society of Petroleum Engineers, Richardson, TX. SPE/IADC paper 16136.

Two approved methods of isolating the producing interval from the surface have been used in the Long Beach Unit of Wilmington field, California. The methods used and approved by the State of California are: (1) installation of a 20-ft Cement Expanded Formation Packer set in a competent shale above the completion interval; and (2) rotating a selected interval of casing during conditioning and cementing. Both methods have been successfully used in high-angle wells in the field. Engineering details are given in the paper.

1072. Webster, R. L. (1984). Petroleum source rocks and stratigraphy of the Bakken formation in North Dakota. in Woodward, J., F. F. Meissner, J. L. Clayton (editors), *Hydrocarbon Source Rocks of the Greater Rocky Mountain Region,* pp. 57-81. Rocky Mountain Association Of Geologists, Denver.

Not examined.

1073. Weeden, Scott L. (March 1, 1989). Baker-Hughes aims beyond drilling - all the way to completion. The Oil Daily, 6.

Baker-Hughes and its subsidiaries provide complete services for drilling and completing wells. Subsidiaries are listed. No information about horizontal drilling is given.

1074. Weeden, Scott L. (March 1, 1989). Lateral holes have 100- year history. The Oil Daily, 7.

A very brief history of drainhole drilling from 1929 (Big Lake Oil Co.) to 1981. The first patent for horizontal drilling using a flexible shaft was granted. The first true lateral hole was drilled in 1929 at Texon, Texas, by Robert E. Lee of Coleman, Texas. Lee sold out to John Eastman in 1941. John Zublin was another pioneer tool designer and driller who drilled the first drainhole in 1944.

1075. Weeden, Scott L. (March 1, 1989). Small business starts off horizontal. The Oil Daily, p. 7.

The newly formed Horizontal Drilling Production Corp. has participated in an Austin Chalk horizontal well as part of a project of Ray Holifield & Associates in Lee County, Texas. The company hopes to become a top horizontal drilling company and to conduct seminars in drilling horizontal wells.

1076. Weeden, Scott L. (March 1, 1989). Short vs long-radius wells: Which is best? The Oil Daily, p. 7.

Long-radius wells are those with an angle build rate of two to six degrees per 100 ft drilled; short-radius wells are those with build rates of 1.5 to 3 degrees per 100 ft drilled. Horizontal boreholes of 5,000 ft are predicted to be possible.

1077. Weirich, J. B., T. E. Zaleski, and P. M. Mulcahy (1987). Perforating the horizontal well: Designs and techniques prove successful. Proceedings 1987 SPE Annual Technical Conference & Exhibition, Production Operations and Engineering volume, September 27-30, 1987, Dallas, pp. 503-508. Society of Petroleum Engineers, Richardson, TX. SPE paper 16929. Also in Proceedings 1988 Rocky Mountain Regional Meeting, Casper, Wyoming, May 11-13, 1988, pp. 129-134.

Perforating horizontal wells is usually done with tubing-conveyed perforation guns. In some applications the perforating is done in an underbalanced condition, which may enhance production by allowing enhancement of flow and surge characteristics. Details of the equipment and operations are given in the paper.

1078. Wenande, B. (1987). A review of the first horizontal well - Denmark. Paper presented at the 1987 European Offshore Drilling Conference, December 2-3, 1987, London.

Not examined.

1079. Werniuk, Jane (1986). Gravity 'Gusher': Devran Petroleum & Shell Canada mining oil in SW Ontario. Canadian Mining Journal 107 (12), 24-25, 27-28. Abstract in The Engineering Index Annual (1987), 86, Part IV, 6531.

Not examined.

1080. Wesenberg, D. L. and J. Polito (1987). A new innovation in directional only MWD systems. Paper presented at the 1987 SPE/IADC Drilling Conference, March 15-18, 1987, New Orleans. Society of Petroleum Engineers, Richardson, TX. SPE/IADC paper 16059.

Directional-only MWD systems have been developed to reduce the cost of drilling. The Sperry-Sun MWD is discussed in this paper. The tool provides directional surveys and steering information. Details of the tool are given in the paper.

1081. West, L. W. and F. M. Doyle-Read (1988). A synergistic evolution of the Norman Wells reservoir description. Journal of Canadian Petroleum Technology 27 (2), 96-103.

Norman Wells field has undergone significant development since 1981, with 237 wells drilled. Geologists and engineers have used a multidisciplinary approach in developing a permeability model for the Devonian reef reservoir. The reservoir is a naturally fractured carbonate, with a fracture system orientation of N21E. Fracture frequency decreases with depth in the reservoir; the vertical fractures do not contribute to interzonal communication. Editor's Note: The development program at Norman Wells includes horizontal and extended-reach wells.

1082. Westerhausen, H. (1991). Wytch Farm - das grobte Onshore- Erdolfeld Westeuropas [Wytch Farm - Largest Onshore Oilfield in Western Europe]. Erdol Erdgas Kohle 107 (4), 148-152. In German.

Discussion of the geology and development of Wytch Farm field, U.K.

1083. Western Oil World (February, 1986). Lateral drilling technique testing in Utah. Western Oil World 43 (2), 21-22.

Texas Eastern's Skyline Oil Co. has completed 68 lateral drainholes from 18 wells in Grassy Trail Creek field, Utah. The drainholes were drilled with a radius-of-curvature of 20-40 ft, using Eastman Whipstock equipment. Grassy Trail Creek field produces from fractured Moenkopi formation. The open, natural fractures are 0.5 to 1.5 inches wide. The Moenkopi in the field is a tight siltstone with 4-5% porosity and 0.01 md permeability. The lateral drainholes average 250 ft in length, with a maximum of 500 ft. Drainhole drilling caused six previously non-producing wells to become significant producers.

1084. Western Oil World (1989). Targeting the Bakken Booty. Western Oil World 46 (5), 16-18.

Review of the Bakken horizontal drilling play with the emphasis on Samuel Gary Jr. & Associates Inc.

1085. Western Oil World (1990). CER undertakes horizontal gas well. Western Oil World 47 (6), 11-12.

CER Corp. under Department of Energy contract is drilling a horizontal well to recover coalbed methane in the Piceance basin, Colorado. The well will be drilled at the DOE Multiwell Experiment Site near Rifle. The Cozzette formation coalbeds of the Mesaverde Group are the drilling target. The well has a projected total depth of 9,625 ft. with a TVD of 7,840 ft.

1086. Western Oil World (1990). Meridian tests new technology. Western Oil World 47 (6), 13.

Meridian Oil Inc. is applying horizontal drilling technology to the Bakken formation, Williston basin, Austin Chalk, Texas, and coal-bed methane in the San Juan basin.

1087. Western Oil World (1990). Niobrara draws horizontal drillers. Western Oil World 47 (7), 9.

Horizontal drilling targets in the Denver-Julesburg basin now include the Niobrara and Codell formations. The Niobrara play was opened by Cowan Oil Co. in Silo field, Wyoming. Other companies drilling or planning horizontal wells in the basin are Basin Exploration Inc. and Union Pacific Resources Co. Oryx Energy Co. and Wolverine Exploration Co. are exploring the Niobrara in the Sand Wash basin. Other Niobrara players are Meridian Oil Inc., Snyder Oil Co., and Wainoco Oil Corp.

1088. Western Oil World (1990). Horizontal wells test Pierre. Western Oil World 47 (5), 24, 29.

Ensign Oil & Gas Co. has drilled three horizontal wells to test the Pierre formation. The Pierre formation is expected to produce from "stressed" zones similar to the Bakken formation and the Austin Chalk. The Ensign wells were drilled in Florence-Canyon City field, Colorado.

1089. Western Oil World (1990). Horizontal drilling heats up in Rockies. Western Oil World 47 (8), 10-11.

Review of horizontal drilling activity in the Rocky Mountain area, including Silo field, East Salt Creek field, and Thompson Creek field in Wyoming, and Salt Creek field, Montana. Wells in Colorado are being drilled in Florence field, Curtis field, Wattenberg field, and Loveland field. One Entrada sandstone horizontal well has been drilled in the San Juan basin, New Mexico.

1090. Western Oil World (1990). Powder River Basin: Drilling sideways in Thompson Creek. Western Oil World 47 (8), 24-25.

Karlton Terry Oil Co. has completed one horizontal well, and plans more horizontal and vertical wells, in Thompson Creek field, Wyoming. Thompson Creek produces 20-gravity oil from the Muddy formation. The horizontal producer tested 1,200 barrels per day. Typical vertical wells produce 12 barrel per day. Production from the horizontal well will probably be 50 barrel per day. The well has 1,000 ft of slotted-liner in the reservoir. Cost of future horizontal wells is estimated at $240,000.

1091. Western Oil World (1990). UPRC brings in horizontal winner in Silo field. Western Oil World 47 (10), 15.

Union Pacific Resources Co. has completed a horizontal well in the naturally-fractured Niobrara reservoir in Silo field, Wyoming. The 8,500 ft well has a horizontal displacement of 2,529 ft. The horizontal section was drilled 2,037 ft into the third bench of the Niobrara Chalk. The horizontal hole crossed six or seven major fractures and 100 smaller fractures. Cowan Resources Co. successfully completed a horizontal well in the Silo field that tested 2,000 BOPD and 500 MCFGPD. The Cowan well now produces 240 barrel of oil per day on pump.

1092. Western Oil World (1990). Four Corners: Horizontal wells test San Juan Mancos. Western Oil World 47 (10), 25.

Horizontal wells are being drilled in the Mancos formation, San Juan basin, by Samuel Gary, Jr. & Associates Inc. and BASF Corp. The Mancos is a naturally fractured Cretaceous formation. Samuel Gary Jr. & Associates are drilling in Rio Puerco field; BASF is drilling in Verde field. Another planned horizontal drilling project in the San Juan basin is a well in Puerto Chiquito field, New Mexico, that well will be drilled by Benson-Montin-Greer Drilling Corp.

1093. Western Oil World (1990). Bakken play: The best is yet to come? Western Oil World 47 (10), 29-31.

U.S. Geological Survey and North Dakota Geological Survey geologists believe that the Bakken play will live up to early expectations. The Bakken may have reserves of 100-150 billion barrels of oil. A brief discussion of the play is presented.

1094. Western Oil World (1990). Williston Basin: Williston settles into comfortable pace. Western Oil World 47 (10), 22, 24-27.

By early September, 1990, the rig count in the Bakken play was down to nine from a high of 13. Attempts to extend the Bakken play outside the fairway have failed. Bakken fracturing is more random than previously believe, making the play more risky. Comparisons are made to the Austin Chalk play in Texas, and drilling projects in the Bakken are discussed in the article.

1095. Western Oil World (1990). Forwest turnkeys horizontal wells. Western Oil World 47 (12), 14.

Forwest Inc. has a turnkey contract to drill horizontal wells for Union Pacific Resources Co., in the Bakken play of North Dakota.

1096. Western Oil World (1991). UPRC scores third sideways producer in Silo field, Wyo. Western Oil World 48 (1), 20.

Union Pacific Resources Co. has completed its third producing horizontal well in Silo field, Wyoming. The well tested 767 BOPD and 326 MCFGPD.

1097. Western Oil World (1991). American Hunter horizontal well scores. Western Oil World 48 (2), 14.

American Hunter Exploration Ltd. has completed a Bakken well in North Dakota. The well was completed, on pump, for 22 BOPD. The well is 17 miles south southeast of a 1990 Bakken discovery by American Hunter.

1098. Western Oil World (1991). Hondo solves Piceance mud-ring problem. Western Oil World 48 (3), 26.

Report of a meeting of Piceance basin/Sand Wash basin operators to discuss drilling problems and solutions. Hondo engineers offered a solution to cuttings/mud ring buildup problems in boreholes through underpressured formations.

1099. White, Cameron (December 3, 1990). Horizontal wells - 6. Formation characteristics dictate completion design. Oil & Gas Journal 88 (49), 58-62, 64.

Design of completions of horizontal wells should be tailored to formation characteristics. Early completions of horizontal wells in fractured reservoirs were restricted to open-hole or slotted-liner completions. Currently, horizontal completions are still not routine, and require careful planning and cooperation between geoscientists and production, reservoir, and drilling personnel. Some of the physical considerations in completion design are: radius of curvature of the borehole; run-in wear of tool and centralizers; tail pipe loads due to gravity; hydrostatic head; hole roundness; liner tops; and solids in the hole. Generally, tubing-conveyed tools should be used, and tools should be hydraulically operated. Completion types discussed include open-hole, open-hole packer, slotted-liner, and cemented liner.

1100. White, C. W. (1989). Drilling and completion of a horizontal Lower Spraberry well including multiple hydraulic fracture treatments. Proceedings 1989 SPE Annual Technical Conference & Exhibition, Production Operations and Engineering volume, October 8-11, 1989, San Antonio, pp. 205-210. Society of Petroleum Engineers, Richardson, TX. SPE paper 19721. Also presented at the 1990 SPE Permian Basin Oil & Gas Recovery Conference, pp. 83-88. Abstract in The Engineering Index Monthly (June, 1990), 28 (6), 672, and (January, 1990), 28 (1), 663.

A horizontal well was drilled by Mobil E & P in the Lower Spraberry of Midland County, Texas. The well was drilled in the direction of the least principle stress, so that vertical fractures would radiate out at a right angles to the horizontal wellbore, when hydraulic fracture treatment was done. The well was drilled as a medium-radius well. The well was drilled to 10,048 ft MD with 1,619 ft of horizontal section. Liner was run partially into the horizontal section, and stuck at 9,905 ft., where it was cemented. The horizontal section of the well was hydraulically fractured in five places. Details of the drilling and completion of the well are given in the paper.

1101. White, S. M. and M. F. Bartle (1987). Slant drilling in Morecambe field, Irish Sea, U.K. Proceedings 1987 SPE/IADC Drilling Conference, March 15-18, 1987, New Orleans, pp. 893- 902. Society of Petroleum Engineers, Richardson, TX. SPE/IADC paper 16151.

Morecambe gas field, U.K. Irish Sea, was discovered by Hydrocarbons Great Britain Ltd. as a result of re-evaluation of logs from two wells drilled and abandoned by another operator. The top of the South Morecambe reservoir is at 2,300 ft; the base of the gas column is at 3,750 ft. Deviated wells may have as much as 2,000 ft of hole exposed to the reservoir. The use of slant rigs, for drilling development wells, required only three platforms instead of seven as required by conventional rigs. The slant drilling technique is suitable for drilling shallow reservoirs in shallow water, and is not suitable for deep water fields. The wells have been drilled in 55 days from rig up to testing. The extra cost of the slant wells was more than offset by the requirement for four fewer platforms. Details of the project are given in the paper.

1102. Whitson, C. D. and G. F. Sandison (1985). Long reach drilling in the Gulf of Mexico. Proceedings 1985 SPE/IADC Drilling Conference, March 5-8, 1985, New Orleans, pp. 433-437. Society of Petroleum Engineers, Richardson, TX. SPE/IADC paper 13473.

A well drilled in Main Pass block 30, Gulf of Mexico, in 1983, had a lateral reach of 11,255 ft. The well was drilled at an average angle of 71 degrees. The measured depth was 14,828 ft and the TVD was 7,164 ft. The long-reach well was drilled in order to reach a target under a shipping fairway. The only new technologies used in drilling the well were MWD, straight hole motors, and drillstring-conveyed logging tools. Details of the project are in the paper.

1103. Whitson, C. D. and G. F. Sandison (1985). Long-reach well successfully drilled. Drilling Contractor 41 (4), 28-29.

Texaco Inc. has successfully drilled an extended-reach well in the Gulf of Mexico. The well was spudded in Breton Sound Block 56, and extended into Main Pass Block 30, to reach a target under the shipping fairway. The horizontal reach is 11,255 ft with a TVD of 7,164 ft. The measured depth of the well is 14,828 ft. A brief account of the drilling is given in the article.

1104. Whitson, C. H. (1985). Discussion of fracturing in chalk completions. Journal of Petroleum Technology 37 (9), 1498

Critical discussion of paper by Hartley, R. and M. G. R. Bosma (1985), Fracturing in chalk completions, Journal of Petroleum Technology 37 (1), 73-80. See also "Author's reply...", Journal of Petroleum Technology 37 (9), 1498.

1105. Whittaker, Alun and Michael J. Kushuba (1987). Realtime logging: Combining MWD with surface measurements for complete well evaluation. Drilling 48 (1), 14-17.

General discussion of MWD directional and formation evaluation capabilities.

1106. Wieseman, Ted R. (1945). Horizontal drilling for oil. Trilobite 1 (8), 2-4.

Not examined.

1107. Wilkirson, J. P., Jay H. Smith, T. D. Stagg,and D. A. Walters (1988). Horizontal drilling techniques at Prudhoe Bay, Alaska. Journal of Petroleum Technology 40 (11), 1445-1451. SPE paper 15372 presented at the 1986 SPE Annual Technical Conference and Exhibition, New Orleans, October 5-8.

Horizontal drilling was used in Prudhoe Bay field to increase ultimate recovery of oil. Three horizontal wells were drilled in areas of the field where water and gas coning made conventional wells uneconomical. Before the first Prudhoe Bay horizontal well was drilled, a horizontal test well was drilled in South Texas. The Texas well showed that MWD was indispensable for directional surveys and formation identification using gamma-ray and resistivity sensors. The three Horizontal (extended-reach) wells in Prudhoe Bay were completed through slotted liners of 1,575, 1,637, and 1,163 feet (480, 499, and 345 meters). Improvements in technology and operating procedures over the three well program resulted in a 40% reduction in cost per foot drilled. Production rates were up to 300% of that for conventional wells. A detailed discussion of each well drilled is given in the paper.

1108. Williams, Bob (August 23, 1982). Kern River's "hotplate" project launched. Oil & Gas Journal 80 (34), 51-52, 54.

An enhanced EOR project using radial drainholes and cyclic steam injection in Kern River heavy oil field will get underway in September, 1982. The project will be operated by Fenix & Scisson Inc. of Tulsa, Oklahoma. The EOR project is expected to recover 75% of the oil-in-place. The EOR process, called HOPCO (Heavy Oil Process), was developed by Barber Oil Inc. and two professors from the University of Texas. The process was acquired by Cornell Oil Inc. in 1981. Participants in the joint venture, called Cornell Heavy Oil Process Inc., are Robert E. Glaze, Holly Corp., Lyco Energy, Placid Oil Co., Sedco Inc., Ladd Petroleum Corp., and Encore Petroleum. Details of the project are given in the article.

1109. Williams, E. T. and J. Kikani (1990). Pressure transient analysis of horizontal wells in a naturally fractured reservoir. Proceedings 1990 SPE Technical Conference & Exhibition, Formation Evaluation and Reservoir Geology volume, September 23-26, 1990, New Orleans, pp. 877-891. Society of Petroleum Engineers, Richardson, TX. SPE paper 20612.

Pressure transient analysis of horizontal wells in the Bakken formation of the Williston basin have provided useful information about reservoir permeability and pressure. The Bakken formation produces from naturally fractured reservoirs. Pressure tests have shown that the Bakken reservoirs are anisotropic and that wellbore orientation affects productivity. These tests show that the preferred orientation of the wells to maximize production can be predicted. The Bakken pressure build up tests indicate that the contribution of the matrix is negligible relative to the fracture contribution.

1110. Williams, N. (1989). Engineering significance of recent production and development projects in Australia, New Zealand and Papua New Guinea. The APEA Journal 29 (2), 76-86.

This article reviews upstream operations in Australia, New Zealand and Papua New Guinea for the Year 1987. Three horizontal wells were reported drilled in the South Pepper and North Herald field, offshore Australia.

1111. Williamson, P. E., C. J. Pegram, J. B. Colwell, A. S. Scherl, K. L. Lockwood, and J. C. Branson (1985). Pre-Eocene stratigraphy, structure, and petroleum potential of the Bass basin. The APEA Journal 25 (1), 362-381.

This article offers a good summary of the geology of the Bass Basin, Australia, where extended-reach and horizontal drilling has been used to 'redevelop' offshore fields.

1112. Wilson, B. L. (1990). Electrical submersible pumps in horizontal wells. Proceedings of the Thirty-Seventh Annual Meeting Southwestern Petroleum Short Course, April 18-19, 1990, Lubbock, Texas, 418-424. Southwestern Petroleum Short Course Association, Inc., Lubbock, Texas.

Production from horizontal wells where fluid flow is low, or bottomhole pressure is low, may develop problems due to "slugging." Electric submersible pumps may be used to increase fluid flow in the horizontal section of the wells and therefore eliminate or reduce the 'slugging' problem.

Specially constructed electric submersible pumps have been built for installation and operation in horizontal wells. This article describes methods of installation and operation of these pumps.

1113. Wilson, K. M. (1990). Practical approach to a fractured chalk play: Untapped reservoirs in the Cretaceous Niobrara formation. A.A.P.G. Bulletin 74 (5), 792.

Abstract. Silo field, Wyoming, is used as an example of the incorrect way to develop fractured reservoirs. An integrated exploration approach is recommended for the location and development of fractured reservoirs in the Niobrara chalk.

1114. Wilson, M. A. and F. L. Sabins (1987). A laboratory investigation of cementing horizontal wells. in 62nd Annual SPE Technical Conference and Exhibition, September 27-30, Dallas, Texas. Society of Petroleum Engineers, Richardson, TX. SPE paper 16928. Also in SPE Drilling Engineering (1988), 3 (3), 275-280.

A large scale laboratory model was used to evaluate cementing in horizontal wells in low-permeability rocks. The experiments showed that for optimal cement performance in horizontal wells, the following practices are important: (1) cement slurries should be designed to contain no more than a trace of free water; (2) design of spacer systems and flow rates is important; (3) hole cleaning is also very important for preventing mud channeling on the low side of the annulus.

1115. Wilson, Mark A. (1989). Cementing horizontal well in preparation for stimulation. World Oil 209 (4), 72, 74-76, 78, 82, 84. Abstract in The Engineering Index Monthly (June, 1990), 28 (6), 668.

Cementing casing in horizontal wells is required if the well is to be hydraulically fractured or otherwise stimulated or worked over. Cementing horizontal or high angle wells requires that cement slurries be modified to prevent free-water from forming high-side channels. Drilling fluids should be designed for complete circulation in the annulus to prevent regions of gelled, immobile fluid. Compaction of solids in the wellbore can cause low-side channeling. Channeling can allow communication between zones that should be kept isolated. Proper centralization of the liner or casing is necessary for a good cement job in horizontal wells.

1116. Wilson, M. A. and F. L. Sabins (1989). Laboratory investigation of cementing horizontal wells. Proceedings 6th Middle East Oil Show, March 11-14, 1989, Manama, Bahrain, pp. 23-33. Society of Petroleum Engineers, Richardson, TX. SPE paper 17925. Abstract in The Engineering Index Monthly (January, 1990), 28 (1), 660.

A successful cement job is especially important in completing a horizontal well. Achieving good mud displacement is difficult in high-angle and horizontal wells. This article reports on laboratory experiments in cementing horizontal wells. The experiments show that good cement jobs require that the cement slurries have no more than a trace of free water. Spacer systems and flow rate are very important in cementing horizontal wells. Hole cleaning and casing centralization is also important for a good cement job. Details of these experiments and results are given in the article.

1117. Wilson, R. C. and D. N. Willis (1986). Successful high-angle drilling in the Statfjord field. Paper presented at the 61st Annual Technical Conference and Exhibition, October 5-8, 1986, New Orleans. Society of Petroleum Engineers, Richardson, TX. SPE paper 15465.

High-angle drilling from offshore platforms allows a much larger reservoir area to be tapped. Drilling at angles of 80 degrees allows an area of 320 km2 to be covered from a single platform, 30 times the area covered by 40 degree drilling. High angle drilling has been used to reduce development costs in Statfjord field, Norwegian North Sea, by reducing the number of platforms required for field development. The project overcame several drilling problems such as the stabilization of unstable formations, hole cleaning at high angles, use of Top Drive systems, and casing design problems for high angle holes.

1118. Winters, W. J., F. DeLucia, and R. Johnson (1988). Drilling a large medium-radius horizontal hole in hard rock. Paper presented at the 1988 SPE Rocky Mountain Regional Meeting, May 11-13, 1988, Casper, Wyoming. Society of Petroleum Engineers, Richardson, TX. SPE paper 17528. Paper not printed in the proceedings.

Not examined.

1119. Withjack, E. M. (1985). Analysis of naturally fractured reservoirs with bottomwater drive: Nido A and B fields, offshore northwest Palawan, Philippines. Journal of Petroleum Technology 37 (9), 1481-1490. SPE paper 12019 presented at the SPE Annual Technical Conference and Exhibition, San Francisco, October 5-8, 1983.

Computer simulations of the naturally fractured Nido A and B fields, Philippines, relate reservoir performance to fracture porosity, matrix-block size, relative permeability, and absolute permeability. Large fracture porosity results in long duration of water-free production before water breakthrough. After breakthrough, absolute permeability and matrix-block size are the main influences.

1120. Wittrisch, C. and Jourdan (1983). Horizontal well logging by Simphor. in Transactions of the VIII European Formation Evaluation Symposium, March 14-15, 1983, London, England, Paper J. London Chapter of the Society of Professional Well Log Analysts, London Chapter.

The SIMPHOR logging system is designed for logging horizontal wells. The SIMPHOR tools are attached to the drill string and logging is done while moving the drill string up or down. Logging tools available for this system are induction, resistivity (SFL), dual laterolog, gamma ray, sonic (BHC), 4-arm caliper, cement bond log, and perforating gun. SIMPHOR has been successfully used to log horizontal wells in Lacq field, France, and Rospo Mare field, Italy. SIMPHOR is a registered trademark.

1121. Woodhouse, R., D. N. Greet and C.R. Mohundro (1984). Induction log vertical resolution improvement in vertical and deviated wells using a practical reconvolution filter. Journal of Petroleum Technology 36 (7), 993-1001. SPE paper 11855 presented at the SPE Rocky Mountain Regional Meeting, Salt Lake City, May 23-25, 1983.

A deconvolution filter was developed to increase the vertical resolution of induction logs in vertical and deviated wells. The filter was tested on synthetic induction logs derived from formations of known resistivity. The filter operates on digitally recorded data at 1 ft increments.

1122. Wu, Jiang and Hans C. Juvkoram-Wold (1990). Discussion of Tubing and casing buckling in horizontal wells. Journal of Petroleum Technology 42 (8), 1062-1063. SPE paper 21308.

Critical discussion of Chen, Yu-Che, Yu-Hsu Lin, and John B. Cheatham (1990), Journal of Petroleum Technology 42 (2), 140-141, 191.

1123. Wu, Jiang, Ping Chen, and Hans Juvkam-Wold (1991). Casing centralization: Centralization of casing in horizontal wells. Popular Horizontal 1 (2), 14-16, 18-21.

The three-moments equation is used to analyze casing centralization and deflection in horizontal wellbores.

1124. Wu, Jiang and Hans C. Juvkam-Wold (April 29, 1991). Drag and torque calculations for horizontal wells simplified for field use. Oil & Gas Journal 89 (17), 49- 53, 56.

Simplified calculation methods are presented for calculating sliding drag and torque in horizontal wells. Equations are given for: drag pulling out of the hole; for drag running in the hole; and torque in the build section.

1125. Yager, D. (January, 1985. Esso drills World's longest horizontal well. The Roughneck.

Not examined.

1126. Yeung, W. S. (September, 1987). The utilization of horizontal wells for the production of convention petroleum - A study of the theory and an analysis of the potential applications in Canada. M. Engineering thesis, University of Calgary, Calgary, Canada.

Not examined.

1127. Yew, C. H. and Y. Li (1988). Fracturing of a deviated well. SPE Production Engineering 3 (4), 429-437. SPE paper 16930 presented at the 1987 SPE Annual Technical Conference and Exhibition, Dallas, September 27-30.

Three-dimensional elasticity theory was used to predict the fracture planes in a hydraulically fractured, deviated well. The axis of a deviated well are not aligned with the direction of the in-situ stress components. The off-plane stress components are important and influence the direction and magnitude of the hydraulically induced fractures and the magnitude of the breakdown pressure. Formulae are derived for the stress components around the wellbore.

1128. Yew, C. H. and J. H. Schmidt (1989). On fracture design of deviated wells. Proceedings SPE Annual Technical Conference & Exhibition, Production Operations and Engineering volume, October 8-11, 1989, San Antonio, pp. 211-224. Society of Petroleum Engineers, richardson, TX. SPE paper 19722. Abstract in The Engineering Index Monthly (January, 1990), 28 (1), 683.

Experimental results of hydraulic fracturing of deviated wells are given in this article. Previously derived equations were modified to analyze the results of plastic model experiments. The following conclusions are reached in this paper: (1) a zone exists around the wellbore surface in which the circumferential stress is tensile; (2) mini-fractures initiated in the wellbore turn perpendicular to the direction of the minimum in-situ stress; (3) criteria for linking two mini-fractures can be established.

1129. Yokoyama, Y. and E. Arima (1989). Pilot development of tight limestone reservoir in the Khafji field. Proceedings 1989 SPE Asia-Pacific Conference, September 13-15, 1989, Sydney, Australia, pp. 185-197. Society of Petroleum Engineers, Richardson, TX. SPE paper 19488.

Khafji field, Saudi Arabia, produces from a low-transmissivity limestone reservoir. A pilot development project in Khafji field consisted of one vertical well and one horizontal well, which were subjected to a variety of treatments including acidizing and selective hydraulic fracturing to determine which methods were best for developing the field. The horizontal well was drilled with two drainholes. The first horizontal segment was drilled using short-radius methods. The horizontal segment was kicked-off at 4,218 ft and drilled to 4,274 ft; the second horizontal segment was kicked-off at 4,208 ft and drilled to 4,975 ft. The two drainholes were treated and tested separately. The horizontal wells showed enhanced productivity. Details of the treatment of each of the wellbore/drainholes are given in the article.

1130. Yost, A. B., II (1980). Results of an experimental directionally drilled well in the Devonian shale. West Virginia University Petroleum Engineering 394.

Not examined.

1131. Yost, A. B., II (1986). Risks and rewards of well drilling in the Devonian shale. Paper presented at the SPE Unconventional Gas Technology Symposium, May 18-21, 1986, Louisville, KY. Society of Petroleum Engineers, Richardson, TX. SPE paper 15232.

An analysis of the economics of drilling in the Devonian shales of Ohio/Kentucky, and West Virginia. The primary conclusions are that Devonian shale wells can be drilled economically, in the study area, with rates of return of 4.8% to 32%. Much of the area is economically marginal using conventional technology. The horizontal drilling was not considered in this study.

1132. Yost, A. B., II, W. K. Overby, and R. S. Carden (1987). Drilling a 2,000 feet horizontal well in the Devonian shale. Proceedings 1987 SPE Annual Technical Conference & Exhibition, Drilling volume, September 27-30, 1987, Dallas, pp. 291-302. Society of Petroleum Engineers, Richardson, TX. SPE paper 16681.

The world's first air-drilled gas well was drilled in a fractured Devonian shale reservoir of the Wilsondale Pool in Wayne county, West Virginia. The well was drilled to 3,400 ft TVD with a 2,000 ft horizontal section. Total length of the hole was 6,020 ft. The well survey methods used were not effective in an air/mist drilling environment. Directional drilling was done with a 7.75 inch low-speed motor, a 2.5 degree bent sub, and an electromagnetic MWD system. Hole cleaning was one of the major problems encountered. The first two motor runs were terminated with the bit plugged with cuttings. The horizontal (90-92 degrees) section was cored using standard 30 ft core barrels. Hole drag below 5,400 ft was the most serious problem encountered and was solved by using additional drill collar at 1,500 ft. The well was logged using drillpipe-conveyed tools, including a borehole television camera, gamma-ray, density, caliper, dual-induction, and temperature logging tools. The temperature log proved best for locating fractures in the Devonian shale. The downhole television camera detected more than 250 fractures. Four and one-half inch diameter casing was run to a depth of 6,020 ft.

1133. Yost, A. B., W. K. Overbey, S. P. Salamy, C. O. Okoye, and M. S. Saradji (1987). Devonian shale horizontal well: Rationale for wellsite selection and well design. Proceedings 1987 SPE/DOE Symposium on Low Permeability Reservoirs, May, 1987, Denver, pp. 207-216. Society of Petroleum Engineers, Richardson, TX. SPE/DOE paper 16410.

A U.S. Department of Energy sponsored horizontal well was drilled in Wilsondale field, West Virginia. The well was drilled to test the effects on productivity of a horizontal well in a naturally fractured, Devonian shale reservoir. Site selection for the well was made after study of the Devonian shale in West Virginia, Kentucky, and Ohio. Wilsondale field site was judged to have the best probability of having fracture permeability in the Devonian shale. Wilsondale field had 38 wells, 17 of which had long production histories. Production histories were also used from 18 wells outside of the study area. The drilling site within the field was selected based on the production histories of the wells and on calculations of the probable productivity of the unknown wells in the field. Theses data were used to simulate the productivity of horizontal wells at five locations within the field. The drilling site was selected based on the reservoir simulation. Well design was based on known fracture orientations in the field, and on the results of a high-angle well drilled in Grand Canyon National Park for the installation of a waterline (see Index). Details of the well design and site selection are given in the article.

1134. Yost, A. B., II, W. K. Overbey, Jr., D. A. Wilkins, C. D. Locke (1988). Hydraulic fracturing of a horizontal well in a naturally fractured reservoir: Gas study for multiple fracture design. Proceedings Gas Technology Symposium: Gas Supply - A critical Balance of Politics, the Market Place, and Technology, June 13-15, 1988, Dallas, pp. 451-460. Society of Petroleum Engineers,

Richardson, TX. SPE paper 17759. Abstract in The Engineering Monthly Index (July, 1989), 27 (7), 550.

A horizontal well was hydraulically fractured in a naturally fractured, low-permeability, low pressure reservoir in Devonian shale. Modeling was done to select a stimulation fluid that would have limited capillary retention. Nitrogen gas, carbon dioxide gas, and nitrogen gas foam were used in the simulation. All three fluids were used to treat several intervals of the horizontal borehole. Production before treatment was 35 MCFGPD, a rate three times that of a vertical well in the same field. Fracturing with carbon dioxide resulted in initially higher production rates than treatment with nitrogen foam, but production returned to baseline soon after treatment. The nitrogen foam treatment retained a higher production rate after treatment. During fracturing, low injection rates 'propagated' the natural fractures that had low angle to the principal stress orientation. New fractures were induced at higher injection rates. Proppants were needed to retain open fractures for more than thirty days. Multiple hydraulic fractures may be made in a horizontal wellbore with a single pumping operation. Details of the operations and results are given in the article.

1135. Yost, Albert B, II (1988). Horizontal gas well promises more Devonian production. The American Oil & Gas Reporter 31 (7), 29-32.

BDM Corp., under contract to the Department of Energy, air-drilled a horizontal well into a fractured Devonian shale gas reservoir in Wayne County, West Virginia. The unstimulated well is expected to produce three times better than a stimulated vertical well in the same reservoir. Details of the drilling and engineering of the well are given in the article. Also mentioned is the Grand Canyon Directional Drilling and Waterline Project.

1136. Yost, A. B.and W. K. Overbey, Jr. (1989). Production and stimulation analysis of multiple hydraulic fracturing of a 2000 ft horizontal well. Proceedings 1989 SPE Gas Technology Symposium, June 7-9, 1989, Dallas, pp. 321-334. Society of Petroleum Engineers, Richardson, TX. SPE paper 19090.

The Department of Energy contracted with BDM Corp. to drill high-angle test wells in naturally-fractured Devonian shale in Wayne county, West Virginia. In 1986, a well (RET #1) was drilled with 2,000 ft of horizontal wellbore in the Devonian shale. Eight zones in the horizontal wellbore were hydraulically treated to open, and hold open, the natural fractures. This treatment resulted in a several fold increase in productivity. Details of the hydraulic fracture treatment and well tests are given in the article.

1137. Younger, June (1991). Heyco, GLC (sic) Energy measure Niobrara potential and horizontal drilling in Powder River basin. Western Oil World 48 (1), 11.

Harvey E. Yates Co. (HEYCO) is preparing to drill a horizontal well near East Fork field in the Powder River basin. The target will be the Niobrara formation. GLG Energy L. P. plans four horizontal exploratory wells near Salt Creek field.

1138. Younger, June (1991). Horizontal drilling: Utah strikes its first horizontal success. Western Oil World 48 (5), 33-34.

Columbia Gas Development Corp. has completed the first horizontal well in the Cane Creek Zone of the Paradox formation, Utah. Columbia Gas' partners in the venture are Exxon Co. USA and Enserch Exploration Partners Ltd. The well potential is 914 BOPD and 290 MCFGPD through a 10/64-inch choke with a flowing tubing pressure of 3,460 psi.

1139. Younger, June (1991). Horizontal probes test more than the Bakken. Western Oil World 48 (5), 38-39.

Brief review of horizontal drilling activity in Wyoming (Union Pacific Resources Co., Snyder Oil Corp., Exxon Co. USA, Gerrity Oil & Gas Corp, GLG Energy L.P., and Meridian Oil Inc. - Niobrara formation), New Mexico (Merrion Oil & Gas Corp - Entrada formation, Ojo Encino field), Montana (Pennzoil Exploration & Production Co. - Madison formation, Ridgelawn field; Columbia Gas Development - Bakken formation), and North Dakota (Pennzoil - North Branch field; Exxon Co. USA - Elkhorn Ranch North field; Meridian Oil Inc.; Conoco Inc.).

1140. Zagalai, B. M. (1989). Reservoir simulation of horizontal wells in the Helder field. Paper presented at the 1989 European Offshore Conference: Offshore Europe 89, September 5-8, 1989, Aberdeen. Society of Petroleum Engineers, Richardson. SPE paper 19296.

The results of a simulation to predict the effect of horizontal wells on water cut in Helder field, Netherlands North Sea, are presented.

1141. Zaleski, T. E., Jr. and Edward Spatz (May 2, 1988). Horizontal completions challenge for industry. Oil & Gas Journal 86 (18), 58, 62, 64, 67-68, 70. Abstract in The Engineering Index Annual (1988), 87, Part III, 5996.

Horizontal wells have been used in consolidated, fractured reservoirs and in sandstone reservoirs to increase the production by exposing more of the reservoir to the wellbore. Horizontal wells have also been used to improve sweep efficiencies of waterfloods, accelerate and improve drainage of reservoirs, and to reduce water- and gas-coning. Horizontal wells are classified on the basis of build rates as long-radius (build rate of 1 to 8 degrees/100 ft), medium-radius (build rates of 9 to 40 degrees/100 ft), and short-radius (build rates over 40 degrees/100 ft). Horizontal completions require good hole cleaning and cement slurry control to be successful if casing or a liner is to be used. If the well is to be perforated, tubing conveyed perforation guns are needed to place the perforations properly in a horizontal wellbore. Sand control methods need special planning in horizontal wellbores.

1142. Zaleski, T. E., Jr. (1989). Innovations in completion technology for horizontal wells. Offshore 49 (2), 34-37.

A general discussion of gravel packing, perforating, and liners and casing in horizontal wells.

1143. Zaleski, Theodore E., Jr. and Jefferson P. Ashton (1990). Gravel packing feasible in horizon tal well completion. Oil & Gas Journal 88 (24), 33-37.

Gravel-packed completions are a solution to sand-control problems in horizontal wells in unconsolidated sediments. In formations with low sand-producing tendencies and formations that are coarse-grained, slotted liners are preferred. Wire-wrapped screens can be substituted for slotted liners in medium- and coarse-grained formations. Prepacked liners are useful for most formations, flow rates, and formation strengths. In formations with substantial clay and fines and high flow capacity when maximum performance is sought, gravel packing is preferred. Diametrical clearance of screens or liners should be approximately 40 mm in non-gravel packed wells. In gravel packed wells radial clearance should be approximately 40 mm; centralizers are recommended.

1144. Zammerilli, A. M. (1989). A procedure for relative ranking of sites selected for horizontal wells in tight, naturally fractured reservoirs. Proceedings 1989 SPE Eastern Regional Conference, October 24-27, 1989, Morgantown, pp. 21-30. Society of Petroleum Engineers, Richardson, TX. SPE paper 19059. Also in Proceedings SPE Gas Technology Symposium, Dallas, June 7-9, 1989, pp. 63-72. Abstract in The Engineering Index Monthly (January, 1990), 28 (1), 613.

Ranking of proposed sites for horizontal wells may be done using a methodology combines risk analysis with a reservoir simulator. Risk analysis was calculated using success ratios and expected monetary value. The analysis is applied to the fractured reservoirs in the Devonian shale of West Virginia. The region under study was partitioned by counties. The analytical procedure is outlined. The analysis shows that three of the partition areas with a total of 10 counties should be profitable with unstimulated horizontal wells.

1145. Zammerilli, A. M. (1989). A simulation study of horizontal, high-angle, and vertical wells in eastern Devonian shale. Proceedings SPE Joint Rocky Mountain Regional/Low Permeability Reservoirs Symposium & Exhibition, March 6-8, 1989, Denver, pp. 641-650. Society of Petroleum Engineers, Richardson, TX. SPE paper 18998. Abstract in The Engineering Index Monthly (January,1990), 28 (1), 613.

Simulations show that horizontal drilling is economically more effective than vertical or high-angle drilling in the Eastern Devonian Shales in West Virginia. The most favorable infill drilling using horizontal wells is where the Devonian shales are 70-100 ft thick. The presence of faulting may significantly alter the effectiveness of this technique and the drainage patterns.

1146. Zammerilli, Anthony (1991). Ranking horizontal-well sites in tight, naturally fractured reservoirs. Journal of Petroleum Technology 43 (1), 89-95. SPE paper 19059 first presented at 1989 SPE Gas Technology Symposium, Dallas, TX, June 7-9, 1990.

Expected Monetary Value and cash flow models are used to rank locations for horizontal wells in the naturally fractured, Devonian shales of the Appalachian basin. Variables used in the evaluations include reservoir parameters (thickness, pressure, success ratios, etc.) and economic variables (gas price and drilling costs). The resulting analysis ranks areas in West Virginia for their favorable economics.

1147. Zamora, Mario and Paul Hanson (1991). More rules of thumb to improve high-angle hole cleaning. Petroleum Engineer International 63 (2), 22, 24, 26-27.

Eighteen 'Rules of Thumb' for hole cleaning in high-angle wells (see previous Zamora article for the first ten rule).

1148. Zamora, Mario and Paul Hanson (1991). Rules of thumb to improve high-angle hole cleaning. Petroleum Engineer International 63 (1), 44-46, 48, 51.

Ten rules of thumb given for hole cleaning in high-angle wells.

1149. Zimmerman, Chris and Donny Winslow (1989). How to select the right tools for stimulating horizontal wells. World Oil 209 (5), 53-54, 56-57. Abstract in The Engineering Index Monthly (February, 1990), 28 (2), 637.

Most horizontal wells being drilled today are medium- and long-radius wells. The well geometry limits the selection of completion tools that can be used. Important factors in stimulation tool selection include number of zone to be treated, borehole environment, and type of treatment. Although many standard tools may be used in horizontal wells, some tools have been developed especially for horizontal completions, such as pressure operated bridge plugs, pressure-operated resettable packers, and tubing-conveyed service tools.

1150. Zimmerman, J. C., D. W. Winslow, R. L. Hinkie, and R. R. Lockman (1990). Selection of tools for stimulation in horizontal cased hole. Proceedings 1990 SPE Permian Basin Oil & Gas Recovery Conference. Positioning for the 90's: Improving Profitability Through Todays Technology, March 8-9, 1990, Midland, pp. 15-26. Society of Petroleum Engineers, Richardson, TX. SPE paper 18995. Also published in the Proceedings California Regional Meeting & Exhibition: Challenges of a New Decade; Ventura, California, April 4-6, 1990, pp, 9-20; and SPE Joint Rocky Mountain Regional/Low Permeability Reservoirs Symposium and Exhibition, Denver, March, 6-8,

1989, pp. 629-640. Abstract in The Engineering Index Monthly (June, 1990), 28 (6), 669 and (January, 1990) 28 (1), 651.

Geometry of a horizontal well may limit the selection and number of tools that can be run in the well. Eight case histories are presented to illustrate the need for planning and the limitation imposed by well geometry.

1151. Zurdo, C., C. Georges, and M. Martin (1986). Mud and cement for horizontal wells. Paper was presented at the 61st SPE Annual Technical Conference and Exhibition, October 5-8, 1986, New Orleans. Society of Petroleum Engineers, Richardson, TX. SPE paper 15464.

Bench tests have been conducted to test the following cementing parameters in horizontal and high angle boreholes: casing-hole eccentricity, drilling fluid rheology behavior, hole geometry, spacer design, and flow rate. Similar tests were conducted to study cuttings transport in horizontal boreholes. Results of these tests are reported in this article. Horizontal wellbores in poor quality reservoirs require drilling fluids that do not further degrade the reservoir. Other important considerations in horizontal wellbores are the hole cleaning capabilities of the drilling mud, fluid balance, and lost circulation plugging, particularly in fractured reservoirs.

1152. Zwicker, Denise Allen (1991). Drilling sideways. The Lamp 73 (1), 8-11.

This is a general article on horizontal drilling activities of Exxon Co. and its affiliates. Brief mention of Austin Chalk and Canadian heavy-oil projects.

Addendum: Articles added after 1 September 1991

1153. Santos, O. L. A. (1991). Well-control operations in horizontal wells. SPE Drilling Engineering 6 (2), 111- 117. SPE paper 21105, presented at the 1990 SPE Latin American Petroleum Engineering Conference, October 14-19, Rio de Janeiro.

This paper presents procedures and mathematical models for: (1) prediction of pressure behavior in the annulus during a gas-kick; (2) procedure for establishing the drillpipe-pressure schedule during replacement of old mud by kill mud; and (3) discussion of kick tolerance in horizontal wells.

1154. Hansen, William B., editor (1991). 1991 Guidebook to Geology and Horizontal Drilling of the Bakken Formation. Montana Geological Society, Billings, Montana. 196 pp.

This volume contains sixteen papers and abstracts on the geology and horizontal drilling of the Bakken Formation, Williston basin, United States and Canada. Of the sixteen abstracts and papers, six have been published previously. This volume also includes production statistics for all wells producing from the Bakken.

1155. Brooks, Dean C. (1991). Case history of an outstanding horizontal Bakken well: Slawson Sidewinder 1- 7, Billings County, North Dakota. in Hansen, William B. (editor) (1991), 1991 Guidebook to Geology and Horizontal Drilling of the Bakken Formation, pp. 105-110, Montana Geological Society, Billings, Montana.

The Slawson Exploration Co. Sidewinder 1-7 well was drilled in Billings County, North Dakota in 1989/1990. The well is 12,656 ft measured depth; 10,351 ft TVD. The lateral section of the well is 2052 ft of which 1814 ft is in the Bakken and 267 ft is in the Mississippian Lodgepole Limestone. The well tested 1362 BOPD through a 18/64-inch choke, and had a flowing tubing pressure of 1050psi. After 21 days the well stabilized at 600 BOPD and 400 psi flowing tubing pressure. This paper describes the drilling and completion methods used on the Sidewinder 1- 7.

1156. Hansen, William B. and George I. W. Long (1991). Bakken production and potential in the U.S. and Canada: Can the Fairway be defined. in Hansen, William B. (editor) (1991), 1991 Guidebook to Geology and Horizontal Drilling of the Bakken Formation, pp. 69-88, Montana Geological Society, Billings, Montana. The paper was presented at the Montana Geological Society Bakken Workshop, Billings, March 27, 1990.

Since Bakken production began in 1953, a Bakken fairway has been defined in North Dakota and Montana. Most of the Bakken production in the fairway is in North Dakota, with scattered production in Montana. Bakken production in Canada is restricted to three fields. Bakken exploration, in Canada, has been limited. One well in western Montana shows that the Bakken exists outside of the Williston basin, although the potential of that area is not known. Horizontal wells improve the possibility of making an economic well.

1157. Hansen, William B. and George I. W. Long (1991). Criteria for horizontal and vertical prospects in the Bakken Formation, Williston basin. in Hansen, William B. (editor) (1991), 1991 Guidebook to Geology and Horizontal Drilling of the Bakken Formation, pp. 151-163, Montana Geological Society, Billings, Montana. This paper was presented at the A.A.P.G.-Rocky Mountain Sectional Meeting, Denver, September 18, 1990.

In Bakken wells, fractures may be identified using as criteria: structure; mud invasion profiles on dual laterolog/microspherically focused resistivity logs; high shut-in pressures on drillstem tests; and high potentials. Details of the study are given in the article. The study was based on data from Roosevelt field, Beaver Creek field, and Elkhorn Ranch field, North Dakota, and Spring Lake field, Montana.

1158. LeFever, Julie A. (1991). History of oil production from the Bakken Formation, North Dakota. in Hansen, William B. (editor) (1991), 1991 Guidebook to Geology and Horizontal Drilling of the Bakken Formation, pp. 3-17, Montana Geological Society, Billings, Montana. Paper presented at the Montana Geological Society Bakken Workshop, Billings, March 27, 1990.

A history of oil production from the Bakken Formation. The first Bakken production was from a Stanolind well in Antelope field, North Dakota. The second Bakken producer (Elkhorn Ranch field) was drilled by Shell in 1960. In 1977, Gulf Oil opened a period of Bakken development using vertical wells. The current horizontal drilling phase of Bakken production started in 1987, when Meridian Oil Inc. completed a horizontal well.

1159. Martiniuk, C. D. (1991). Regional geology and petroleum potential of the Bakken Formation. in Hansen, William B. (editor) (1991), 1991 Guidebook to Geology and Horizontal Drilling of the Bakken Formation, pp. 43-67, Montana Geological Society, Billings, Montana. This paper was presented at the Montana Geological Society Bakken Workshop, Billings, March 27, 1990. This paper is reprinted from "Regional Geology and Petroleum Potential of the Bakken Formation, Southwestern Manitoba" by C. D. Matiniuk, 1988; Manitoba Energy and Mines, Petroleum Open File Report POF 8-88.

The Bakken Formation in Manitoba differs from the Bakken in the United States portion of the Williston basin. The chief difference is that the lower member of the Bakken is absent, due to nondeopsition, in most of Manitoba. Details of the geology of the Bakken in Manitoba are given in the paper. Oil has been discovered and produced from Daly field, in what appear to be stratigraphic traps.

1160. Meissner, Fred F. (1991). Petroleum geology of the Bakken Formation Williston Basin, North Dakota and Montana. in Hansen, William B. (editor), (1991), 1991 Guidebook of Geology and Horizontal Drilling of the Bakken Formation, pp. 19-42, Montana Geological Society, Billings, Montana. This paper is reprinted from the 1978 Montana Geological Society Guidebook: Economic Geology of the Williston Basin.

The Bakken Formation upper and lower shale members are mature source rocks, in the deeper parts of the Williston basin. Description of the petroleum geology of the Bakken and of the fracturing by formation fluid overpressure, to produce oil reservoirs. Regional fracturing mechanisms are also discussed.

1161. Shepard, Warren (1991). Tectonics and thermal maturity of the Bakken Formation of the Williston Basin, with comments on Landsat analysis applications. in Hansen, William B. (editor) (1991), 1991 Guidebook to Geology and Horizontal Drilling of the Bakken Formation, pp. 167-177, Montana Geological Society, Billings, Montana. This paper was presented at the Montana Geological Society Bakken Workshop, Billings, March 28, 1990.

A plate tectonic models presented to explain the thermal maturity and fracturing of the Bakken Formation in the Williston Basin. Landsat image interpretations is used to support the model.

1162. Coudeyre, Michel, Bernard LeGoc, and Jean Paul Lombez (1 July, 1991). Guiding a horizontal well using geological monitoring. World Oil 212 (7), 59-60, 62, 64, 66.

Total-Exploration and its partners used "geological piloting" to guide the drillbit into and maintain it in a thin productive zone in the Villeperdue field, France. The technique uses gas monitoring, cuttings analysis, fluorescence, gamma-ray logs, calcimetry, and drilling parameters. In the example cited, gas monitoring using the Anadrill Gaslogger and Geoservices Advanced Logging System were the most usefull tools. Editor's Note: Geological monitoring, in the sense of this article, is commonly used, in the horizontal Austin Chalk play of South Texas, to guide the drillbit in the productive zones of the Austin Chalk. In the Austin Chalk play, cuttings analysis, fluorescence, and gas monitoring can be critical to the success of the well.

1163. Wilson, Douglas and Mark F. Barrilleaux (July 8, 1991). Water-packing techniques successful in gravel packing high- angle wells. Oil & Gas Journal 89 (27), 30- 32, 34-37.

Description of gravel packing techniques used by Exxon on wells drilled in Mississippi Canyon Block 280, Gulf of Mexico.

1164. Montgomery, Scott (1991). Horizontal drilling in the Austin Chalk. Part 2: Patterns of production. Petroleum Frontiers 7 (4), 1-58.

An overview of the petroleum geology of the Austin Chalk/Selma Chalk trend is presented, along with a discussion of Austin Chalk productivity, with examples, and a summary of drilling and completion methods. The Austin Chalk/Selma Chalk trend extends from Alabama on the east to the Rio Grande River on the west. Productive areas are localized by fracturing, which may have a variety of causes. Non-fracture-related production is alluded to but not discussed.

1165. Renard, Gerard and J. M. Dupuy (1991). Formation damage effects on horizontal-well flow efficiency. Journal of Petroleum Technology 43 (7), 786-789. For critical discussion see: Gilman, J. R. (1991), Journal of Petroleum Technology 43 (7), 870; SPE paper 23526.

Flow efficiency, the ratio of actual productivity index of a well to ideal productivity index, is a better index of the efficiency of a well. This article gives equations for the determination of flow efficiency in horizontal wells.

1166. Gilman, J. R. (1991). Discussion of formation damage effects on horizontal-well flow efficiency. Journal of Petroleum Technology 43 (7), 870.

Critical discussion of the article by Renard and Dupuy (item 1165 in this bibliography).

1167. Coll, David (September 17, 1990). Manitoba offers royalty holiday for horizontal wells. Oilweek 41 (31), 3.

The Province of Manitoba is offering a tax holiday on freehold prodution taxes and Crown royalties on oil produced from horizontal wells or horizontal drainholes drilled prior to January 1, 1992. The tax incentives cover the first 7,500 cubic meters of oil produced from a new well and 1,500 cubic meters of produced from drainholes drilled from pre-existing vertical wells.

1168. Coll, David (September 17, 1990). Horizontal program paying dividends. Oil week 41 (31), 13-14.

CS Resources Ltd. is using horizontal wells to develop Pelican Lake field. Devran Petroleum Ltd. has a 23% interest in the project. The Wabiskaw sandstone is the reservoir at Pelican Lake. CS Resources is also active in Winter field, Saskatchewan, with Saskoil, Lasmo Canada, and Devran as partners. The five horizontal wells in Winter field have horizontal sections of 2,500 ft and initial production rates of 200 BOPD.

1169. Coll, David B. (October 15, 1990). Teachers put lessons to work. Oilweek 41 (35), 4.

North Continental Energy Ltd. has completed a horizontal well in Hastings field, Alberta. The well was completed in the Frobisher limestone.

1170. Coll, David B. (December 10, 1990). Austin Chalk: Fabled formation rewards Ridgeway. Oilweek 41 (43), 5-6.

Ridgeway Petroleum Corp., a Canadian company, is participating in the Austin Chalk play of South Texas. Ridgeway also owns leases in the Sturgeon Lake area of Alberta that have horizontal drilling potential. Ridgeway has participated in two Austin Chalk wells, the first well flowed 1,968 BOPD (36 gravity API) with a water cut of 0.07%.

INDEX

A

B

C

D

E

F

G

H

I

J

K

L

M

N

O

P

Q

R

S

T

U

V

W

Y

Z

Additional copies of this book with soft cover may be obtained from SOUTHWEST GEOSERVICES, 11023 Fuqua, Box 279, Houston, Texas, 77089, (713) 481-6007, for $50 (Texas residents add $4.16 sales tax). If shipping is to an address outside the United States, please add $5 for postage and handling.